Springer
Tokyo
Berlin
Heidelberg
New York
Barcelona
Hong Kong
London
Milan
Paris
Singapore

K. Shimamoto (Ed.)

Molecular Biology of Rice

With 48 Figures, Including 7 in Color

Springer

Ko Shimamoto
Professor
Laboratory of Plant Molecular Genetics
Nara Institute of Science and Technology
8916-5 Takayama, Ikoma, Nara 630-01, Japan

ISBN 4-431-70215-6 Springer-Verlag Tokyo Berlin Heidelberg New York

Library of Congress Cataloging-in-Publication Data
Molecular biology of rice / K. Shimamoto, (ed.).
p. cm.
Includes bibliographical references and index.
ISBN 4-431-70215-6 (alk. paper)
1. Rice--Breeding. 2. Rice--Genetics. I. Shimamoto, Ko.
SB191. R5 M575 1998
633.1'82--dc21

98-33426
CIP

Printed on acid-free paper

Printed in Tokyo

Typesetting: From the authors' electronic files
Printing & binding: Obun, Japan

SPIN: 10656065

Preface

Rice (*Oryza sativa* L.) has been the subject of molecular biological studies for the last 10 years, mainly as a result of advances in genome analysis and methods for transformation. The realization that we may face a major food problem in the coming century has stimulated research in the molecular biology of rice not only in Asia but in the United States and Europe. Large-scale international collaboration also has been a major force to advance the molecular biology of rice.

The aim of this book is to summarize the current understanding of rice from the view of molecular biology. Part 1 deals with the genome analysis of rice. The genome of rice is now well characterized with respect to its sequences. Our next goal is to learn the functions of 20,000 to 40,000 rice genes. To this end, promising methods using transposons are being developed and they will become essential tools of rice research in the future. Currently, we know little about how the form of the rice plant is determined. From the three papers in Part 2 we can envision that genes which play important roles in rice development will be found in the near future with the use of the wealth of morphological mutants available in rice. Better understanding of rice development undoubtedly will help us to modify rice for better productivity.

Many key biological steps in rice are regulated by networks of genes at several levels. Part 3 presents our current understanding of four such networks—in particular, those related to seed development and disease resistance. It is clear from these papers that we are beginning to understand the regulation of important biological processes at the molecular level. Two areas of rice molecular biology in which many efforts have been made in the last 10 years are the improvement of transformation methods and the introduction and expression of economically important genes. The clear success in these efforts is shown in Part 4. We now have more than one reliable transformation method for rice, and the list of transgenic rice showing improved agricultural traits is rapidly growing. The enormous potential in molecular improvement of rice in the future can be seen in these papers.

Is rice becoming a model for plant molecular biologists? This book, I hope, gives a clear answer to this question. Ten years from now, many fundamental questions in plant biology will be addressed by using rice as an experimental system. At the same time, genetically engineered rice will help provide a sustained supply of food in many countries worldwide.

Ko Shimamoto
Nara
August 1998

Contents

Part 1

Genome analysis and gene isolation

1
Current status of and future prospects for genome analysis in rice

Takuji Sasaki

Rice Genome Research Program (RGP), NIAR/STAFF, 1-2, Kannondai 2-chome, Tsukuba, Ibaraki 305-8602, Japan

1 Introduction

The genome is the set of genes by which all the genetic characters of a species are defined. In the case of rice, the genome is composed of 12 chromosomes per haploid, with a total size of 430 Mb. Today we know that DNA is the material responsible for encoding genetic information, but before molecular biology, genetics depended solely on the study of phenotypes, such as the well-known studies of peas by Gregor Mendel. The formulation of important principles of genetics, such as linkage and crossing over, also depended on phenotypes. Although polymorphism within DNA can be used to explain Mendelian laws, precise and detailed analysis of the genome is needed to completely explain inheritance as a molecular event. Technological innovation in the analysis of DNA has produced a new concept, genomics, which is the study of the genome.

Rice is an important staple food for people in Asia and Africa and has a long history of improvement by breeding. In 1893, the Japanese government launched the breeding of rice by the National Institutes. As well as improvements in rice quality and yield, genetic knowledge of many rice phenotypes has since been accumulated.

Mutant phenotypes are used for making conventional linkage maps with about 190 loci (Kinoshita 1995). Although the genetic distance between the loci is not necessarily reliable, a linkage map is the only resource for locating genes corresponding to rice phenotypes. To investigate and understand the rice genome further, we have to thoroughly analyze the DNA, which controls all the phenotypic

Key words. genome, genomics, linkage analysis, RFLP (restriction fragment length polymorphism), physical map, YAC (yeast artificial chromosome), cDNA, EST (expressed sequence tag), map-based cloning, disease resistance gene, QTL (quantitative trait loci), rice (*Oryza sativa* L.)

traits. Since the milestone established by Watson and Crick (Watson and Crick 1953), genetics has developed in more and more detail. Linkage analysis is now based on polymorphisms within the sequences of DNA (Botstein et al. 1980), and the accuracy of linkage maps has improved so much as to allow them to be used to isolate a gene corresponding to a phenotype. Researchers also use physical mapping to reconstruct a genome from contiguously overlapping fragments of DNA and to catalogue genes to understand the genome better.

This article describes the current status of rice genomics, or the understanding of the rice genome based on DNA, and indicates the prospects for the ultimate goals of genome analysis, genome sequencing, and function analysis.

2 Linkage analysis using DNA markers

The first step in determining the anatomy of the rice genome is to generate many DNA markers within it. Any differences in nucleotide sequences at allelic loci are used for linkage analysis. Restriction fragment length polymorphism (RFLP) (Botstein et al. 1980), simple sequence repeats (SSR) (Hearne et al. 1992), and cleaved amplified polymorphic sequences (CAPS) (Lyamichev et al. 1993) are the three major types of codominant markers, and amplified fragment length polymorphism (AFLP) (Vos et al. 1995) and random-amplified polymorpic DNA (RAPD) (Williams et al. 1990) are the two major types of polymorphisms showing dominant characteristics. Because codominant characters give more accurate information about allelic fragments than dominant ones in detection systems such as electrophoresis, linkage maps based on codominant markers are preferred, and can also be used for identifying unambiguous markers for physical map construction.

There are several rice linkage maps based on polymorphisms in DNAs (McCouch et al. 1988; Saito et al. 1991; Kurata et al. 1994; Causse et al. 1994; Harushima et al. 1998). The latest one contains 2275 markers within 12 chromosomes, with a total genetic distance of 1521.6 cM (Harushima et al. 1998):

- Almost all the markers are sequenced, although not fully.
- About 70% of the mapped markers are rice cDNAs.
- About 30% show significant sequence similarity to genes with known function.
- The centromeric regions of all 12 chromosomes are indicated.
- Chromosome orientation, such as short arm at the top and long arm at the bottom, is decided using rice secondary and telotrisomic mutants (Singh et al. 1996).

The map, including the sequences of markers and in some cases the identified gene names of the markers, is available on the Worldwide Web at <http://www.dna.affrc.go.jp:84/publicdata/geneticmap98/geneticmap98.html>.

All five main maps in use were constructed using independent mapping populations. However, it is desirable to share the markers and genotype information for wider use in rice genetics and molecular biology. So far, about 3500 markers have been produced, mainly as RFLP markers. Another 150 loci based on SSRs are known (Akagi et al. 1996; McCouch et al. 1997). SSRs are easier than RFLPs to apply to large numbers of samples because they are detected by PCR. A map integrating the main maps is now under construction (S. McCouch, personal communication) to increase the opportunities to obtain DNA markers more closely related to the targeted trait for tagging or gene isolation. So far, our knowledge is limited (Yoshimura et al. 1997), and more extensive work of tagging the traits by DNA markers is required.

RFLP markers can be used to accurately identify genotypes, but Southern blot analysis, used to detect RFLPs, needs a lot of DNA, many experimental steps, and expensive reagents such as radioisotopes. To overcome these restrictions, it is desirable to convert RFLP markers to PCR-based markers, or CAPS (Lyamichev et al. 1993). The conversion of RFLP to CAPS requires information on RFLP marker sequences, especially unique sequences within markers. For this conversion, introns in an expressed gene seem to give a chance to produce polymorphisms for a restriction enzyme. In the case of the RGP map, 150 of 400 mapped rice cDNAs tried have been converted to CAPS (M. Yano, personal communication). CAPS markers are powerful tools for examining many rice plants quickly in a map-based cloning strategy.

3 Physical mapping using large DNA fragments

To determine the structure of a genome, we have to reconstruct the genome from fragments of DNA, just like assembling a jigsaw puzzle. The completed puzzle is called a physical map. Unlike a jigsaw puzzle, nobody knows the original picture, but knowing even part of the sequence of each piece helps.

DNA markers, discussed in the previous chapter, are distinguished by differences in the order of the bases A, T, G, and C within a specific genomic region. In RGP, these markers are used to screen pieces of rice DNA. This strategy, called the chromosomal landing method (Tanksley et al. 1995), is useful when many DNA markers are available on the genetic linkage map, because the positions of picked DNA fragments within the genome are defined. However, it cannot determine whether adjacent pieces overlap or the direction of each piece.

For cloning of large pieces of genomic DNA, several types of vectors based on artificial chromosomes in microbes are used, such as yeast artificial chromosomes (YACs) (Burke et al. 1987) and bacterial artificial chromosomes (BACs and PACs) (Shizuya et al. 1992; Ioannou et al. 1994). RGP uses YACs. Its constructed library is composed of about 7000 clones of 350 kb average size (Umehara et al. 1995). To screen YACs for DNA markers on the linkage map, high-density YAC filters are prepared. On each filter, 1536 YACs are dotted and grown as colonies. Colony hybridization is performed using each DNA marker as a probe, and positive clones are used for further Southern analysis to confirm the identity of patterns with the same restriction enzyme as used for RFLP analysis. In some cases, the positive

Table 1. Status of YAC library screening in RGP by DNA markers on RGP linkage map (12)

Chromosome	Number of contigs	Genetic distance covered by contigs (cM)	Number of isolated YACs
1	40	102.7	31
2	23	53.5	18
3	26	66.4	17
4	23	54.3	21
5	26	38.0	23
6	21	50.2	15
7	24	37.0	26
8	17	24.9	9
9	11	19.8	19
10	14	24.7	23
11	12	15.4	28
12	7	8.0	39
Total	244	494.9	269

YACs give a different pattern from that used for mapping the probe markers. These YACs are grouped as "locus unknown YACs."

Other methods for screening are also used to complement colony hybridization. One uses PCR; another is fingerprinting. Among the markers on our linkage map, about 60 were generated by RAPD and were converted to sequence tagged site (STS) markers. Three-dimensional pools of YACs were made and used as PCR templates. YACs carrying the tags were identified by dissecting the pools. Fingerprinting was detected by repetitive sequences widely distributed within the rice genome or as AFLP by a fluorescent dye. For several practical reasons, however, we could not get satisfactory results from fingerprinting to complement the colony hybridization and PCR methods.

In 1996–1997, RGP reported the results of screening of the YAC library using 1400 DNA markers on the linkage map reported in 1994 (Umehara et al. 1996; Saji et al. 1996; Wang et al. 1996; Shimokawa et al. 1996; Antonio et al. 1996; Koike et al. 1997; Tanoue et al. 1997; Umehara et al. 1997). In all, 2100 YACs were assigned to loci, and the total coverage of the rice genome was estimated to be 50%. Continuation of screening after these reports with about 1000 new markers could add 400 YACs to the previous physical map, increasing the degree of coverage to 65%. The fraction of chimeric YACs is at most 12%. Table 1 summarizes the results.

We have also identified 244 YAC contigs composed of more than two YACs. These have a total genetic distance of 495 cM. Because of the limitations of the chromosome landing method, about 270 YACs have been identified as single, partly because of the lack of markers common to adjacent YACs. One method to determine the presence of overlapping between adjacent YACs is to map YAC-end clones to each other. This method could decrease the number of island YACs, even using the same number of markers on the linkage map.

Our rice physical map is the only one of which we know. It needs to be completed. To do this, we have to increase the number of markers so that we can pick up more YACs to fill the gaps, or prepare another type of genomic DNA library composed of clones that are more easily handled than YACs. These points are discussed in the following chapters.

4 Rice expressed gene catalogue

Besides genetic and physical mapping, we need information about the number and population of expressed genes to complete the rice genome. It is time-consuming and laborious to make full-length cDNAs and sequence them completely, and so partial sequencing of incomplete cDNAs is used. These sequence data are used mainly for cataloguing rice expressed genes and for tagging the genes for mapping. The partial sequence data of such cDNAs are called expressed sequence tagged sites (ESTs) and are now available in the public database dbEST, held by the National Center for Biotechnology Information (NCBI) in the United States, at <http://www.ncbi.nlm.nih.gov/dbEST/index.html>. Rice occupies seventh place in the list of species in dbEST; about 80% of these EST data are from RGP. The rest are mainly from Korea (http://bioserver.myongji.ac.kr/ricemac.html).

RGP has partially sequenced about 40 000 cDNA clones, mainly from their 5'-ends (Sasaki et al. 1994; Yamamoto and Sasaki 1997). More than 1000 clones have been analyzed in each of 14 cDNA libraries. The sources of these libraries were callus grown under several different conditions: young roots, green shoots, etiolated shoots, and panicles at different growing stages. The total number of genes in the rice genome is roughly estimated as 20 000 to 40 000. The grouping of 40 000 sequences based on their overlapping parts using software called the TIGR assembler revealed 16 000 independent sequence groups. Although this is the maximum estimate, about half the rice genes are thought to be registered in the EST catalogue of RGP. These cDNAs are now being sequenced from their 3'-ends to design clone-specific PCR primers (see below). These new ESTs will be used for more accurate estimation of redundancy of analyzed clones and for a tissue-specific expression profile of multiple genes.

The EST catalogue is used to identify the putative function of gene products by similarity search against public protein databases. About 30% of the translated amino acid sequences (12 000 cDNAs) show significant similarity to known sequences from other species, including bacteria, plants, and animals (Sasaki 1996). The huge amounts of data from the large-scale sequencing of many species have been incorporated into the protein databases, and some match the rice ESTs. For instance, the completed yeast genome sequence indicated unexpectedly similar homologues between rice and yeast, although their functions have not yet been identified. A higher degree of similarity has been found between plants, such as rice and *Arabidopsis*. When 12 300 rice ESTs were compared with 29 200 *Arabidopsis* ESTs using the nucleotide-based Basic Local Alignment Search Tool (BLASTN) algorithm, 1700 groups of rice ESTs showed significant similarities to *Arabidopsis* ESTs. This result suggests a high probability of finding the rice homologues of *Arabidopsis* genes. Recently, the development of *Arabidopsis* genomics has enabled the isolation of genes by map-based cloning or T-DNA tagging. In almost all cases, rice-homologous sequences were found in the catalogue. This fact strongly supports the utility of rice ESTs as a resource for gene hunting in conjunction with genetic analysis (Sasaki et al. 1996) or transposon tagging of phenotypic traits (Hirochika 1997).

ESTs or cDNAs used as RFLP markers have added molecular information to the genetic information. These markers, used as probes for YAC screening, give information on the physical location of expressed genes on each picked YAC (Agyare et al. 1997). By increasing the EST markers on a physical map, we can increase our knowledge of the location of genes and might be able to identify the gene location after working out the long-range genome sequences. The next chapter discusses this in more detail.

5 EST map

One of the ultimate goals of mapping is to know the exact location of target genes within the genome. Genetic mapping based on linkage analysis is one strategy. However, this mapping strategy is valid only when polymorphism is observed among siblings. To overcome this restriction, the direct landing method for mapping is used. In linkage analysis, the accuracy of the distance between mapped genes is defined by the population size used for mapping; for example, 0.25 cM by 200 rice plants (400 gametes) used to construct a linkage map in RGP. The physical distance corresponding to 0.25 cM in rice is about 70 kb, based on the total genetic distance of 1520 cM for the total rice genome size, 430 Mb. This means that the physical distance between the markers mapped on YACs is accurate only to 70 kb. On the other hand, direct mapping of ESTs on YACs can in principle identify only the YACs carrying the EST sequences. However, where the YACs have already been overlapped by many markers, the area for the location of the ESTs can be narrowed. In addition, the position of each EST is unambiguously determined on the genome when the EST-specific sequence is used for mapping. Southern hybridization depends on "hybridization with hybridizable sequences," which does not mean the strict matching of the sequences. The mapped EST, on the other hand, will allow the placement of the open reading frames for the flat genomic sequence.

Another important role of mapped ESTs is as markers to fill the gaps between isolated YACs or YAC contigs (Fig. 1). To identify the position of a genomic fragment, an accurate physical map is required. In the RGP physical map of chromosome 6, which was extensively studied to make a complete contig, 36 contigs and

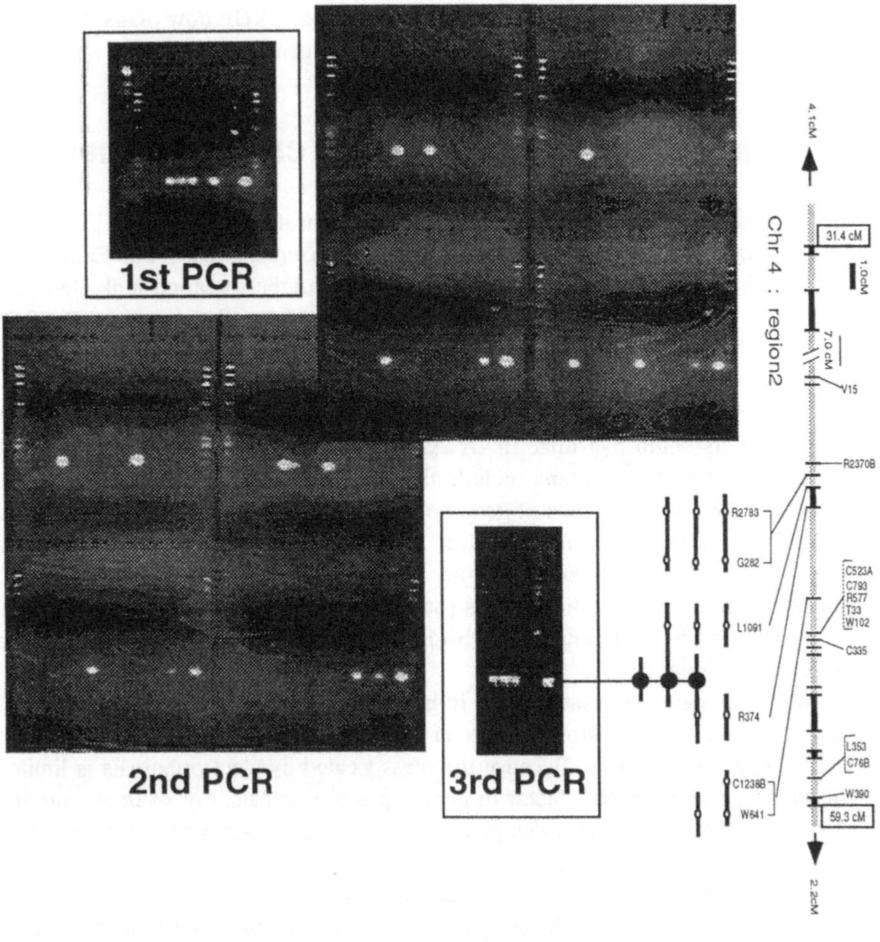

Fig. 1. Example of an EST mapping of a rice cDNA clone by PCR-based screening of the YAC library. A PCR-amplified DNA band using 3'-end primers was detected in gels during the three-step PCR. This clone was finally mapped onto three YAC clones (*vertical bars with filled circle*) in region 2 of chromosome 4 (Koike et al. 1997), reducing the gaps between the YAC contigs.

YACs are still to be joined into 1 contig as a completely reconstructed chromosome. New YACs identified by ESTs are needed to make bridges between these islands. The complete YAC physical map with many EST markers is a prerequisite to starting genome sequencing to make a new sequence-ready physical map.

To make an EST map, each cDNA-specific sequence, or sequence within its 3'-untranslated region, is used to design PCR primers. These primers are used to amplify the target sequence within 7000 YACs. The surveying procedure uses a 3-dimensional pooling method (Wu et al. 1997), and so far more than 2000 EST markers have been generated for the 12 chromosomes. RGP now plans the generation of 10 000 EST markers within the whole rice genome.

6 Colinearlity with other cereal crop genomes

For the researchers of the genomes of maize and wheat in the USA and Europe, the genome size of these species is a serious problem (Arumuganathan and Earle 1991). The size is in the range 10^9-10^{10} bp per haploid, and the so-called junk DNA may hinder the extraction of useful genetic information. But as the genome size of rice is the smallest among the important cereal crops, the results from the rice research can be used to facilitate understanding of other crops.

The first step in RGP's comparison of crop genomes was to clarify whether rice DNA markers could hybridize to DNAs of other crops, such as wheat or maize. Factors affecting the outcome include the degree of similarity of sequences with each other and the difference of genome sizes among these crops. The former was examined by comparing the nucleotide sequences of cDNAs with the same gene function. The rice cDNAs show a higher sequence similarity to those from monocot plants than to those from dicot plants (Sasaki 1996). This means that when cDNAs are used as probes for Southern hybridization, their counterparts in other cereal crops can be identified.

When the detection system used in hybridization depends on enhanced chemiluminescence, the signal strength is determined by the molar concentration of DNAs shown by electrophoresis. Because the mass treated by electrophoresis is limited, the genome size is another factor in getting positive signals. For example, probing rice with wheat DNA is promising, but the converse is disappointing. In this case, radiolabeled probes were replaced.

Collaborative mapping was conducted among rice, wheat, foxtail millet, maize, sorghum, and barley by exchanging probes; this comparison clarified the synteny, or colinearity of genes along the chromosomes. Mike Gale and Graham Moore of the John Innes Centre, UK, devised wonderful graphs to show syntenic relationships among crops (Moore et al. 1995; Devos and Gale 1997). We can learn a lot through comparative research of cereal crops to complement each other's work. In addition, synteny strengthens the importance of rice genome sequencing as basic knowledge because this crop has the smallest genome size among cereals.

It is becoming recognized that colinearity of genes exists between rice and *Arabidopsis*, a model plant for dicots. This information is obtained not by cross-hybridization but by a direct comparison of positions of similar genes along the chromosomes. The positional information comes from the rice linkage map and the *Arabidopsis* genome sequence. We can expect more precise comparison between both species as rice genome sequencing progresses.

7 Lesson of gene isolation by map-based cloning

One of the dreams for molecular geneticists and one of the goals of genome research is to clarify the gene controlling the phenotype. However, there are very few examples. In the case of the common pea, for example, only two traits—wrinkled seed (Bhattacharyya et al. 1990) and short internode (Martin et al. 1997; Lester et al. 1997)—of the seven that Mendel used to establish his rule of inheritance have been identified. The genes for five remaining traits are still unknown more than 100 years after Mendel's discovery.

At present, we have two main methods for identifying the genes behind phenotypes: map-based cloning and gene disruption. The former uses fine linkage and YAC physical maps, and the latter uses endogenous retrotransposons (Hirochika 1997). So far, only two rice genes, *Xa1* (Yoshimura et al. 1998) and *Xa21* (Song et al. 1995), both genes for resistance to bacterial blight disease, have been isolated by map-based cloning. In both cases, ordinal fine-scale mapping, physical mapping using BAC (*Xa21*) or YAC (*Xa1*), and complementary testing with the candidate gene have been used with much effort to identify the genes. Additional information about the structure of genes for plant disease resistance has helped. A fragment of the *Xa21* gene was accidentally used as an RFLP marker cosegregating with *Xa21*, and the partial nucleotide sequence of one of the cosegregating cDNAs with *Xa1* showed significant homology to the known motif in disease resistance genes.

Surprisingly, although both present resistance against the same pathogen, *Xanthomonas oryzae* pv. *oryzae*, the motifs characterizing the protein structure are completely different. The XA1 protein carries characteristics common to the best-known type of plant disease resistance gene—nucleotide-binding sites (NBSs) and leucine-rich repeats (LRRs). This fundamental structure is found in more than 10 resistance genes, such as *N6* from flax (Lawrence et al. 1995) and *RPS2* from *Arabidopsis* (Bent et al. 1994; Mindrinos et al. 1994). In XA1, the LRR is a novel type with an almost completely identical repeat of 93 amino acids. The XA21 protein has a unique structure carrying both a receptor kinase-like domain and an LRR motif. The serine-threonine receptor kinase domain was recognized in the first identified plant disease resistance gene, *Pto* in tomato (Martin et al. 1993), by map-based cloning. The typical LRR motif, with 24 amino acids, is observed in *Cf9* (Jones et al. 1994) and *Cf2* (Dixon et al. 1996) in tomato. XA21 is the only example carrying two motifs in one protein molecule. These findings suggest that XA1 and XA21, having different structures, present resistance to different races of *Xanthomonas oryzae* pv. *oryzae*. They might thus be a model for understanding the race-specificity of resistance.

The advantage of map-based cloning is its accuracy based on genetic analysis. However, this requires accurate genetic analysis to start with. Repeated backcrossing and genotyping of progenies using DNA markers can target traits controlled by more than two quantitative trait loci (QTLs) when these QTLs are isolated as single Mendelian factors (Tanksley 1993). In RGP, the heading date of rice, which is determined by at least two factors—photoperiod sensitivity and degree of vegetative growth—was analyzed as QTLs (Yano et al. 1997). So far, at least 10 QTLs of heading date have been detected in *japonica* (Nipponbare) and *indica* (Kasalath) crosses. Four have been identified as loci for photoperiod sensitivity. Each of these loci was confined within near-isogenic lines with a Nipponbare genome background. Fine mapping could be used to select a few YACs carrying each photoperiod sensitivity gene. Trials are in progress to narrow the candidate region by screening a cosmid library.

Recently, a new gene disruption method using a rice endogenous retrotransposon, Tos17, has been developed. It is being used to identify genes causing changes in phenotype (Hirochika et al. 1996). Two methods—tagging the phenotype by Tos17 and searching the expressed gene (cDNA) disrupted by Tos17—are used, and both can benefit from the many cDNAs partially sequenced in RGP. If a mutation can be caused by disrupting an unknown gene, the function of its product might be elucidated. This could confirm whether a single Tos17 disrupts a single gene.

8 How can we further understand the rice genome?

Comparing the situations before and after the big wave of genome analysis during the past decade, we can see that enormous amounts of information and molecular resources are now available. However, to understand the rice plant, we need much more information. Sequencing the entire genome is the most basic and ultimate goal.

We have two choices for cloning the DNA to sequence: the two-step cloning method and the one-step whole genome shotgun and cloning method. The latter is used for sequencing microorganisms (Fleischmann et al. 1995) and eukaryote genomes. To apply this method to a genome of more than 100 Mb requires improvements in sequencing and the development of intelligent computer software for editing raw sequence data. Therefore rice genome sequencing uses the first method, the conventional strategy.

For the cloning of large-sized DNA fragments (~100 kb), BAC (Shizuya et al. 1992) or PAC (Ioannou et al. 1994) is chosen as a vector. After a BAC or PAC contig is made by such methods as fingerprinting, overlapping sequences of BAC or PAC ends, or landing DNA markers on the BAC or PAC, the most reliably and economically assembled clones are selected, broken into pieces by sonication, and subcloned into a plasmid vector. Sequencing these subclones to cover the whole region several times is needed to complete the sequence.

Considering the importance of rice as a reference plant for cereal crops, sequencing of the rice genome needs to be completed as soon as possible, and the data should be widely disseminated. To achieve this goal, representatives from Japan, the United States, the European Union (EU), China, and Korea organized an international cooperative consortium. Other countries will join in the near future. Members have discussed how to sequence the rice genome rapidly and accurately. RGP's

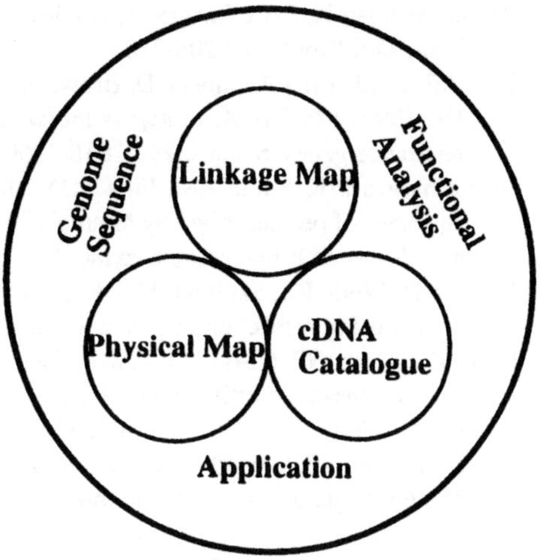

Fig. 2. Conceptual structure of rice genomics. Three main subjects studied in the past decade—linkage map, physical map, and cDNA catalogue—will be expanded to the next three subjects—genome sequence, functional analysis, and application—to complete the rice genome database.

japonica variety Nipponbare has been chosen as a common DNA resource. Details can be found at <http://www.staff.or.jp>.

All the data on rice plants have to be integrated for the information to be used effectively. Two other rice databases exist: RiceGenes (http://probe.nalusda.gov:8300/index.html) and RiceDB (http://bank.dna.affrc.go.jp/htbin/browse/RiceDB). Both are constructed mainly from linkage maps linked to sequences of DNA markers or YACs identified by the markers. In future, a user-friendly relational database will be needed.

The results of rice genome research will be important for adapting rice to grow in a wider range of conditions. Because of the expected shortage of staple foods worldwide in the 21st century, the practice of using DNA markers to accelerate breeding should be more enthusiastically embraced (Ribaut and Hoisington 1998). The value of the establishment of rice genomics is undoubtedly high for all cereal crops, which was the main reason why rice was chosen as a priority research target (Phillips and Freeling 1998). Figure 2 shows the conceptual structure of the integrated information and materials obtained through rice genome research. Full use of these subjects will help us to understand the rice plant better.

References

Agyare FD, Lashkari DA, Lagos A, Namath AF, Lagos G, Davis RW, Lemieux B (1997) Mapping expressed sequence tag sites on yeast artificial chromosome clones of *Arabidopsis thaliana* DNA. Genome Res 7:1-9

Akagi H, Yokozeki Y, Inagaki A, Fujimura T (1996) Microsatellite DNA markers for rice chromosomes. Theor Appl Genet 93:1071-1077

Antonio BA, Emoto M, Wu J, Umehara Y, Kurata N, Sasaki T (1996) Physical mapping of rice chromosomes 8 and 9 with YAC clones. DNA Res 3:393-400

Arumuganathan K, Earle ED (1991) Nuclear DNA content of some important plant species. Plant Mol Biol Rep 9:208-218

Bent AF, Kunkel BN, Dahlbeck D, Brown KL, Schmidt R, Giraudat J, Leung J, Staskawicz BJ (1994) *RPS2* of *Arabidopsis thaliana*: a leucine-rich repeat class of plant disease resistance genes. Science 265:1856-1860

Bhattacharyya MK, Smith AM, Ellis THN, Hedley C, Martin C (1990) The wrinkled-seed character of pea described by Mendel is caused by a transposon-like insertion in a gene encoding starch-branching enzyme. Cell 60:115-122

Botstein D, White RL, Skolnick M, Davis RW (1980) Construction of a genetic linkage map in man using restriction fragment length polymorphisms. Am J Hum Genet 32:314-331

Burke DT, Carle GF, Olson MV (1987) Cloning of large segments of exogenous DNA into yeast by means of artificial chromosome vectors. Science 236:806-813

Causse MA, Fulton TM, Cho YG, Ahn SN, Chuneongse J, Eu K, Xiao J, Yu Z, Ronald PC, Harrington SE, Second G, McCouch SR, Tanksley SD (1994) Saturated molecular map of the rice genome based on an interspecific backcross population. Genetics 138:1251-1274

Devos KM, Gale MD (1997) Comparative genetics in the grasses. Plant Mol Biol 35:3-15

Dixon MS, Jones DA, Keddie JS, Thomas CM, Harrison K, Jones JDG (1996) The tomato *Cf-2* disease resistance locus comprises two functional genes encoding leucine-rich repeat proteins. Cell 84:451-459

Fleischmann RD, Adams MD, White O, Clayton RA, Kirkness EF, Kerlavage AR, Bult CJ, Tomb JF, Dougherty BA, Merrick JM, McKenny K, Sutton G, FitzHugh W, Fields C, Gocayne JD, Scott J, Shirley R, Liu LI, Glodek A, Kelley JM, Weidman JF, Phillips CA, Spriggs T, Hedblom E, Cotton MD, Utterback TR, Hanna MC, Nguyen DT, Saudek DM, Brandon RC, Fine LD, Fritchman JL, Fuhrmann JL, Geoghagen NSM, Gnehm CL, McDonald LA, Small KV, Fraser CM, Smith HO, Venter JC (1995) Whole-genome random sequencing and assembly of *Haemophilus influenzae* Rd. Science 269:496-512

Harushima Y, Yano M, Shomura A, Sato M, Shimano T, Kuboki Y, Yamamoto T, Lin SY, Antonio BA, Parco A, Kajiya H, Huang N, Yamamoto K, Nagamura Y, Kurata N, Khush GS, Sasaki T (1998) A high-density rice genetic linkage map with 2275 markers using a single F_2 population. Genetics 148:1-16

Hearne CM, Ghosh S, Todd JA (1992) Microsatellites for linkage analysis of genetic traits. Trends Genet 8:288-294

Hirochika H, Sugimoto K, Otsuki Y, Tsugawa H, Kanda M (1996) Retrotransposons of rice involved in mutations induced by tissue culture. Proc Natl Acad Sci USA 93:7783-7788

Hirochika H (1997) Retrotransposons of rice: their regulation and use for genome analysis. Plant Mol Biol 35:231-240

Ioannou PA, Amemiya CT, Garnes J, Kroisel PM, Shizuya H, Chen C, Batzer MA, de Jong PJ (1994) A new bacteriophage P1-derived vector for the propagation of large human DNA fragments. Nat Genet. 6:84-89

Jones DA, Thomas CM, Hammond-Kosack KE, Balint-Kurti PJ, Jones JDG (1994) Isolation of the tomato *Cf-9* gene for resistance to *Cladosporoium fulvum* by transposon tagging. Science 266:789-793

Kinoshita T (1995) Report on gene symbolization. Rice Genet Newsl 12:9-153

Koike K, Yoshino K, Sue N, Umehara Y, Ashikawa I, Kurata N, Sasaki T (1997) Physical mapping of rice chromosomes 4 and 7 using YAC clones. DNA Res 4:27-33

Kurata N, Nagamura Y, Yamamoto K, Harushima Y, Sue N, Wu J, Antonio BA, Shomura A, Shimizu T, Lin SY, Inoue T, Fukuda A, Shimano T, Kuboki Y, Toyama T, Miyamoto Y, Kirihara T, Hayasaka K, Miyao A, Monna L, Zhong HS, Tamura Y, Wang ZX, Momma T, Umehara Y, Yano M, Sasaki T, Minobe Y (1994) A 300 kilobase interval genetic map of rice including 883 expressed sequences. Nat Genet 8:365-372

Lawrence GJ, Finnegan EJ, Ayliffe MA, Ellis JG (1995) The L6 gene for flax rust resistance is related to the *Arabidopsis* bacterial resistance gene *RPS2* and the tobacco viral resistance gene *N*. Plant Cell 7:1195-1206

Lester DR, Ross JJ, Davies PJ, Reid JB (1997) Mendel's stem length gene (Le) encodes a gibberellin 3 beta-hydroxylase. Plant Cell 9:1435-1443

Lyamichev V, Brow MAD, Dahlberg JE (1993) Structure-specific endonucleolytic cleavage of nucleic acids by eubacterial DNA polymerase. Science 260:778-783

Martin GB, Brommonschkel SH, Chunwongse J, Frary A, Ganal MW, Spivey R, Wu T, Earle ED, Tanksley SD (1993) Map-based cloning of a protein kinase gene conferring disease resistance in tomato. Science 262:1432-1436

Martin DN, Proebsting WM, Hedden P(1997) Mendel's dwarfing gene: cDNAs from the Le alleles and function of the expressed proteins. Proc Natl Acad Sci USA 94:8907-8911

McCouch SR, Kochert G, Yu ZH, Wang ZY, Khush GS, Coffman WR, Tanksley SD (1988) Molecular mapping of rice chromosomes. Theor Appl Genet 76:815-829

McCouch SR, Chen X, Panaud O, Temnykh S, Xu Y, Cho YG, Huang N, Ishii T, Blair M (1997) Microsatellite marker development, mapping and applications in rice genetics and breeding. Plant Mol Biol 35:89-99

Mindrinos M, Katagiri F, Yu GL, Ausubel FM (1994) The *A. thaliana* disease resistance gene *RPS2* encodes a protein containing a nucleotide-binding site and leucine-rich repeats. Cell 78:1089-1099

Moore G, Devos KM, Wang Z, Gale MD (1995) Cereal genome evolution, grasses, line up and form a circle. Curr Biol 5:737-739

Phillips RL, Freeling M (1998) Plant genomics and our food supply: an introduction. Proc Natl Acad Sci USA 95:1969-1970

Ribaut JM, Hoisington D (1998) Marker-assisted selection: new tools and strategies. Trends Plant Sci 3:236-239

Saito A, Yano M, Kishimoto N, Nakagahra M, Yoshimura A, Saito K, Kuhara S, Ukai Y, Kawase M, Nagamine T, Yoshimura S, Ideta O, Ohsawa R, Hayano Y, Iwata N, Sugiura M (1991) Linkage map of restriction fragment length polymorphism loci in rice. Jpn Breed 41:665-670

Saji S, Umehara Y, Kurata N, Ashikawa I, Sasaki T (1996) Construction of YAC contigs on rice chromosome 5. DNA Res 3:297-302

Sasaki T (1996) Rice cDNAs as a model for expressed genes in plants. In: Heslop-Harrison JS (ed) Symposia of the Society for Experimental Biology, Number L: Unifying Plant Genomes. The Company of Biologists Limited, Cambridge, UK, pp 11-15

Sasaki T, Song J, Koga-Ban Y, Matsui E, Fang F, Higo H, Nagasaki H, Hori M, Miya M, Murayama-Kayano E, Takiguchi T, Takasuga A, Niki T, Ishimaru K, Ikeda H, Yamaamoto Y, Mukai Y, Ohta, I, Miyadera N, Havukkala I, Minobe Y (1994) Toward cataloguing all rice genes: large-scale sequencing of randomly chosen rice cDNAs from a callus cDNA library. Plant J 6:615-624

Sasaki T, Yano M, Kurata N, Yamamoto K (1996) The Japanese rice genome research program. Genome Res 6:661-666

Shimokawa T, Kurata N, Wu J, Umehara Y, Ashikawa I, Sasaki T (1996) Assignment of YAC clones spanning rice chromosomes 10 and 12. DNA Res 3:401-406

Shizuya H, Birren B, Kim UJ, Mancino V, Slepak T, Tachiiri Y, Simon M (1992) Cloning and stable maintenance of 300-kilobase-pair fragments of human DNA in *Escherichia coli* using an F-factor-based vector. Proc Natl Acad Sci USA 89:8794-8797

Singh K, Ishii T, Parco A, Huang N, Brar DS, Khush GS (1996) Centromere mapping and orientation of the molecular linkage map of rice (*Oryza sativa* L.). Proc Natl Acad Sci USA 93:6163-6168

Song WY, Wang GL, Chen L, Kim HS, Pi LY, Gardner J., Wang B, Holsten T, Zhai WX, Zhu LH, Fauquet C, Ronald P (1995) A receptor kinase-like protein encoded by the rice disease resistance gene *Xa21*. Science 270:1804-1806

Tanksley SD (1993) Mapping polygenes. Annu Rev Genet 27:205-233

Tanksley SD, Ganal MW, Martin GB (1995) Chromosomal landing: a paradigm for map-based cloning in plants with large genomes. Trends Genet 11:63-68

Tanoue H, Shimokawa T, Wu J, Sue N, Umehara Y, Ashikawa I, Kurata N, Sasaki T (1997) Ordered YAC clone contigs assigned to rice chromosomes 3 and 11. DNA Res 4:133-140

Umehara Y, Inagaki A, Tanoue H, Yasukochi Y, Nagamura Y, Saji S, Otsuki Y, Fujimura T, Kurata N, Minobe Y (1995) Construction and characterization of a rice YAC library for physical mapping. Mol Breed 1:79-89

Umehara Y, Tanoue H, Kurata N, Ashikawa I, Minobe Y, Sasaki T (1996) An ordered yeast artificial chromosome library covering over half of rice chromosome 6. Genome Res 6:935-942

Umehara Y, Kurata N, Ashikawa I, Sasaki T (1997) Yeast artificial chromosome clones of rice chromosome 2 ordered using DNA markers. DNA Res 4:133-140

Vos P, Hogers R, Bleeker M, Reijans M, van de Lee T, Hornes M, Frijters A, Pot J, Peleman J, Kuiper M, Zabeau M (1995) AFLP: a new technique for DNA fingerprinting. Nucleic Acids Res 23:4407-4414

Wang ZX, Idonuma A, Umehara Y, Van Houten W, Ashikawa I, Minobe Y, Sasaki T (1996) Physical mapping of rice chromosome 1 with yeast artificial chromosomes (YACs). DNA Res 3:291-296

Watson JD, Crick FHC (1953) A structure for deoxyribose nucleic acid. Nature 171:737-738

Williams JGK, Kubelik AR, Livak KJ, Rafalski JA, Tingey SV (1990) DNA polymorphisms amplified by arbitrary primers are useful as genetic markers. Nucleic Acids Res 18:6531-6535

Wu J, Simokawa T, Maehara T, Koike K, Umehara Y, Matsumoto T (1997) Mapping rice ESTs onto YAC clones—toward the establishment of a comprehensive rice expression map. Rice Genome 6:2-3

Yamamoto K, Sasaki T (1997) Large-scale EST sequencing in rice. Plant Mol Biol 35:135-144

Yano M, Harushima Y, Nagamura Y, Kurata N, Minobe Y, Sasaki T (1997) Identification of quantitative trait loci controlling heading date in rice using a high-density linkage map. Theor Appl Genet 95:1025-1032

Yoshimura A, Ideta O, Iwata N (1997) Linkage map of phenotype and RFLP markers in rice. Plant Mol Biol 35:49-60

Yoshimura S, Yamanouchi U, Katayose Y, Toki S, Wang ZX, Kono I, Kurata N, Yano M, Iwata N, Sasaki T (1998) Expression of *Xa1*, a bacterial blight-resistance gene in rice, is induced by a bacterial inoculation. Proc Natl Acad Sci USA 95:1663-1668

2
Genome analysis and breeding

David J. Mackill

USDA-ARS, Department of Agronomy & Range Science, University of California, Davis, CA, USA

1 Introduction

To appreciate the recent advances made in rice genome analysis, one must only reflect on the state of the genetic map 15 years ago. The linkage map of 1985, developed over a half century of work, contained only 119 genes (Kinoshita 1986). A scientist desiring to locate a new gene on the map was faced with the unattractive prospect of crossing its source to a collection of genetic tester stocks representing all the linkage groups. The alternative of using trisomics to determine the chromosomal location of the gene did not offer much improvement. Furthermore, few breeders could see any relevance of the genetic map to applied breeding programs. Most of the mapped genes were morphological mutants of little direct agronomic value. Also, the map did not offer any substitution for phenotypic selection in breeding programs.

In 1980, the novel concept that differences in restriction sites or insertions/deletions between adjacent sites could be used as markers to construct genetic maps was first proposed (Botstein et al. 1980). These restriction fragment length polymorphism (RFLP) markers were soon applied to genetic mapping in plants, and the first RFLP map of rice was constructed in the lab of Steven Tanksley at Cornell University (McCouch et al. 1988). RFLPs had many advantages over previous markers such as isozymes. Most importantly, they were abundantly available for all the 12 rice chromosomes, and, with only one cross, genes of economic importance could be mapped by linkage to RFLP markers.

Before the advent of genome analysis, rice was regarded as an undesirable species for genetic studies due to the difficulty of visualizing its minute chromosomes compared with crops like wheat or maize. (A single wheat chromosome contains

Key words. genetics, molecular markers, gene mapping, QTL, RFLP, RAPD, AFLP, microsatellite, *Oryza sativa*

more DNA than the entire 12 chromosomes of the rice genome.) What was previously considered a weakness has now propelled rice to the forefront of plant molecular genetics. After the minuscule weed *Arabidopsis*, rice is the primary model species for plant molecular biology. The synteny observed among the chromosomes of the cereals has stimulated widespread interest in rice as a model grass species for genome studies (Moore et al. 1995). Its importance as a food for nearly half the world's population is an additional incentive for using rice as a model. The suitability of rice for molecular marker studies has led to tremendous growth in information on genetic characterization and gene mapping of important economic traits. Rice also represents the first example in monocot species of successful cloning of a gene (the bacterial blight resistance gene *Xa21*) through map position (Song et al. 1995).

This chapter aims to review the progress in application of genome studies to rice breeding. The various molecular marker systems available in rice are introduced. The use of markers in characterization of rice germplasm and mapping of loci controlling economic traits is described, and the progress to date in mapping important genes is reviewed. The potential application of genome maps for cloning genes is briefly discussed, and future prospects for marker-assisted selection and isolation of agronomically important genes is summarized in light of current genome initiatives.

2 Molecular marker systems for rice

Initial genome maps in rice were constructed with RFLP markers, and these maps continue to be used extensively for gene tagging. The principal RFLP maps were developed at Cornell University (Causse et al. 1994) and at the Rice Genome Program in Japan (Kurata et al. 1994). Fortunately, both groups have made their probes available to plant geneticists, and this has greatly facilitated the mapping of economically important traits.

RFLPs are generally very reliable markers for gene mapping. They are codominant, which improves the power of mapping quantitative traits. Since their map position is usually well established, loci mapped with RFLPs on different populations can be easily compared. A major disadvantage with RFLPs is that their detection involves the laborious process of southern blotting, and requires at least 50 μg of high-quality DNA per individual for mapping. Another disadvantage is that the researcher must acquire and maintain a sufficient number of probes from one of the sources mentioned above.

Markers based on the polymerase chain reaction (PCR) have been developed to overcome some of the disadvantages of RFLPs. The use of these markers does not require the researcher to maintain a large number of probes that have been mapped by one of the large mapping projects. Instead, primers can be ordered from one of several suppliers, and there is no need for southern blotting and labeling probes. Very little DNA is required for PCR-based markers. This opens up the potential to assay markers with DNA extracted from small amounts of tissue.

Random amplified polymorphic DNA (RAPD) markers (Williams et al. 1990) were the first PCR-based markers to be widely applied in gene mapping and germplasm characterization. Shortly after the first report of this technology, RAPDs were detected in rice accessions (Zheng et al. 1991, Fukuoka et al. 1992) and applied to gene mapping (Ronald et al. 1992). Only one RAPD map has been published (Redoña and Mackill 1996b). For RAPDs, genomic DNA is subjected to PCR with a random 10-bp primer, followed by electrophoresis, usually on an agarose gel, and staining with ethidium bromide. Markers are usually scored as presence or absence of a specific fragment. The simplicity of detection and abundance of RAPDs have made them attractive markers for application to rice genetic studies, but they also have several disadvantages. They can be somewhat unreliable, especially when the same primers are compared in different laboratories (Skroch and Nienhuis 1995). Their dominant nature gives them less power in mapping quantitative traits.

Another PCR-based marker, amplified fragment length polymorphism (AFLP), was recently developed by the company Keygene (Vos et al. 1995). AFLPs are detected by cutting genomic DNA with two restriction enzymes, a rare cutter and a frequent cutter, and ligating different adapters to the restriction sites. These adapters serve as templates for amplifying a subset of the fragments through PCR. The primer pairs used for amplifying AFLPs have either a 2- or 3-bp overhang whose sequence determines the specific spectrum of amplified fragments. Most work on higher plants involves two primers with 3-bp overhangs, but small genome crops can employ +2/+3 combinations, which can produce 100 or more fragments per reaction. The fragments are detected through radioactive or nonradioactive labeling followed by electrophoresis on a polyacrylamide sequencing gel. Although the process of detecting AFLPs is somewhat laborious, they are more reliable than RAPDs, and a large number of fragments can be detected for each primer combination. They are particularly suitable for fine-scale mapping and saturating a particular region to obtain markers tightly linked to a gene of interest (Meksem et al. 1995). AFLPs are abundant in rice (Cho et al. 1996, Mackill et al. 1996), and Maheswaran et al. (1997) produced an AFLP map containing 208 markers using only 20 +3/+3 primer pairs. The level of polymorphism of AFLPs may not be as high as RAPDs or RFLPs on a per fragment basis (Mackill et al. 1996, Maheswaran et al. 1997), but the abundance of fragments detected can overcome this limitation. Like RAPDs, AFLPs are usually scored as codominant markers; however, software is being developed that, combined with image analysis or automated sequencing equipment, could allow codominant scoring. AFLPs require relatively high quality DNA to ensure that it is completely digested by the two restriction enzymes.

Microsatellite markers are based on short (2-4 bp) DNA sequences that are tandemly repeated from 10 to more than 30 times. These markers, sometimes referred to as simple sequence repeats (SSR) or simple sequence length polymorphism (SSLP), can be amplified by PCR with primers flanking the repeats. Polymorphisms are detected as length variations of fragments, usually on polyacrylamide sequencing gels. Initial studies in rice indicated that these markers were highly polymorphic (Zhao and Kochert 1992, Wu and Tanksley 1993, Zhao and Kochert 1993b, Yang et al. 1994). While RFLPs usually only exhibit 2 or 3 alleles per locus, as many as 25 alleles have been observed at a single microsatellite locus among a diverse set of cultivars (Yang et al. 1994). Unfortunately, microsatellite markers have not been abundantly available due to the expense of identifying them, sequencing the nonrepeated flanking regions, and mapping the loci. Recently, several groups have reported sequences and map positions for microsatellite markers in rice (Akagi et al. 1996, Miyao et al. 1996, Panaud et al. 1996), and a map of 120 markers has been published (Chen et al. 1997, McCouch et al. 1997). As additional microsatellite markers are discovered, they are likely to become the marker of choice for gene mapping and genetic characterization of rice cultivars.

3 Germplasm characterization

3.1 Characterization of *Oryza* species

The *Oryza* genus consists of over 20 species organized into four complexes (Vaughan 1994). The sativa complex, consisting of diploids with the AA genome, contains the two cultivated species: Asian rice *O. sativa*, which originated from the wild species *O. rufipogon*, and African rice *O. glaberrima*, which was domesticated from *O. barthii*. *Oryza* species had been previously classified using morphology, hybridization, and isozymes. Wang et al. (1992) used 15 RFLP probes with one restriction enzyme to study 71 accessions representing 19 species of *Oryza*. Cluster analysis performed on genetic distance estimates revealed a similar classification to the conventional one based on previous methods, with the four species complexes forming the major clusters. This study demonstrated the high polymorphism and large number of alleles present among wild accessions, and also illustrated the potential for using DNA markers to follow the introgression of traits from wild into cultivated rice.

Tandem or interspersed repeated sequences often exhibit pronounced conservation within species; however, they can be highly specific to individual species or genomes. Repeats have been identified that are specific for various rice genomes (Zhao et al. 1989, Wu and Wu 1992, Reddy et al. 1993, Zhao and Kochert 1993a, Kiefermeyer et al. 1995, Nakajima et al. 1996). Other types of repeated elements, such as the intergenic spacer from rDNA (Cordesse et al. 1992), p-SINE1 (short interspersed element) repeats (Mochizuki et al. 1993, Motohashi et al. 1997), and retrotransposons (Noma et al. 1997), have been used to study rice genomes. The isolation of repeated sequences specific to non-AA genomes or non-*sativa* AA genomes has been suggested as a means to monitor the introgression of alien sequences in interspecific crosses (Aswidinnoor et al. 1991).

3.2 Characterization of cultivated rice accessions

The ability to classify rice cultivars into genetically similar groups has important applications in breeding programs. Classification of the germplasm into subspecies or intrasubspecific groups assists the breeder in selecting parents for specific crosses. Assignment of cultivars to heterotic groups would greatly enhance the ability to identify parents for hybrid rice breeding programs (see next section). The ability to reliably distinguish through DNA fingerprinting methods closely related cultivars that are grown in the same region and derive from similar parents has additional implications. Plant variety protection (PVP) issues are becoming more important as companies or research institutions attempt to protect their investments in varietal research. Ensuring cultivar purity for quality control purposes is another motivation for being able to fingerprint cultivars. Eliminating potential duplicate accessions from germplasm collections or from advanced stages of evaluation in breeding programs would reduce the costs of germplasm enhancement programs. The ability to isolate DNA from seed (McDonald et al. 1994, Donini et al. 1997) makes the application of molecular markers to varietal identification more feasible.

A major difficulty with DNA fingerprinting is that most cultivars are actually collections of closely related genotypes. Variability within accessions is expected to be the rule rather than an exception. There have been few studies where individuals within cultivars have been tested for genetic variability. Early work with RFLPs indicated approximately 25% of cultivars or wild accessions tested were variable among three to five individuals (Wang and Tanksley 1989, Wang et al. 1992). A reliable marker should be uniform within but variable between cultivars.

The important study of Glaszmann (1987) showed that isozymes could be used to classify rice cultivars into genetically similar groups. Glaszmann assigned 1688 rice cultivars into six isozyme groups, of which groups 1 and 6 represented the two main subspecies *indica* and *japonica*, respectively. Similarly, Wang and Tanksley (1989) classified 70 rice cultivars using 10 RFLP probes with 5 restriction enzymes. They found general agreement with the results of Glaszmann (1987). Isozyme groups 1 to 5 were classified into a large "indica" cluster, although the individual isozyme groups formed smaller clusters within this larger one. The main divergence was between this large indica cluster and the japonica (group 6) cultivars. Similar RFLP studies by other authors confirmed their effectiveness in differentiating the *indica* and *japonica* subspecies (Nakano et al. 1992, Zhang et al. 1992, Ishii et al. 1993, Ishii et al. 1995).

In addition to the labor and equipment involved, RFLPs have the disadvantage of limited variation within closely related germplasm. Most rice gene mapping experiments have been performed in intersubspecific (i.e., indica × japonica) populations. Unfortunately, breeders rarely use these crosses because they are characterized by spikelet sterility and poor agronomic characteristics. RFLP maps have been used to locate genes controlling quantitative traits in crosses between two indica cultivars (Lin et al. 1996). This has been more difficult for crosses between two japonica cultivars, where RFLP variation is lower than in indica rice. Kajiya et al. (1996) used three rounds of subtractive hybridization to obtain clones that showed polymorphism between two japonica cultivars. However, for most purposes, the frequency of RFLPs between two japonicas is too low for gene mapping.

Due to their ease of use, RAPD markers have proven very popular in studies of rice germplasm. Fukuoka et al. (1992) classified 16 rice accessions into indica and japonica clusters based on 36 10-bp primers. Yu and Nguyen (1994) could clearly differentiate upland cultivars of the tropical japonica group from indica upland and lowland cultivars. Virk et al. (1995) similarly showed that indicas and japonicas were clearly differentiated with RAPDs. Mackill (1995) employed 10 RAPD primers to classify a group of 134 mostly japonica cultivars. Again, indica and japonica cultivars formed the two main clusters. The japonica cluster consisted of two subgroups, the temperate types (the typical short- or medium-grain cultivars) and tropical types (javanica, upland, or USA long-grain cultivars). The abundance of RAPD polymorphisms detected between temperate and tropical japonicas was used to construct a RAPD map of a cross between the two types (Redoña and Mackill 1996b). An interesting feature of this map was the abundance of markers clustered on chromosomes 10 and 11, which have comparatively few markers in RFLP maps. The differences for chromosome 10 were particularly striking; 18% of the RAPDs mapped to chromosome 10 in their japonica map compared to 1% to 5% in three RFLP maps between indica and japonica crosses and one interspecific cross (Redoña and Mackill 1996b). This suggests that genes on these chromosomes may be important in the temperate-tropical divergence within the *japonica* subspecies. RAPD primers have also been used to classify rice species (Ishii et al. 1996, Martin et al. 1997) and weedy strains of *O. sativa* (Suh et al. 1997).

On a per marker basis, AFLPs appear to have lower polymorphism than RAPDs or RFLPs, but it is still possible to obtain a large number of polymorphic fragments with relatively few primer combinations (Mackill et al. 1996). AFLP markers revealed very similar classification schemes to RAPDs based on a sample of 14 cultivars (Mackill et al. 1996).

Due to their high allelic diversity, microsatellite markers should provide a very powerful tool for DNA fingerprinting of rice cultivars (McCouch et al. 1997). Wu and Tanksley (1993) observed 5 to 11 alleles for 10 microsatellite markers applied to 20 rice accessions. The increase in polymorphism over that for RFLPs was especially pronounced in the within-japonica comparisons. Yang et al. (1994) assessed variability of 10 microsatellites, 8 of which were used in the experiment of Wu and Tanksley (1993), among 238 accessions of *O. sativa* landraces and cultivars. The number of alleles per locus ranged from 3 to 25. Interestingly, the high level of diversity within cultivars was associated with reduced differentiation between the two subspecies. In contrast, Mackill et al. (1996) observed distinct differences between alleles present in two indica accessions compared to 12 japonica accessions. Within the japonica group, five California medium-grain temperate japonica cultivars were nearly identical at 14 microsatellite loci, whereas two temperate japonicas from Korea and Italy and four tropical japonicas possessed different alleles at many loci. Akagi et al. (1997) observed that a sample of 59 genetically similar japonica cultivars showed multiple alleles for 17 of 20 microsatellite markers.

3.3 Classification of heterotic groups

Due to the limited potential for expansion of the cultivated rice area, future yield increases are imperative to keep up with the expanding population. At the present time, exploitation of heterosis appears to be the most promising approach for significantly increasing rice yield potential. In China, hybrid rice is grown on approximately 18 million ha, or over half the total area, with an estimated yield advantage of 15%-20% over conventional cultivars (IRRI 1997). Hybrid rice production is rapidly spreading outside China, and research to develop commercial hybrids has accelerated worldwide. Identification of appropriate parental combinations for achieving high F_1 hybrid yield or heterosis is one of the most laborious aspects of varietal development.

The suitability of RFLP markers for germplasm classification generated considerable interest in applying them to predicting heterosis in hybrid crops such as maize. Evidence of their usefulness in this respect has been ambiguous. In some germplasm samples, there appear to be correlations between heterozygosity for molecular markers and hybrid yields (Lanza et al. 1997). However, in crosses between heterotic groups, which are of most interest to breeders, marker heterozygosity is often not predictive (Boppenmaier et al. 1993, Dhillon et al. 1993). Direct identification of quantitative trait loci (QTL) involved in yield heterosis may be a more promising approach (Stuber et al. 1992, Schön et al. 1994, Ajmone-Marsan et al. 1995). Zhang et al. (1994, 1995b) studied heterosis in a diallele cross of eight indica hybrid rice parents. They found positive correlations for heterozygosity of specific markers and midparent heterosis for yield and other agronomic traits. They indicated that correlations between molecular markers and heterosis were high for specific markers that were linked to QTL involved in the traits. A subsequent study indicated that correlations between marker heterozygosity and heterosis for yield and yield components were low within both an indica and a japonica set of intercrosses (Zhang et al. 1996). Xiao et al. (1996a) found positive correlations between yield heterosis and genetic distance based on RAPD and microsatellite markers in a sample of 15 indica × indica and 6 japonica × japonica crosses.

Most of the studies mentioned above were based on a relatively few parental varieties. It is likely that more information on the relationship between overall heterozygosity and heterosis will become available in the future, and this might produce a strategy for employing markers in classifying germplasm into heterotic groups and identification of the most productive hybrid combinations. Markers associated with chromosomal segments involved directly in yield heterosis will probably be the most useful for screening germplasm.

4 Mapping loci for economic traits

4.1 Procedures for gene mapping

The prospect of using molecular markers to locate genes of economic importance, especially those underlying quantitative traits, has attracted enormous interest since the first molecular maps were published in the late 1980s. Identification of convenient tags that can be used to screen for desirable alleles is an obvious enticement for undertaking gene mapping studies. Placing genes on a molecular map, however, has far more advantages than the employment of the markers in selection among breeding populations. The ability to rapidly map genes of economic importance has provided an effective method for identifying the uniqueness of new genes. For example, rapid mapping of blast resistance genes identified at IRRI (Yu et al. 1991) facilitated allelism tests with genes previously described in Japan (Inukai et al. 1994, McCouch et al. 1994). In addition to clarifying the relationship among genes from different sources, map position of promising alleles is the first requirement for isolating the genes through map-based cloning.

Rice mapping populations usually consist of plants derived from a single cross between two parents differing for the traits of interest. F_2 or F_2-derived F_3 populations are frequently used. The disadvantage of using F_2 populations is that each plant is a unique genotype, precluding replication except through clonal propagation. F_3 populations can be replicated to some extent, but will be segregating within families. Recombinant inbred lines (RIL) are developed by single seed descent of random plants from an F_2 population. By the F_6 generation, most genes are homozygous. RILs are thus immortal populations, and can be propagated indefinitely through seed (so long as outcrossing is avoided). Doubled haploids (DH) are similar to RILs, except that homozygotes are developed directly from the F_1 plant through anther culture, and the repeated crossovers that can occur with each generation of selfing are avoided. Genetic distance should thus be expanded in DH populations, and more closely spaced markers may be needed. On the other hand, the repeated rounds of meiosis that accompany the generation of RILs may lead to high levels of distorted segregation in indica × japonica crosses (Wang et al. 1994). Because of the ability to generate a large seed supply, RILs have been used to map many QTL in rice (Wang et al. 1994, Tsunematsu et al. 1996, Xiao et al. 1996b, Nandi et al. 1997). DH populations have also been used (Lorieux et al. 1996, Yadav et al. 1997), but not as extensively due to the difficulty in generating a sufficient number of DH plants through anther culture. Backcrosses and other types of crosses have not been widely used in rice gene mapping.

Mapping major genes involves identifying markers linked to the trait of interest. Near-isogenic lines (NIL), if available, provide a rapid means of detecting putative associations between markers and interesting genes (Yu et al. 1991). Once associations are detected, linkage must be confirmed in segregating populations. Bulk segregant analysis (BSA), a powerful method to rapidly identify marker-trait associations, was described by Michelmore et al. (1991). In this approach, DNA is extracted from approximately 10 F_2 plants of each of the two extreme phenotypes and bulked to form two DNA samples. The two samples are screened with PCR-based markers such as RAPDs and AFLPs. Once positive markers are identified, they can be confirmed on individual plants from the segregating population to determine linkage with the trait, and then mapped on a separate population. In practice, it is often more efficient to use two DNA pools for each phenotype (four total samples) to minimize false positives. With high-volume markers such as AFLPs, the BSA approach is an effective means to rapidly assay thousands of markers in a matter of weeks.

Identifying markers linked to QTL is more complicated than for major gene traits. A considerable amount of work is required to identify and map markers that cover the entire 12 chromosomes, and approximately 100 or more well-spaced markers are needed. Codominant markers such as RFLPs or microsatellites are preferred, because they improve the power of detecting QTL. The simplest approach to identifying marker-QTL associations involves a single marker analysis, where ANOVA or regression is used to determine each marker's effect on a trait. The use of interval analysis was advocated as a more powerful method of detecting QTL (Lander and Botstein 1989). The computer program Mapmaker-QTL (Paterson et al. 1988) has been used extensively for interval mapping of QTL. Conventional interval mapping suffers from the inability to take into account the effects of multiple QTL on a trait. Composite interval mapping approaches that account for the effects of multiple QTL (Jansen and Stam 1994, Zeng 1994) have become more popular, and computer programs have been developed to perform these analyses in various mapping populations.

The problem of determining thresholds for QTL detection has also received considerable attention. The objective for a statistical test of QTL significance is to maximize the rejection of false QTL (reduced type I error) while minimizing the rejection of true QTL (reduced type II error). The level of significance is usually measured as the log-odds ratio (LOD). LOD values of 2.0 to 2.5, which have been commonly used for detecting QTL in rice, have been thought by some to present a high risk of false positives. A permutation test was suggested as a means to minimize the chance that a single false QTL for a particular trait will be declared to below 0.05 (Churchill and Doerge 1994). This will typically increase the threshold LOD values above 3 or 4. Unfortunately, this also increases the chance that true QTL will be rejected. In many cases, this may not be a serious problem, because the QTL with the largest effects are the primary interest of plant breeders. QTL of borderline significance or small effects are unlikely targets for use in breeding programs. However, there are examples where such borderline QTL have been shown to have significant effects when they are in a more uniform genetic background in which QTLs of larger effect are not segregating (Paterson 1998).

Ahn et al. (1993) used QTL analysis for cooked kernel elongation; however, they were dealing with a trait controlled by only one locus. The first rice QTL study was reported by Wang et al. (1994), who used an indica × japonica RIL population to identify loci conferring partial resistance to blast. They identified ten chromosomal segments from the upland japonica Moroberekan conferring reduced lesion number. Yano and Sasaki (1997) summarized the results of 13 QTL reports that involved over 20 plant traits. The number of QTL mapping papers in rice has more than doubled since that publication. As more mapping studies are conducted in different populations, it becomes possible to identify QTL that are expressed in several populations. Preliminary results indicate that different QTL are often identified in different populations. A single QTL experiment may not be a very reliable gauge of what QTL are the most important for a particular trait.

4.2 Candidates for marker-assisted selection

Several studies on the theoretical aspects of marker-assisted selection (MAS) have been reported, most focusing on quantitative traits. These show that MAS is generally more efficient than selection based on phenotype alone (Lande and Thompson 1990, Gimelfarb and Lande 1994, Whittaker et al. 1995). Detection of QTL for traits controlled by many genes with low heritability requires population sizes larger than most experiments employ (Lande 1992); therefore, MAS is likely to be most effective when a few QTL control a significant portion of the variability. This is commonly observed for many quantitative traits. As with QTL, MAS for major genes is an effective means of recovering the recurrent parent in backcrossing schemes. Flanking markers can be employed to select for recombination on either side of the target gene in alternate backcrossing cycles when there is close linkage to undesirable genes in the exotic donor (Tanksley et al. 1989).

Marker-assisted backcrossing is a promising approach for introducing QTL from exotic cultivars or wild species into elite breeding lines. Simulation studies show that markers can provide a gain of about two generations over conventional backcrossing (Hospital et al. 1992, Visscher et al. 1996, Hospital and Charcosset 1997), which can be a very significant advantage in terms of rapid development and release of new cultivars. In all cases of MAS for QTL, close linkage of the marker with the QTL is important to maintain the marker-trait association throughout the various rounds of selection (Edwards and Page 1994). Flanking markers are highly desirable to help prevent recombination between the marker and the QTL (Visscher et al. 1996).

In conventional QTL mapping studies that utilize exotic cultivars or wild species as parents, identification of favorable QTL is complicated by a generally inferior phenotype of the donor, which is associated with a large number of unfavorable loci compared to favorable loci in unadapted germplasm. The advanced backcross QTL (ABQTL) method has been proposed as a means to identify and simultaneously transfer QTL from wild species donors into highly adapted lines (Tanksley and Nelson 1996). The advantages of this approach result from the fact that wild species, which are inferior to the elite cultivars in economic traits, often possess alleles that positively influence these traits. These alleles are exposed by reducing the percent of exotic alleles in the population through repeated backcrossing. This is followed by molecular marker analysis to identify the exotic alleles that positively impact the traits of interest. Initial results in rice indicate that the ABQTL method has potential for exploitation of wild rice relatives in the improvement of agronomic traits such as yield (Xiao et al. 1996c). ABQTL populations have been developed at several locations, and the effectiveness of this method should become clearer in the near future.

Considering the expense of molecular marker technology and the facilities required, one must carefully consider the advantages of MAS compared to conventional phenotypic selection for each trait of interest. In conventional breeding programs, large population sizes (up to 10,000 F_2 plants per cross) are typically used to increase the chance that desirable recombinants can be recovered. In most cases, employment of markers in breeding programs will not be suitable for traits that have a high heritability and can be screened easily in the seeding stage. For example, most disease resistance genes can be detected in seedlings, and thousands of plants can be rapidly screened in disease nurseries. The breeder's ability to select plants combining resistance with other desirable features easily surpasses the efficiency possible through MAS with molecular markers. An exception is when two or more genes that express a similar phenotype need to be transferred into a cultivar. In the case of rice blast, combining genes that confer resistance to different lineages of the pathogen into a single cultivar is thought to be a means of increasing the durability of resistance. Since the resistance genes often express a similar phenotype with one or a mixture of pathogen races, it is very difficult to determine which lines contain both genes without screening with differential races of the blast pathogen. Markers can thus be an effective means of combining disease resistance genes in a breeding program (Huang et al. 1997).

MAS is likely to be more advantageous compared to phenotypic selection for traits that are difficult to measure accurately or traits that are expressed at a very late growth stage. For example, resistance to some diseases such as stem rot is laborious to measure and requires repeated seasons of data for accurate evaluation. It is thought that this trait is mostly controlled by a single locus (Mackill et al., in preparation). Resistance to stresses at the reproductive stage such as salinity or low temperature are difficult to measure, and cannot be identified before flowering. MAS is probably most attractive for traits that are expressed in progeny of crosses, as these take effectively two generations to measure in addition to a laborious process of making multiple test crosses. Examples include restorer genes for cytoplasmic male sterility (cms) (Ichikawa et al. 1997) and the wide compatibility allele S_5^n (Williams et al. 1997). In the case of restorer genes, plants must be crossed with the cms source, and the fertility of the resulting progeny indicates the presence of the restorer allele. In the case of the wide compatibility allele, the plants must be crossed with either an indica or japonica tester, and the fertility of the progeny measured. In both these cases, MAS would be an exceedingly attractive alternative to phenotypic selection.

With advances in technology we can expect that the cost of DNA marker analysis will be reduced. PCR-based markers such as AFLPs and microsatellites lend themselves well to automation. In addition, DNA extraction methods have been improved, and it is relatively easy to isolate sufficient quantities for PCR-based markers such as microsatellites. Rapid extraction methods that work with rice have been reported (Williams and Ronald 1994, Zheng et al. 1995), and it is even possible to isolate DNA from half-seeds so as to assay marker types before planting (Chunwongse et al. 1993, Wang et al. 1993). In the future, we can also expect new types of markers that do not require electrophoresis (Gu et al. 1995).

4.3 Progress in gene mapping

4.3.1 Resistance to disease and insect pests

Disease resistance genes have figured prominently in the application of molecular markers to gene mapping, and the first report of RFLP mapping in rice involved resistance to blast (Yu et al. 1991). A number of major genes have been tagged with markers, and this has made it possible to determine potential allelic relationships among newly identified genes. While location in a region of the map that is devoid of existing loci indicates that it is a new gene, location near previously described loci is not necessarily proof of allelism. Crosses must still be performed to determine if the genes are allelic or closely linked. For example, Inukai et al. (1994) determined that *Pi-1* is linked with *Pi-k* on chromosome 11 at about 6.1 cM. On the other hand, *Pi-2* appears to be allelic to Pi-z on chromosome 6. Blast resistance genes seem to be clustered on chromosomes 4, 6, 11, and 12 (McCouch et al. 1994). As with major genes, partial resistance to blast was the first subject of QTL mapping in rice (Wang et al. 1994). Partial or quantitative resistance is thought to be an important component of durable resistance, being less likely to be overcome by new races. Bacterial blight is another widespread and destructive rice disease, and a number of major genes have been mapped (Ronald et al. 1992, Yoshimura et al. 1992). As with blast, chromosome 11 appears to be an important source of resistance genes. PCR-based markers that are closely linked to several blast (Hittalmani et al. 1995, Inukai et al. 1996, Naqvi and Chattoo 1996) and bacterial blight (Yoshimura et al. 1995, Huang et al. 1997) resistance genes have been developed for application in MAS, making it possible now to pyramid specific genes into a desired genetic background. A gene for resistance to rice tungro spherical virus (RTSV) has recently been mapped on chromosome 4 (Sebastian et al. 1996), and it is hoped this will facilitate the transfer of this trait into superior cultivars.

For insect resistance, gene mapping has focused on gall midge and brown plant hopper (BPH). Two major gall midge resistance genes, *Gm2* and *Gm4*, have been mapped (Mohan et al. 1994, Mohan et al. 1997), and PCR-based markers have been developed for MAS (Nair et al. 1995, Nair et al. 1996). Likewise, two BPH resistance genes have been mapped with molecular markers (Ishii et al. 1994, Hirabayashi and Ogawa 1995).

4.3.2 Resistance to abiotic stresses

Drought is one of the most widespread constraints in vast areas of South and Southeast Asia as well as in West Africa and South America, and drought resistance is likewise one of the most intractable breeding objectives due to the complexity of the trait and paucity of good donors. Molecular markers have been applied to identify QTL associated with component traits of drought resistance. Drought avoidance through improved root systems is thought to be the most promising resistance mechanism, and QTL identified for traits such as root thickness and depth have been associated with drought resistance scores (Champoux et al. 1995). QTL studies have focused on three crosses between lowland indica cultivars and upland japonicas as sources of resistance: a RIL population from CO39 × Moroberekan (upland) (Champoux et al. 1995), a DH population from IR64 × Azucena (upland) (Yadav et al. 1997), and an F_2 population from Bala × Azucena (upland) (Price and Tomos 1997, Price et al. 1997). These three populations have allowed some comparison of QTL identified for root parameters in different studies. Yadav et al. (1997) listed QTL for five root traits that were identified in both the CO39/Moroberekan RILs and in the IR64/Azucena DH lines. Loci on chromosomes 1, 7, and 9 appeared to be involved in several of these traits. A QTL for improved osmotic adjustment, contributed by the drought-susceptible parent CO39, was mapped on chromosome 8 (Lilley et al. 1996). There is evidence that loci associated with root thickness also contribute to greater ability of roots to penetrate an obstruction (Ray et al. 1996), an important trait for rainfed lowland situations.

Submergence tolerance, the ability to survive short (<2 weeks) periods of inundation, is more simply inherited than drought resistance. Most of the variation for this trait is controlled by a single locus that maps near the centromere of chromosome 9 and is designated *Sub1* (Xu and Mackill 1996). Nandi et al. (1997) identified four additional QTL that influenced submergence tolerance. Both the *Sub1* locus and the QTL derive ultimately from the highly tolerant cultivar FR13A.

For adverse soils, a locus conferring tolerance to salinity was mapped on chromosome 7 (Zhang et al. 1995a), and QTL on chromosomes 1 and 2 were associated with tolerance to iron toxicity (Wu et al. 1997). While tolerance of low soil fertility is not usually considered an important breeding objective, there has been interest in improving tolerance to low nitrogen by facilitating association of roots with N_2-fixing microorganisms. Wu et al. (1995) identified four QTL that control this trait.

4.3.3 Morphological and agronomic traits

Important major genes for morphological and agronomic traits that have been mapped with RFLP markers include the semidwarf gene *sd1* (Ogi et al. 1993), the photoperiod sensitivity gene *Se-1* (Mackill et al. 1993), and the shattering gene *sh-2* (Ogi et al. 1993). While these traits are not strong candidates for MAS due to their ease of visual evaluation, their map position is important in the eventual isolation of these genes. As mentioned previously, more urgent targets for MAS include genes that are difficult or time-consuming to evaluate, two examples being the fertility-restoring genes and the wide compatibility genes. The fertility-restoring gene for the BT cms system, which is used in some japonica hybrids, has been mapped on chromosome 10, and selectable markers developed (Ichikawa et al. 1997); markers for *Rf-3*, which is involved with fertility restoration for the more prevalent WA cms, have been identified on chromosome 1 (Zhang et al. 1997). WA cms usually requires two restorer genes for complete fertility restoration, so two loci must be selected for in MAS.

Most agronomic traits are controlled by QTL systems, and they have figured prominently in mapping studies. There are several problems, however, in extracting useful information from these types of studies. Most mapping experiments are limited to indica × japonica crosses which are rarely, if ever, used for yield improvement in rice breeding programs. If loci are identified in such crosses that have sufficient additive effect that they could improve performance when transferred into one or the other subspecies background, than MAS has potential for facilitating such a transfer. This is because visual selection for these traits in segregating populations is problematic due to their low heritability. The recent interest in developing indica × japonica hybrids could benefit by the identification of loci that are associated with intersubspecific heterosis (Xiao et al. 1995). The promise of the ABQTL approach described above has generated considerable excitement, especially with the identification of two QTL from *O. rufipogon* on chromosomes 1 and 2 that each contributed 17%-18% yield advantage (Xiao et al. 1996c).

Another complication with agronomic traits is that it appears certain that QTL will be cross-specific, and even environment-specific, within a cross. For example, very few QTL identified for seedling vigor were expressed in two different crosses and even under two temperature regimes within the same cross (Redoña and Mackill 1996a, b). Zhuang et al. (1997), on the other hand, detected little variation for QTL for agronomic and yield traits in three populations of the same cross grown in different environments, although not all QTL were expressed in all environments.

With the accumulation of additional QTL studies, it will become clearer how reliable and robust are individual QTL for agronomic traits. Those QTL that have the highest effects and are detected in several crosses would seem to be the most promising for MAS and eventually for map-based cloning. As an example, QTL for number of spikelets per panicle and grain weight identified in several studies are placed on a reference map in Fig. 1. While a number of QTL were only observed in one experiment, it is clear that some chromosome regions appear prominent in several studies. For spikelets per panicle, loci on chromosomes 3, 4, 6, and 8 were observed in at least two studies. For grain weight, loci on chromosomes 4 and 5 were particularly prominent. Since QTL can be fine-mapped with the appropriate populations (Paterson et al. 1990), it is likely that QTL controlling important yield components will be the subject of map-based cloning in the future.

4.3.4 Grain quality

Grain quality is becoming a more important breeding objective in many countries, and in some breeding programs takes complete precedence over yield. Molecular markers have not proven extremely useful in improving quality-related characters, because (a) these characters are often under complex genetic control, (b) the chemical determinants of improved cooking quality have not been fully elucidated, and (c) the two subspecies have distinct market requirements and meaningful evaluation of quality must be done within subspecies (and within temperate or tropical japonica). The low level of polymorphism within subspecies, especially within temperate and tropical japonica cultivars, makes it difficult to map genes in these types of crosses. As an example, M-401 and M-202 are two California japonica

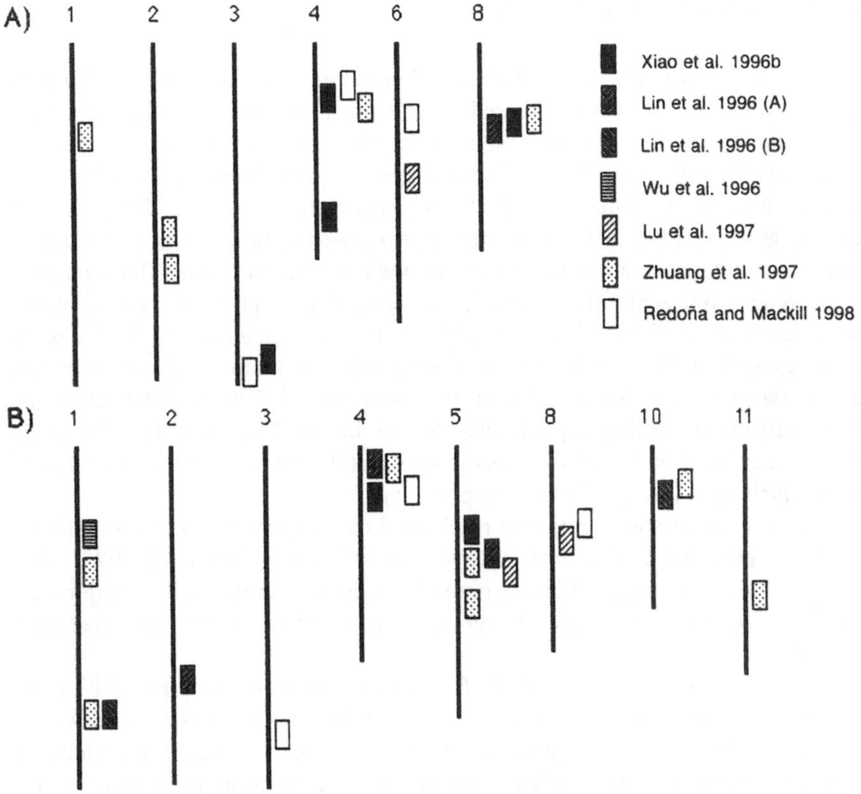

Fig. 1. Approximate location of QTL identified for spikelets per panicle (**A**) and grain weight (**B**) in six studies. Markers are placed in the relative position of the closest RFLP markers in the map of Causse et al. (1994). Indica × japonica crosses were used by Xiao et al. (1996b), Wu et al. (1996), Lu et al. (1997), and Redoña and Mackill (1998). Lin et al. (1996) used two indica × indica crosses, designated (**A**) and (**B**). Zhuang et al. (1997) used the (**A**) population studied by Lin et al. but in different environments and with different generations.

cultivars that are more than 90% identical for RAPDs and AFLPs (Mackill et al. 1996), but M-401 has much improved cooking quality over M-202. Amylose content, which has a major effect on grain quality, is largely determined by alleles from the *wx* gene, which has been cloned. This trait, however, can be measured rapidly through NIR equipment; therefore, it is not a likely candidate for MAS. Similarly, aroma, which is required in some specialty rices, is determined by a locus mapped to chromosome 8 (Ahn et al. 1992). This trait can also be evaluated by the easily detected aroma of leaf tissue or grains. The identification of a locus associated with cooked kernel elongation on chromosome 8 (Ahn et al. 1993) may offer the prospect of MAS for this more difficult to measure trait.

5 Map-based cloning

One of the more exciting applications of rice genome analysis is the isolation of the genes themselves through map-based or "positional" cloning. Cloning genes that were only known through their phenotypic effect has been relatively routine in small microbial genomes, and in the small genome plant *Arabidopsis*, but has only recently become possible in the genomes of agricultural species. The process of cloning genes involves determining the approximate position of a gene through the usual mapping procedures, saturating the region of interest with DNA markers, developing a fine-scale map with a large segregating population, and isolating a large insert clone (from a YAC or BAC library) that encompasses the locus. In most cases, the BAC or YAC clones must be subcloned into smaller vectors, such as lambda or cosmid clones, and contigs are constructed. The candidate clones are then analyzed for coding regions that may be involved in the trait. Ultimately, these clones must be transformed into plants that do not possess the trait to determine which gene confers the desired phenotype.

The first successful example of map-based cloning in rice was the isolation of the *Xa-21* gene that confers resistance to bacterial blight (Song et al. 1995). Several map-based cloning projects are currently under way in rice, with a major focus on disease resistance genes (Yoshimura et al. 1996, Monna et al. 1997, Nakamura et al. 1997).

The cloning of agronomically useful genes opens up exciting possibilities for rice improvement. As transformation methods become more routine, it will be possible to introduce cloned genes directly into a range of elite cultivars which can be released directly as new varieties. In conventional breeding programs, attempting to select for several genes in a segregating population is quite cumbersome, even if markers are available. Once the most important genes are cloned, they can be placed on a single construct and introduced back into the rice plant. These multiple genes can then be selected through conventional breeding or MAS as a single Mendelian unit (Michelmore 1995).

6 Conclusions

In the 1970s, most breeders would be hard-pressed to answer the question, How does one apply the science of genetics to plant breeding? Such is no longer the case. The tremendous progress in mapping agronomically useful genes over the last decade offers the enticing prospect of directly applying these molecular genetic tools in plant breeding. However, with all the recent emphasis on plant genomics it must be stressed that the conventional methods of plant breeding are still an indispensable component of cultivar development. With current technology, the application of MAS will depend on a number of factors that include the genetic control, heritability, and ease of measurement of a trait, as well as the availability of appropriate markers and equipment. At the present time, it seems that MAS will be most appropriate for intersubspecific or interspecific transfer of traits that have been difficult to improve using conventional methods. Examples might include the transfer of durable blast resistance or drought tolerance from upland japonica cultivars into indica cultivars, the transfer of cold tolerance from japonicas into indica cultivars, the transfer of restorer genes for WA cytoplasm into japonica cultivars, and the transfer of submergence tolerance from indica cultivars into japonica cultivars. It will also be useful for pyramiding resistance genes into a single line. Routine use of MAS for agronomic or yield-related characters in typical breeding populations still seems problematic.

It is estimated that the genome of *Arabidopsis* will be completely sequenced by the year 2000, and rice, a major crop with the smallest genome, will shortly follow. This means that we will soon have within reach the complete sequence of all the 40,000 or so genes that make up the rice plant. It is expected that the detailed molecular maps with associated major genes and QTL, which are currently being developed worldwide, will serve as a basis to assign functions to the genes that will be identified through sequencing. Knowing the complete sequence naturally opens up many new opportunities for plant breeding. However, obtaining the sequence of a single cultivar is only part of the picture. The rich diversity of cultivated rice and its wild relatives is derived from a wide range of alleles that will be different from those of the sequenced individual. Unraveling this complex relationship between genes and phenotype and applying this information to developing better crops will be the work of genome scientists and plant breeders for the next century.

References

Ahn SN, Bollich CN, McClung AM, Tanksley SD (1993) RFLP analysis of genomic regions associated with cooked-kernel elongation in rice. Theor Appl Genet 87:27-32

Ahn SN, Bollich CN, Tanksley SD (1992) RFLP tagging of a gene for aroma in rice. Theor Appl Genet 84:825-828

Ajmone-Marsan P, Monfredini G, Ludwig WF, Melchinger AE, Franceschini P, Pagnotto G, Motto M (1995) In an elite cross of maize a major quantitative trait locus controls one-fourth of the genetic variation for grain yield. Theor Appl Genet 90:415-424

Akagi H, Yokozeki Y, Inagaki A, Fujimura T (1996) Microsatellite DNA markers for rice chromosomes. Theor Appl Genet 93:1071-1077

Akagi H, Yokozeki Y, Inagaki A, Fujimura T (1997) Highly polymorphic microsatellites of rice consist of AT repeats, and a classification of closely related cultivars with these microsatellite loci. Theor Appl Genet 94:61-67

Aswidinnoor H, Nelson RJ, Dallas JF, McIntyre CL, Leung H, Gustafson JP (1991) Cloning and characterization of repetitive DNA sequences from genomes of *Oryza minuta* and *Oryza australiensis*. Genome 34:790-798

Bevan M, Ecker J, Theologis S, Federspiel N, Davis R, McCombie D, Martienssen R, Chen E, Waterston B, Wilson R, Rounsley S, Venter C, Tabata S, Salanoubat M, Quetier F, Cherry JM, Meinke D (1997) Objective: the complete sequence of a plant genome. Plant Cell 9:476-478

Boppenmaier J, Melchinger AE, Seitz G, Geiger HH, Herrmann RG (1993) Genetic diversity for RFLPs in European maize inbreds. 3. Performance of crosses within versus between heterotic groups for grain traits. Plant Breed 111:217-226

Botstein D, White RL, Skolnick M, Davis RW (1980) Construction of a genetic linkage map in man using restriction fragment length polymorphism. Am J Hum Genet 32:314-331

Causse MA, Fulton TM, Cho YG, Ahn SN, Chunwongse J, Wu KS, Xiao JH, Yu ZH, Ronald PC, Harrington SE, Second G, McCouch SR, Tanksley SD (1994) Saturated molecular map of the rice genome based on an interspecific backcross population. Genetics 138:1251-1274

Champoux M, Wang G, Sarkarung S, Huang N, Mackill DJ, O'Toole JC, McCouch SR (1995) Locating genes associated with root morphology and drought avoidance in rice via linkage to molecular markers. Theor Appl Genet 90:969-981

Chen X, Temnykh S, Xu Y, Cho YG, McCouch SR (1997) Development of a microsatellite framework map providing genome-wide coverage in rice (*Oryza sativa* L.). Theor Appl Genet 95:553-567

Cho YG, Blair MW, Panaud O, McCouch SR (1996) Cloning and mapping of variety-specific rice genomic DNA sequences: amplified fragment length polymorphisms (AFLP) from silver-stained polyacrylamide gels. Genome 39:373-378

Chunwongse J, Martin GB, Tanksley SD (1993) Pre-germination genotypic screening using PCR amplification of half-seeds. Theor Appl Genet 86:694-698

Churchill GA, Doerge RW (1994) Empirical threshold values for quantitative trait mapping. Genetics 138:963-971

Cordesse F, Grellet F, Reddy AS, Delseny M (1992) Genome specificity of rDNA spacer fragments from *Oryza sativa* L. Theor Appl Genet 83:864-870

Dhillon BS, Boppenmaier J, Pollmer WG, Herrmann RG, Melchinger AE (1993) Relationship of restriction fragment length polymorphisms among European maize inbreds with ear dry matter yield of their hybrids. Maydica 38:245-248

Donini P, Elias ML, Bougourd SM, Koebner RMD (1997) AFLP fingerprinting reveals pattern differences between template DNA extracted from different plant organs. Genome 40:521-526

Edwards MD, Page NJ (1994) Evaluation of marker-assisted selection through computer simulation. Theor Appl Genet 88:376-382

Fukuoka S, Hosaka K, Kamijima O (1992) Use of random amplified polymorphic DNAs (RAPDs) for identification of rice accessions. Jpn J Genet 67:243-252

Gimelfarb A, Lande R (1994) Simulation of marker assisted selection for non-additive traits. Genet Res 64:127-36

Glaszmann JC (1987) Isozymes and classification of Asian rice varieties. Theor Appl Genet 74:21-30

Gu WK, Weeden NF, Yu J, Wallace DH (1995) Large-scale, cost-effective screening of PCR products in marker-assisted selection applications. Theor Appl Genet 91:465-470

Hirabayashi H, Ogawa T (1995) RFLP mapping of *Bph-1* (Brown planthopper resistance gene) in rice. Breeding Sci 45:369-371

Hittalmani S, Foolad MR, Mew T, Rodriguez RL, Huang N (1995) Development of a PCR-based marker to identify rice blast resistance gene, *Pi-2*(t), in a segregating population. Theor Appl Genet 91:9-14

Hospital F, Charcosset A (1997) Marker-assisted introgression of quantitative trait loci. Genetics 147:1469-1485

Hospital F, Chevalet C, Mulsant P (1992) Using markers in gene introgression breeding programs. Genetics 132:1199-1210

Huang N, Angeles ER, Domingo J, Magpantay G, Singh S, Zhang G, Kumaravadivel N, Bennett J, Khush GS (1997) Pyramiding of bacterial blight resistance genes in rice: marker-assisted selection using RFLP and PCR. Theor Appl Genet 95:313-320

Ichikawa N, Kishimoto N, Inagaki A, Nakamura A, Koshino Y, Yokozeki Y, Oka M, Samoto S, Akagi H, Higo K, Shinjyo C, Fujimura T, Shimada H (1997) A rapid PCR-aided selection of a rice line containing the *Rf-1* gene which is involved in restoration of the cytoplasmic male sterility. Mol Breeding 3:195-202

Inukai T, Nelson RJ, Zeigler RS, Sarkarung S, Mackill DJ, Bonman JM, Takamure I, Kinoshita T (1994) Allelism of blast resistance genes in near-isogenic lines of rice. Phytopathology 84:1278-1283

Inukai T, Zeigler RS, Sarkarung S, Bronson M, Dung LV, Kinoshita T, Nelson RJ (1996) Development of pre-isogenic lines for rice blast resistance by marker aided selection from a recombinant inbred population. Theor Appl Genet 93:560-567

IRRI (1997) Rice Almanac. International Rice Research Institute, Manila

Ishii T, Brar DS, Multani DS, Khush GS (1994) Molecular tagging of genes for brown planthopper resistance and earliness introgressed from *Oryza australiensis* into cultivated rice, *O. sativa*. Genome 37:217-221

Ishii T, Brar DS, Second G, Tsunewaki K, Khush GS (1995) Nuclear genome differentiation in Asian cultivated rice as revealed by RFLP analysis. Jpn J Genet 70:643-652

Ishii T, Nakano T, Maeda H, Kamijima O (1996) Phylogenetic relationships in A-genome species of rice as revealed by RAPD analysis. Genes Genet Syst 71:195-201

Ishii T, Terachi T, Mori N, Tsunewaki K (1993) Comparative study on the chloroplast, mitochondrial and nuclear genome differentiation in two cultivated rice species, *Oryza sativa* and *O. glaberrima*, by RFLP analyses. Theor Appl Genet 86:88-96

Jansen RC, Stam P (1994) High resolution of quantitative traits into multiple loci via interval mapping. Genetics 136:1447-1455

Kajiya H, Shimano T, Monna L, Yano M, Sasaki T (1996) Effective production of RFLP markers between closely related varieties of rice by genomic subtraction. Breeding Sci 46:387-391

Kiefermeyer MC, Reddy AS, Delseny M (1995) Characterization of a dispersed repetitive DNA sequence associated with the CCDD genome of wild rice. Genome 38:681-688

Kinoshita T (1986) Standardization of gene symbols and linkage maps in rice. In: Rice Genetics. Proceedings of the International Rice Genetics Symposium, 27-31 May 1985. International Rice Research Institute, Manila, pp 215-228

Kurata N, Nagamura Y, Yamamoto K, Harushima Y, Sue N, Wu J, Antonio BA, Shomura A, Shimizu T, Lin SY, Inoue T, Fukuda A, Shimano T, Kuboki Y, Toyama T, Miyamoto Y, Kirihara T, Hayasaka K, Miyao A, Monna L, Zhong HS, Tamura Y, Wang ZX, Momma T, Umehara Y, Yano M, Sasaki T, Minobe Y (1994) A 300 kilobase interval genetic map of rice including 883 expressed sequences. Nat Genet 8:365-372

Lande R (1992) Marker-assisted selection in relation to traditional methods of plant breeding. In: Stalker HT, Murphy JP (eds) Plant breeding in the 1990s. CAB International, Wallingford, United Kingdom, pp 437-451

Lande R, Thompson R (1990) Efficiency of marker-assisted selection in the improvement of quantitative traits. Genetics 124:743-756

Lander ES, Botstein D (1989) Mapping Mendelian factors underlying quantitative traits using RFLP linkage maps. Genetics 121:185-199

Lanza LLB, Desouza CL, Ottoboni LMM, Vieira MLC, Desouza AP (1997) Genetic distance of inbred lines and prediction of maize single-cross performance using RAPD markers. Theor Appl Genet 94:1023-1030

Lilley JM, Ludlow MM, McCouch SR, O'Toole JC (1996) Locating QTL for osmotic adjustment and dehydration tolerance in rice. J Exp Bot 47:1427-1436

Lin HX, Qian HR, Zhuang JY, Lu J, Min SK, Xiong ZM, Huang N, Zheng KL (1996) RFLP mapping of QTLs for yield and related characters in rice (*Oryza sativa* L.). Theor Appl Genet 92:920-927

Lorieux M, Petrov M, Huang N, Guiderdoni E, Ghesquiere A (1996) Aroma in rice: genetic analysis of a quantitative trait. Theor Appl Genet 93:1145-1151

Lu CF, Shen LS, Tan ZB, Xu YB, He P, Chen Y, Zhu LH (1997) Comparative mapping of QTLs for agronomic traits of rice across environments by using a doubled-haploid population. Theor Appl Genet 94:145-150

Mackill DJ (1995) Classifying japonica rice cultivars with RAPD markers. Crop Sci 35:889-894

Mackill DJ, Salam MA, Wang ZY, Tanksley SD (1993) A major photoperiod-sensitivity gene tagged with RFLP and isozyme markers in rice. Theor Appl Genet 85:536-540

Mackill DJ, Zhang Z, Redona ED, Colowit PM (1996) Level of polymorphism and genetic mapping of AFLP markers in rice. Genome 39:969-977

Maheswaran M, Subudhi PK, Nandi S, Xu JC, Parco A, Yang DC, Huang N (1997) Polymorphism, distribution, and segregation of AFLP markers in a doubled haploid rice population. Theor Appl Genet 94:39-45

Martin C, Juliano A, Newbury HJ, Lu BR, Jackson MT, Ford-Lloyd BV (1997) The use of RAPD markers to facilitate the identification of *Oryza* species within a germplasm collection. Genet Resour Crop Evol 44:175-183

McCouch SR, Chen XL, Panaud O, Temnykh S, Xu YB, Cho YG, Huang N, Ishii T, Blair M (1997) Microsatellite marker development, mapping and applications in rice genetics and breeding. Plant Mol Biol 35:89-99

McCouch SR, Kochert G, Yu ZH, Wang ZY, Khush GS, Coffman WR, Tanksley SD (1988) Molecular mapping of rice chromosomes. Theor Appl Genet 76:815-829

McCouch SR, Nelson RJ, Tohme J, Zeigler RS (1994) Mapping of blast resistance genes in rice. In: Zeigler RS, Leong SA, Teng PS (eds) Rice blast disease. CAB International, Oxon, U.K., pp 167-186

McDonald MB, Elliot LJ, Sweeney PM (1994) DNA extraction from dry seeds for RAPD analyses in varietal identification studies. Seed Sci Technol 22:171-176

Meksem K, Leister D, Peleman J, Zabeau M, Salamini F, Gebhardt C (1995) A high-resolution map of the vicinity of the R1 locus on chromosome V of potato based on RFLP and AFLP markers. Mol Gen Genet 249:74-81

Michelmore RW (1995) Molecular approaches to manipulation of disease resistance genes. Annu Rev Phytopathol 33:393-427

Michelmore RW, Paran I, Kesseli RV (1991) Identification of markers linked to disease-resistance genes by bulked segregant analysis: a rapid method to detect markers in specific genomic regions by using segregating populations. Proc Natl Acad Sci USA 88:9828-9832

Miyao A, Zhong HS, Monna L, Yano M, Yamamoto K, Havukkala I, Minobe Y, Sasaki T (1996) Characterization and genetic mapping of simple sequence repeats in the rice genome. DNA Res 3:233-238

Mochizuki K, Ohtsubo H, Hirano HY, Sano Y, Ohtsubo E (1993) Classification and relationships of rice strains with AA genome by identification of transposable elements at nine loci. Jpn J Genet 68:205-217

Mohan M, Nair S, Bentur JS, Rao UP, Bennett J (1994) RFLP and RAPD mapping of the rice *Gm2* gene that confers resistance to biotype 1 of gall midge (*Orseolia oryzae*). Theor Appl Genet 87:782-788

Mohan M, Sathyanarayanan PV, Kumar A, Srivastava MN, Nair S (1997) Molecular mapping of a resistance-specific PCR-based marker linked to a gall midge resistance gene (*Gm4t*) in rice. Theor Appl Genet 95:777-782

Monna L, Miyao A, Zhong HS, Yano M, Iwamoto M, Umehara Y, Kurata N, Hayasaka H, Sasaki T (1997) Saturation mapping with subclones of YACs: DNA marker production targeting the rice blast disease resistance gene, *Pi-b*. Theor Appl Genet 94:170-176

Moore G, Devos KM, Wang Z, Gale MD (1995) Grasses, line up and form a circle. Curr Biol 5:737-739

Motohashi R, Mochizuki K, Ohtsubo H, Ohtsubo E (1997) Structures and distribution of *p-SINE1* members in rice genomes. Theor Appl Genet 95:359-368

Nair S, Bentur JS, Rao UP, Mohan M (1995) DNA markers tightly linked to a gall midge resistance gene (*Gm2*) are potentially useful for marker-aided selection in rice breeding. Theor Appl Genet 91:68-73

Nair S, Kumar A, Srivastava MN, Mohan M (1996) PCR-based DNA markers linked to a gall midge resistance gene, *Gm4t*, has potential for marker-aided selection in rice. Theor Appl Genet 92:660-665

Nakajima R, Noma K, Ohtsubo H, Ohtsubo E (1996) Identification and characterization of two tandem repeat sequences (*TrsB* and *TrsC*) and a retrotransposon (RIRE1) as genome-general sequences in rice. Genes Genet Syst 71:373-382

Nakamura S, Asakawa S, Ohmido N, Fukui K, Shimizu N, Kawasaki S (1997) Construction of an 800-kb contig in the near-centromeric region of the rice blast resistance gene *Pi-ta*^{2} using a highly representative rice BAC library. Mol Gen Genet 254:611-620

Nakano M, Yoshimura A, Iwata N (1992) Phylogenetic study of cultivated rice and its wild relatives by RFLP. Rice Genet Newsl 9:132-134

Nandi S, Subudhi PK, Senadhira D, Manigbas NL, Sen-Mandi S, Huang N (1997) Mapping QTLs for submergence tolerance in rice by AFLP analysis and selective genotyping. Mol Gen Genet 255:1-8

Naqvi NI, Chattoo BB (1996) Development of a sequence characterized amplified region (SCAR) based indirect selection method for a dominant blast-resistance gene in rice. Genome 39:26-30

Noma K, Nakajima R, Ohtsubo H, Ohtsubo E (1997) *RIRE1*, a retrotransposon from wild rice *Oryza australiensis*. Genes Genet Syst 72:131-140

Ogi Y, Kato H, Maruyama K, Saito A, Kikuchi F (1993) Identification of RFLP markers closely linked to the semidwarfing gene at the *sd-1* locus in rice. Jpn J Breed 43:141-146

Panaud O, Chen X, McCouch SR (1996) Development of microsatellite markers and characterization of simple sequence length polymorphism (SSLP) in rice (*Oryza sativa* L.). Mol Gen Genet 252:597-607

Paterson AH (1998) QTL mapping in DNA marker-assisted plant and animal improvement. In: Paterson AH (ed) Molecular dissection of complex traits. CRC Press, Boca Raton, pp 131-143

Paterson AH, DeVerna JW, Lanini B, Tanksley SD (1990) Fine mapping of quantitative trait loci using selected overlapping recombinant chromosomes in an interspecific cross of tomato. Genetics 124:735-742

Paterson AH, Lander ES, Hewitt JD, Peterson S, Lincoln SE, Tanksley SD (1988) Resolution of quantitative traits into Mendelian factors by using a complete linkage map of restriction fragment length polymorphisms. Nature 335:721-726

Price AH, Tomos AD (1997) Genetic dissection of root growth in rice (*Oryza sativa* L.). 2. Mapping quantitative trait loci using molecular markers. Theor Appl Genet 95:143-152

Price AH, Young EM, Tomos AD (1997) Quantitative trait loci associated with stomatal conductance, leaf rolling and heading date mapped in upland rice (*Oryza sativa*). New Phytol 137:83-91

Ray JD, Yu L, McCouch SR, Champoux MC, Wang G, Nguyen HT (1996) Mapping quantitative trait loci associated with root penetration ability in rice (*Oryza sativa* L.). Theor Appl Genet 92:627-636

Reddy AS, Kiefermeyer MC, Delseny M (1993) Characterization of new variants of a satellite DNA from *Oryza officinalis*, specific for the CC genome of wild rice. Genome 36:750-761

Redoña ED, Mackill DJ (1996a) Mapping quantitative trait loci for seedling vigor in rice using RFLPs. Theor Appl Genet 92:395-402

Redoña ED, Mackill DJ (1996b) Molecular mapping of quantitative trait loci in japonica rice. Genome 39:395-403

Redoña ED, Mackill DJ (1998) Quantitative trait locus analysis for rice panicle and grain characteristics. Theor Appl Genet 96:957-963

Ronald PC, Albano B, Tabien R, Abenes L, Wu K-S, McCouch S, Tanksley SD (1992) Genetic and physical analysis of the rice bacterial blight disease resistance locus, *Xa21*. Mol Gen Genet 236:113-120

Schön CC, Melchinger AE, Boppenmaier J, Brunklausjung E, Herrmann RG, Seitzer JF (1994) RFLP mapping in maize - quantitative trait loci affecting testcross performance of elite European flint lines. Crop Sci 34:378-389

Sebastian LS, Ikeda R, Huang N, Imbe T, Coffman WR, McCouch SR (1996) Molecular mapping of resistance to rice tungro spherical virus and green leafhopper. Phytopathology 86:25-30

Skroch P, Nienhuis J (1995) Impact of scoring error and reproducibility of RAPD data on RAPD based estimates of genetic distance. Theor Appl Genet 91:1086-1091

Song WY, Wang GL, Chen LL, Kim HS, Pi LY, Holsten T, Gardner J, Wang B, Zhai WX, Zhu LH, Fauquet C, Ronald P (1995) A receptor kinase-like protein encoded by the rice disease resistance gene, *Xa21*. Science 270:1804-1806

Stuber CW, Lincoln SE, Wolff DW, Helentjaris T, Lander ES (1992) Identification of genetic factors contributing to heterosis in a hybrid from two elite maize inbred lines using molecular markers. Genetics 132:823-839

Suh HS, Sato YI, Morishima H (1997) Genetic characterization of weedy rice (*Oryza sativa* L.) based on morpho-physiology, isozymes and RAPD markers. Theor Appl Genet 94:316-321

Tanksley S, Young N, Paterson A, Bonierbale M (1989) RFLP mapping in plant breeding: new tools for an old science. Bio/Technology 7:257-264

Tanksley SD, Nelson JC (1996) Advanced backcross QTL analysis: a method for the simultaneous discovery and transfer of valuable QTLs from unadapted germplasm into elite breeding lines. Theor Appl Genet 92:191-203

Tsunematsu H, Yoshimura A, Harushima Y, Nagamura Y, Kurata N, Yano M, Sasaki T, Iwata N (1996) RFLP framework map using recombinant inbred lines in rice. Breeding Sci 46:279-284

Vaughan DA (1994) The wild relatives of rice. International Rice Research Institute, Manila

Virk PS, Newbury HJ, Jackson MT, Ford-Lloyd BV (1995) The identification of duplicate accessions within a rice germplasm collection using RAPD analysis. Theor Appl Genet 90:1049-1055

Visscher PM, Haley CS, Thompson R (1996) Marker-assisted introgression in backcross breeding programs. Genetics 144:1923-1932

Vos P, Hogers R, Bleeker M, Reijans M, van de Lee T, Hornes M, Frijters A, Pot J, Peleman J, Kuiper M, Zabeau M (1995) AFLP: a new technique for DNA fingerprinting. Nucleic Acids Res 23:4407-4414

Wang GL, Mackill DJ, Bonman JM, McCouch SR, Champoux MC, Nelson RJ (1994) RFLP mapping of genes conferring complete and partial resistance to blast in a durably resistant rice cultivar. Genetics 136:1421-1434

Wang GL, Wing RA, Paterson AH (1993) PCR amplification from single seeds, facilitating DNA marker-assisted breeding. Nucleic Acids Res 21:2527-2527

Wang ZY, Second G, Tanksley SD (1992) Polymorphism and phylogenetic relationships among species in the genus *Oryza* as determined by analysis of nuclear RFLPs. Theor Appl Genet 83:565-581

Wang ZY, Tanksley SD (1989) Restriction fragment length polymorphism in *Oryza sativa* L. Genome 32:1113-1118

Whittaker JC, Curnow RN, Haley CS, Thompson R (1995) Using marker-maps in marker-assisted selection. Genet Res 66:255-265

Williams CE, Ronald PC (1994) PCR template-DNA isolated quickly from monocot and dicot leaves without tissue homogenization. Nucleic Acids Res 22:1917-1918

Williams CE, Yanagihara S, McCouch SR, Mackill DJ, Ronald PC (1997) Predicting success of *Indica/Japonica* crosses in rice, based on a PCR marker for the *S-5*^{n} allele at a hybrid-sterility locus. Crop Sci 37:1910-1912

Williams JGK, Kubelik AR, Livak KJ, Rafalski JA, Tingey SV (1990) DNA polymorphism amplified by arbitrary primers are useful as genetic markers. Nucleic Acids Res 18:6531-6535

Wu KS, Tanksley SD (1993) Abundance, polymorphism and genetic mapping of microsatellites in rice. Mol Gen Genet 241:225-235

Wu P, Luo A, Zhu J, Yang J, Huang N, Senadhira D (1997) Molecular markers linked to genes underlying seedling tolerance for ferrous iron toxicity. Plant Soil 196:317-320

Wu P, Zhang G, Huang N (1996) Identification of QTLs controlling quantitative characters in rice using RFLP markers. Euphytica 89:349-354

Wu P, Zhang G, Ladha JK, McCouch SR, Huang N (1995) Molecular marker facilitated investigation on the ability to stimulate N_2 fixation in the rhizosphere by irrigated rice plants. Theor Appl Genet 91:1177-1183

Wu TY, Wu R (1992) A novel repetitive DNA sequence in the genus *Oryza*. Theor Appl Genet 84:136-144

Xiao J, Li J, Yuan L, McCouch SR, Tanksley SD (1996a) Genetic diversity and its relationship to hybrid performance and heterosis in rice as revealed by PCR-based markers. Theor Appl Genet 92:637-643

Xiao J, Li J, Yuan L, Tanksley SD (1996b) Identification of QTLs affecting traits of agronomic importance in a recombinant inbred population derived from a subspecific rice cross. Theor Appl Genet 92:230-244

Xiao JH, Grandillo S, Ahn SN, McCouch SR, Tanksley SD, Li JM, Yuan LP (1996c) Genes from wild rice improve yield. Nature 384:223-224

Xiao JH, Li JM, Yuan LP, Tanksley SD (1995) Dominance is the major genetic basis of heterosis in rice as revealed by QTL analysis using molecular markers. Genetics 140:745-754

Xu K, Mackill DJ (1996) A major locus for submergence tolerance mapped on rice chromosome 9. Mol Breed 2:219-224

Yadav R, Courtois B, Huang N, McLaren G (1997) Mapping genes controlling root morphology and root distribution in a doubled-haploid population of rice. Theor Appl Genet 94:619-632

Yang GP, Maroof MAS, Xu CG, Zhang Q, Biyashev RM (1994) Comparative analysis of microsatellite DNA polymorphism in landraces and cultivars of rice. Mol Gen Genet 245:187-94

Yano M, Sasaki T (1997) Genetic and molecular dissection of quantitative traits in rice. Plant Mol Biol 35:145-153

Yoshimura S, Umehara Y, Kurata N, Nagamura Y, Sasaki T, Minobe Y, Iwata N (1996) Identification of a YAC clone carrying the *Xa-1* allele, a bacterial blight resistance gene in rice. Theor Appl Genet 93:117-122

Yoshimura S, Yoshimura A, Iwata N, McCouch SR, Abenes ML, Baraoidan MR, Mew TW, Nelson RJ (1995) Tagging and combining bacterial blight resistance genes in rice using RAPD and RFLP markers. Mol Breeding 1:375-387

Yoshimura S, Yoshimura A, Saito A, Kishimoto N, Kawase M, Yano M, Nakagahra M, Ogawa T, Iwata N (1992) RFLP analysis of introgressed chromosomal segments in three near-isogenic lines of rice for bacterial blight resistance genes, *Xa-1*, *Xa-3* and *Xa-4*. Jpn J Genet 67:29-37

Yu LX, Nguyen HT (1994) Genetic variation detected with RAPD markers among upland and lowland rice cultivars (*Oryza sativa* L.). Theor Appl Genet 87:668-672

Yu ZH, Mackill DJ, Bonman JM, Tanksley SD (1991) Tagging genes for blast resistance in rice via linkage to RFLP markers. Theor Appl Genet 81:471-476

Zeng ZB (1994) Precision mapping of quantitative trait loci. Genetics 136:1457-1468

Zhang G, Bharaj TS, Lu Y, Virmani SS, Huang N (1997) Mapping of the R*f-3* nuclear fertility-restoring gene for WA cytoplasmic male sterility in rice using RAPD and RFLP markers. Theor Appl Genet 94:27-33

Zhang GY, Guo Y, Chen SL, Chen SY (1995a) RFLP tagging of a salt tolerance gene in rice. Plant Sci 110:227-234

Zhang QF, Gao YJ, Maroof MAS, Yang SH, Li JX (1995b) Molecular divergence and hybrid performance in rice. Mol Breeding 1:133-142

Zhang QF, Gao YJ, Yang SH, Ragab RA, Maroof MAS, Li ZB (1994) A diallel analysis of heterosis in elite hybrid rice based on RFLPs and microsatellites. Theor Appl Genet 89:185-192

Zhang QF, Maroof MAS, Lu TY, Shen BZ (1992) Genetic diversity and differentiation of indica and japonica rice detected by RFLP analysis. Theor Appl Genet 83:495-499

Zhang QF, Zhou ZQ, Yang GP, Xu CG, Liu KD, Maroof MAS (1996) Molecular marker heterozygosity and hybrid performance in indica and japonica rice. Theor Appl Genet 93:1218-1224

Zhao X, Kochert G (1992) Characterization and genetic mapping of a short, highly repeated, interspersed DNA sequence from rice (*Oryza sativa* L.). Mol Gen Genet 231:353-359

Zhao X, Wu T, Xie Y, Wu R (1989) Genome-specific repetitive sequences in the genus *Oryza*. Theor Appl Genet 78:201-209

Zhao XP, Kochert G (1993a) Clusters of interspersed repeated DNA sequences in the rice genome (*Oryza*). Genome 36:944-953

Zhao XP, Kochert G (1993b) Phylogenetic distribution and genetic mapping of a (GGC)n microsatellite from rice (*Oryza sativa* L.). Plant Mol Biol 21:607-614

Zheng K, Subudhi PK, Domingo J, Magpantay G, Huang N (1995) Rapid DNA isolation for marker assisted selection in rice breeding. Rice Genet Newsl 12:255-258

Zheng K-L, Shen B, Qian H-R (1991) DNA polymorphisms generated by arbitrary primed PCR in rice. Rice Genet Newsl 8:134-136

Zhuang JY, Lin HX, Lu J, Qian HR, Hittalmani S, Huang N, Zheng KL (1997) Analysis of QTL x environment interaction for yield components and plant height in rice. Theor Appl Genet 95:799-808

3
Retrotransposons of rice as a tool for forward and reverse genetics

Hirohiko Hirochika

Department of Molecular Genetics, National Institute of Agrobiological Resources, Tsukuba, Ibaraki 305, Japan

1 Introduction

Most plant transposable elements have been isolated and studied as agents causing unstable mutations. These elements (class II elements) comprise well-characterized *Ac/Ds*, *En/Spm*, and *Mu* elements of maize, *Tam* elements of snapdragon, and *dTph1* of petunia. Because of their ability to induce mutations at high frequency and at the same time tag the causative genes molecularly, they have been utilized as molecular genetic tools for cloning genes (transposon-tagging). Many important genes such as genes for transcription factors regulating pigmentation and flower development and a disease resistance gene have been isolated. The development of such a powerful transposon-tagging system in rice has been long awaited. One way to develop the system is to use heterologous transposable elements such as *Ac/Ds* (see Chapter 4).

Another important challenge is to find new active transposable elements of rice, especially new types of elements, which may provide advantages over the characterized elements. Although *Ac/Ds*-, *En/Spm*-, and *Mu*-like elements (Hirochika and Fukuchi 1992, Motohashi et al. 1996, R. Ishikawa and M. Freeling, personal communication) have been identified by molecular analysis in rice, most of them seem inactive. Consistent with this finding, none of the spontaneous *waxy* mutations were induced by transposable elements (Okagaki and Wessler 1988). Recent studies have shown the existence of another type of transposable elements, class I elements, in many plants including rice (for review, see Grandbastien 1998). Because class I elements undergo replicative transposition through RNA as an intermediate, they usually induce only stable mutations. Class I elements are divided into retrotransposons (LTR-retrotransposons) and retroposons (non-LTR

Key words. gene isolation, insertion mutation, retrotransposon, rice, somaclonal variation, transposable elements, transposon tagging, PCR, reverse and forward genetics

retrotransposons). Retrotransposons carry long terminal direct repeats (LTRs) at both ends and an internal domain encoding Gag and Pol polyproteins (Fig. 1). Enhancer-promoter sequences in the LTRs regulate the synthesis of both RNA template for reverse transcription and mRNA required for protein synthesis. The Pol protein is processed to yield proteins with integrase, reverse transcriptase and RNase H activities. Based on the sequence similarity of the Pol proteins, retrotransposons can be further subdivided into two phylogenetically distinct groups, *Ty1-copia* and *Ty3-gypsy* (Doolittle et al. 1989, Xiong and Eickbush 1990). Retroposons do not have LTRs, and thus are also known as non-LTR retrotransposons. In this group, long interspersed nuclear elements (LINEs) and short interspersed nuclear elements (SINEs) are included. Whereas LINEs encode Gag and Pol polyproteins, SINEs do not carry any protein-coding region. Many class I elements have been isolated from different plant species, but only a few retrotransposons of the *Ty1-copia* group have been shown to be active (Grandbastien 1998). Interestingly, these retrotransposons are active only under certain stress conditions, such as protoplast formation (*Tnt1* of tobacco, Grandbastien et al. 1989) and tissue culture (*Tto1* of tobacco, Hirochika 1993).

Recent studies showed that the retrotransposon of rice, named *Tos17*, is highly active only during tissue culture and that the activation of *Tos17* is responsible for

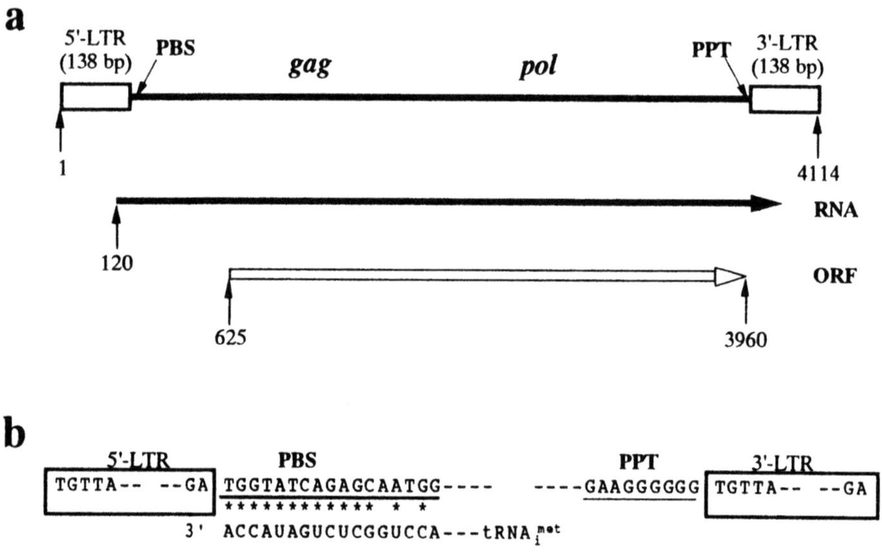

Fig. 1a,b. Structure and expression of *Tos17*. **a** Structure deduced from the complete nucleotide sequence and transcription of *Tos17*. **b** Nucleotide sequences flanking the LTRs of *Tos17*. LTRs are *boxed*; the primer binding site (PBS) and polypurine tract (PPT) are indicated by *thick* and *thin underlining*. The complementarity between the PBS and the 3'-end of the initiator methionine tRNA is shown by *asterisks*. The *thick arrow* indicates the transcript of *Tos17*, and the *open arrow* indicates the longest open reading frame (ORF) starting with the ATG.

tissue culture-induced mutations (Hirochika et al. 1996). Unique features of *Tos17* suggested that it can be used for transposon-tagging and reverse genetic studies. Reverse genetic studies of rice have become extremely important, because a large number of cloned genes with unknown function have been identified by large-scale sequencing of cDNA (Sasaki et al. 1994, Yamamoto and Sasaki 1997). To determine the biological function of a gene, site-selected insertional mutagenesis with transposable elements is a powerful method. In site-selected mutagenesis, mutants of a specific gene are identified in a large mutant population by polymerase chain reaction (PCR) using transposable element- and gene-specific primers. Loss-of-function phenotypes shown by identified mutants can then be studied to determine the biological function of the gene. This method was first developed in *Drosophila* (Ballinger and Benzer 1989, Kaiser and Goodwin 1990) and has since been used in *Caenorhabditis elegans* (Zwaal et al. 1993), petunia (Koes et al. 1995), and maize (Bensen et al. 1995, Das and Martienssen 1995). Although class II elements are used in these systems, it is possible to use other types of insertion elements such as the Ti plasmid (McKinney et al. 1995, Krysan et al. 1996) and retrotransposons. In this paper, the current status of forward and reverse genetic studies of rice using retrotransposons is reviewed.

2 Isolation of active retrotransposons of rice

Thirty two families of retrotransposons of rice have been isolated by using two molecular methods (Hirochika et al. 1992, Hirochika et al. 1996, H. Hirochika et al., unpublished results) and general features of these retrotransposons have been recently reviewed (Hirochika 1997). One of the methods is based on the fact that the sequence of a primer binding site (PBS) is highly conserved among plant retrotransposons. Using an oligonucleotide corresponding to the PBS, the rice genomic DNA library was screened for retrotransposons. Based on this screening, the copy number of retrotransposons in the rice genome was estimated to be 1000 (Hirochika et al. 1992). The existence of about 30 families of retrotransposons was suggested and three of them (*Tos1-Tos3*) were characterized. However, no direct evidence showing transpositional activity was obtained. The second method involves polymerase chain reaction (PCR) based on the conservation of certain amino acid sequences in the reverse transcriptase domain among *Ty1-copia* group retrotransposons. By PCR-amplification of the rice genomic DNA, 12 families (*Tos4-Tos15*) have been isolated (Hirochika et al. 1992, Hirochika et al. 1996). Among these families, *Tos10* alone has been shown to be active only during tissue culture (Hirochika et al. 1996). Interestingly, the activity of *Tos10* was detected in an *indica* variety but not in two *japonica* varieties examined, Nipponbare and Koshihikari. Although the molecular basis of this difference remains unknown, a correlation between transcriptional and transpositional activities was observed, suggesting that the transcription step is rate-limiting.

The tobacco *Tto1* is the first plant retrotransposon whose activity was shown to be regulated by tissue culture (Hirochika 1993). This was isolated by applying the PCR screening method to cDNA prepared from cultured cells. This strategy is based on the fact that most inactive plant retrotransposons are not transcribed and active retrotransposons are transcribed only under certain stress conditions. As has been discussed (Hirochika 1993), this strategy should be applicable to any plant species. This method was used to isolate active retrotransposons of rice (Hirochika et al. 1996). Sequences of four new families of retrotransposons (*Tos17-Tos20*) were amplified by PCR on cDNA prepared from cultured cells. *Tos17* and *Tos19* were shown to be activated by tissue culture, indicating that the cDNA-PCR method is effective in isolating active retrotransposons. Further survey revealed 12 new families (*Tos21-Tos32*), and two of them (*Tos21*, *Tos25*) were shown to be activated by tissue culture (H. Hirochika, unpublished). Among the five active retrotransposons of rice, *Tos17* has been shown to be most active (Hirochika et al. 1996, H. Hirochika, unpublished). As shown in Fig. 1, *Tos17* is 4312 bp long (H. Okamoto and H. Hirochika, unpublished), and carries one open reading frame corresponding to Gag and Pol polyproteins, two identical 138-bp LTRs, and the PBS complementary to the 3'-end of the initiator methionine tRNA.

Eight new retrotransposons (*RIRE1-RIRE8*) including *Ty3-gypsy* as well as *Ty1-copia* group elements have been isolated as highly repetitive elements or insertion elements in another transposable element (Noma et al. 1997, Ohtsubo et al., personal communication). However, none of these elements have been shown to be active.

3 Induction of insertion mutations by tissue culture

Whereas the original copy number of *Tos17* is two in the *japonica* variety Nipponbare, five to 30 transposed *Tos17* copies were detected in all of the plants regenerated from tissue cultures (Fig. 2), including transgenic plants. The transcript of *Tos17* was only detected under tissue culture conditions, indicating that the transposition of *Tos17* is regulated mainly at the transcriptional level as in the case of tobacco retrotransposons such as *Tnt1* (Pouteau et al. 1991, Grandbastien 1998) and *Tto1* (Hirochika 1993, Takeda et al. 1998).

As shown in Fig. 2, at least five new insertions of *Tos17* were induced during the 3- to 9-month culture period. Each regenerated plant showed a different pattern of hybridization with the *Tos17* probe, indicating that independent transposition events occurred in each cell of the tissue culture. In contrast, the patterns are almost the same in the regenerated plants from the 16-month-old culture. This suggests that a specific cell type was selected during prolonged culture. Considering the frequency of transposition, *Tos17* is most likely to be involved in tissue culture-induced mutations. Tissue culture-induced mutations have been reported in many plant species (Larkin and Scowcroft 1981, Bajaj 1990), but their molecular causes are largely unknown. Direct evidence that *Tos17* insertions lead to muta-

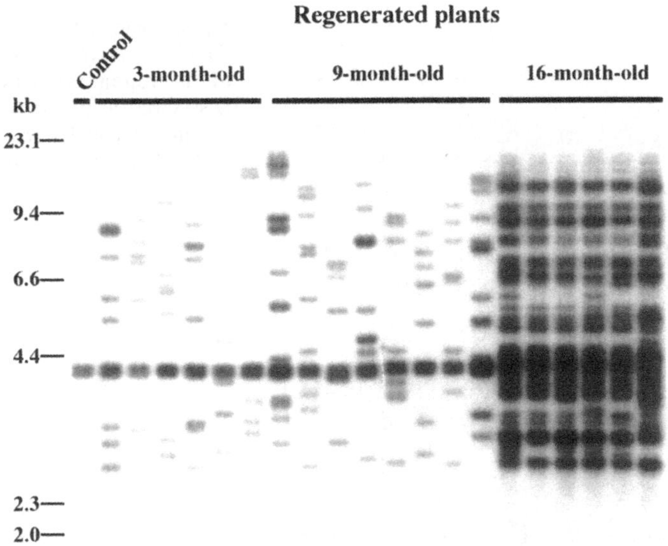

Fig. 2. Activation of *Tos17* by tissue culture. DNAs were prepared from leaves of normally propagated rice plants (*Control*) and plants regenerated from 3-, 9-, and 16-month-old cultures and analyzed by DNA blot hybridization with a *Tos17*-specific probe after digestion with *Xba*I.

tions of genes was obtained by analyzing the target sites of *Tos17* transposition (Hirochika et al. 1996, H. Hirochika, unpublished). Nine target sites were amplified by using inverse PCR and sequenced. Five of the nine target sites showed significant homology to known genes at the amino acid level, indicating that *Tos17* transposed into genes. Because seven out of the nine targets were transcribed, most of the target sites are likely to be in genes. Regenerated plants with *Tos17*-insertions in the phytochrome A gene, the S-receptor kinase-related gene, and the gene encoding a ring-finger protein were identified. Taken together, these results indicate that activation of *Tos17* is an important cause of tissue culture-induced mutations and that regenerated rice plants and their progeny can be used as mutagenized material for forward and reverse genetic studies.

To make rice lines mutagenized with *Tos17* for genetic studies, calli were induced from scutella by culturing dehusked mature seeds and cultured in a liquid medium for 3 to 5 months, as described by Otsuki (1990). Five to ten transposed copies were detected in each plant regenerated from these cultures. By using this culture protocol, it is possible to maintain regeneration ability for more than 1 year. Because culturing of tissue and regeneration from tissue culture are easy for most rice varieties, tissue culture-induced activation of *Tos17* for saturation mutagenesis (see the last section) should be feasible in most rice varieties. It should be noted that *Tos17* was almost not activated by tissue culture in some varieties (H. Hirochika et al., unpublished).

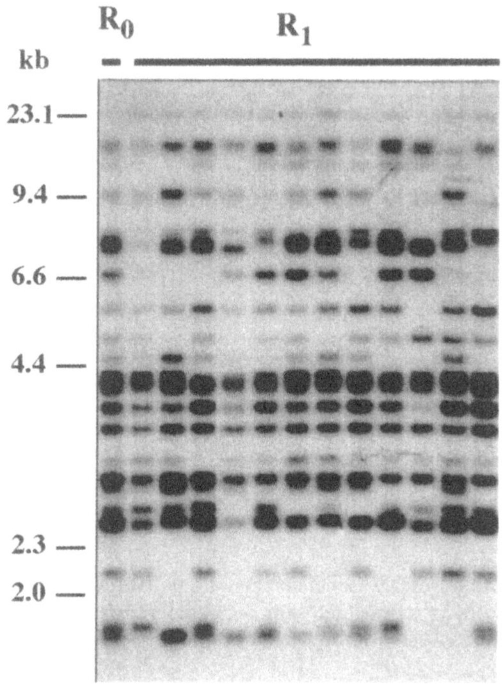

Fig. 3. Segregation of transposed *Tos17* copies in the progeny (R_1) of the regenerated plant (R_0). DNAs were analyzed by DNA blot hybridization with a *Tos17*-specific probe after digestion with *Xba*I.

Although *Tos17* is actively transcribed in tissue culture, no transcript of *Tos17* was detected in plants regenerated from tissue culture, suggesting that transposition is restricted under tissue culture condition. This was confirmed by the following experiments. The copy number of *Tos17* was compared between the regenerated plant and its progeny (Fig. 3). The transposed copies detected in the regenerated plant were stably inherited, and no new transposed copy was detected in the progeny. These findings indicate that the transposition of *Tos17* is restricted under tissue culture conditions and that transposed copies are germinally transmitted in the next generation. Some maize class II elements such as *Ac/Ds* (Peschke et al. 1987) and *En/Spm* (Peschke and Phillips 1991) are also known to be activated by tissue culture. In contrast to *Tos17*, their activity is maintained in regenerated plants and their progeny. Another important result obtained from the same analysis is that each transposed *Tos17* copy was segregated. This indicates that *Tos17* transposes into different loci on the chromosomes. This was further confirmed by mapping transposed *Tos17* copies on the rice chromosomes (Yamazaki et al., unpublished results). Because *Ac/Ds* and *En/Spm* elements preferentially transpose to linked sites (Bancroft and Dean 1993, Gierl 1995), many lines carrying these elements located throughout the genome are needed for saturation insertion mutagenesis. The unique features of *Tos17* described here should be useful when it is used as a tool for genetic studies.

The transpositional activity of the second active *Tos19* was analyzed. To compare the activities of *Tos17* and *Tos19* under the same culture condition, the DNA gel blot used for the hybridization with *Tos17* probe was rehybridized with a *Tos19* probe. A copy number increase was only found in the regenerated plants from the 16-month-old culture. Even in the plants in which an increase was found, the increased copy number was only one to three, indicating that *Tos17* is responsible for most of the insertion mutations induced by the known retrotransposons. Thus, the focus in the following sections is primarily on *Tos17*.

4 Forward and reverse genetic studies using retrotransposons

Features of *Tos17* useful for forward and reverse genetic studies are summarized as follows. (1) Transposition can be regulated: *Tos17* is only active during tissue culture. (2) Highly mutagenic during tissue culture: *Tos17* transposes preferentially into genes, and at least five genes are disrupted in each plant regenerated from 5-month-old culture. (3) Transposition target sites are unlinked with the site of original copy: it is easy to generate a large collection of random insertions for saturation mutagenesis. (4) Induced mutations are stable. (5) Original copy number is quite low: one to four depending on varieties. This is important for identifying the transposed copy responsible for the specific mutation. (6) Endogenous transposon: screening and characterization of mutants in the field are possible without any regulation of recombinant DNA experiments. This must be important, because many agronomically important traits are best characterized in the field.

4.1 Forward genetics: use as a tool for gene-tagging

To demonstrate the feasibility of gene-tagging using *Tos17*, we have first screened R1 generations of regenerated rice for mutants based on the phenotypes in the paddy field. Many mutants have been recognized, and seven mutants showing clear phenotypes (Fig. 4) were picked out and further subjected to co-segregation analysis. Genetic analysis showed that all these mutations are recessive. Using segregating populations of R1 and R2 generations, co-segregation of mutant phenotypes and *Tos17* copies was analyzed. Because the original copy number of *Tos17* is quite low (one to four depending on cultivars) in contrast to other plant retrotransposons such as *Tnt1* (several hundred copies; Grandbastien et al. 1989) and *Tto1* (30 copies; Hirochika 1993), it is possible to identify the specific *Tos17* copy co-segregating with the mutant phenotype by DNA gel blot analysis. In the mutants, five to 30 transposed *Tos17* copies were detected, and perfect co-segregation of one *Tos17* copy was observed with two mutations (viviparous and stripe) using more than 60 segregating individuals. These results strongly suggest that the mutations are caused by a *Tos17* insertion. Genomic sequences corresponding to mutant loci were amplified by TAIL-PCR (Liu and Whittier 1995), cloned, and partially sequenced. Sequence data indicate that these are new genes (Liu et al.,

Fig. 4. Mutants found in the progeny of regenerated rice plants. Viviparous, stripe, dwarf, and brittle culm mutants are shown in the *left* of each panel and the control plant is shown in the *right*. Two independent lesion mimic mutants and one sterile mutant are shown in the *right two panels*. To clearly show the viviparous phenotype, mature seeds of the mutant and the wild type were incubated in high humidity.

unpublished; Yamazaki et al., unpublished). The viviparous mutant was shown to be deficient in abscisic acid (ABA) synthesis (under the induced condition, the mutant level is less than 1% of the wild type). Two genes involved in the biosynthesis of ABA have been recently cloned by transposon-tagging in *Nicotiana plumbaginifolia* (Martin et al. 1996) and maize (Tan et al. 1997). The sequence data indicate that the rice gene is distinct from these two genes.

Further experiments are needed to obtain final proof that the cloned genes are causative genes. One straightforward strategy is molecular complementation: introduction of cloned wild-type genes into mutants to restore the wild-type phenotype. This can be done by using an efficient method of *Agrobacterium*-mediated transformation of rice (Hiei et al. 1994). Another important strategy is to find new alleles of the mutations by using PCR-screening (see the following section). If all the null mutant alleles show an identical phenotype, this would be proof. This strategy is expected to become widely adopted, because of its simplicity.

Although final proof is needed, the above results suggest that transposon-tagging using *Tos17* is quite feasible. However, we must consider one problem; that is, only two of seven mutations examined are due to a *Tos17* insertion. A similar problem was also noted when T-DNA and class II elements were used: only 35%-40% of the mutations are caused by a T-DNA insertion (Azpiroz-Leehan and Feldmann 1997) and six out of nine by a *Ds* insertion (Long et al. 1993, Bancroft et al. 1993). In these cases, non-tagged mutations are thought to be caused by abortive insertion events which result in a rearrangement at the target site without integration of the elements (Azpiroz-Leehan and Feldmann 1997, Bancroft et al. 1993). In the case of mutagenesis with class II elements, a rearrangement caused by imprecise excision would be another important factor, as has been discussed in rice where the *Ac/Ds* system was employed (Izawa et al. 1997). Mutations not tagged with *Tos17* may be due to insertions of unknown transposable elements activated by tissue culture. In addition, mechanisms of tissue culture-induced mutations not related to transposable elements (Larkin and Scowcroft 1981, Brettell et al. 1986, Dennis et al. 1987) may also be responsible for non-tagged mutations. To increase the efficiency of transposon-tagging, it is important to find new transposable elements responsible for tissue culture-induced mutations and/or reduce mutation frequency not related to transposable elements. As discussed, transposition of plant retrotransposons including *Tos17* is mainly regulated at the transcriptional level. Thus, it would be possible to avoid the problems associated with tissue culture by modifying LTRs of *Tos17* so that it can transpose under conditions other than tissue culture.

4.2 Use as a tool for reverse genetic studies

We are using two strategies to screen mutants for reverse genetic studies. One is random sequencing of mutated genes by isolating the sequences flanking transposed *Tos17*. Another important strategy is PCR screening of mutants of the gene of interest.

4.2.1 Systematic screening of mutants by sequencing *Tos17*-insertion sites

As we reported previously, mutants of interesting genes were identified by sequencing only eight sequences flanking *Tos17* insertions (Hirochika et al. 1996). In this case, inverse-PCR was used to amplify the flanking sequences. To carry out large-scale sequencing of the flanking sequences, simpler PCR methods such as TAIL-PCR (Liu and Whittier 1995) and suppression PCR (Siebert et al. 1995) were adopted, and PCR products were directly sequenced using an automated DNA sequencer. Sequences obtained were subjected to similarity search using the BLAST program. Among 194 target sequences examined, 45 and 29 showed a significant

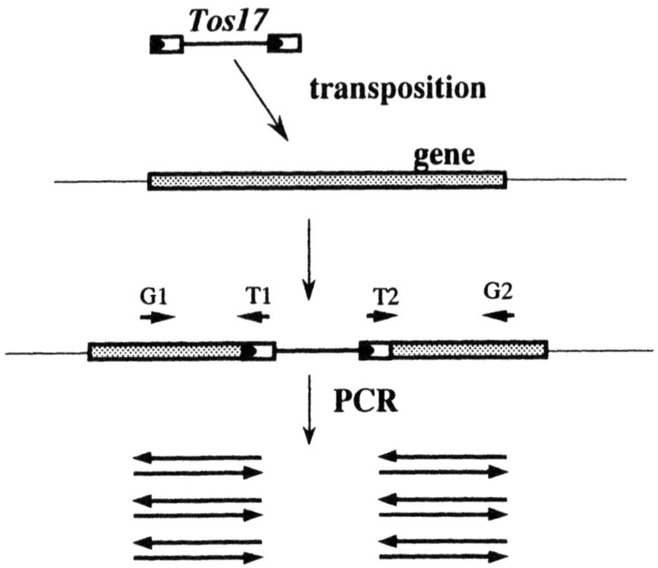

Fig. 5. Use of PCR to detect *Tos17*-insertion mutants of a specific gene. Four combinations of two gene-specific primers (*G1, G2*) and two *Tos17*-specific primers (*T1, T2*) were used in separate reactions to detect *Tos17* insertions irrespective of their position and orientation.

sequence similarity to sequences in data bases at nucleotide and amino acid levels, respectively (A. Miyao et al., unpublished results). Different classes of genes, such as genes for transcription factors, genes involved in the signal transduction, genes with similarity to disease resistance genes, and genes involved in metabolism, have been shown to be disrupted by *Tos17* insertion. In several cases, insertion into a retroposon and retrotransposons has been found. This strategy should be useful not only for identification of mutants but also for discovery of new genes. Thirty-six thousand expressed sequence tags (ESTs) of rice have been sequenced, but these correspond to only one-third to one-half of the rice genes, considering the redundancy (Sasaki 1997). This suggests that at least one-half of the mutant genes identified by random sequencing of the *Tos17*-flanking sequences are new genes.

Mutants identified by this method are being characterized as to their phenotypes. Some showed clear phenotypes. For example, the mutant of a ring finger protein gene showed a strong reduction of lateral root elongation and decreased elongation of many parts of organs such as leaves, stems, and roots (H. Hirochika et al., unpublished data). A similar phenotype was observed with the *tir*3 mutant of *Arabidopsis*, which has reduced auxin transport (Ruegger et al. 1997). Ring finger proteins have been reported in many species (Bouchard and Cote 1993) including plants and appear to be involved in various cellular functions, such as DNA repair, transcriptional regulation, and RNA binding. The ring finger protein gene of rice is most similar to one class involved in transcriptional regulation. These results raise the possibility that the rice ring finger protein gene may be involved in auxin transport by regulating gene expression or gene expression regulated by auxin.

In summary, all these results indicate that the strategy described in this section is feasible and applicable to the genome-wide screening of mutants. Another important result obtained by the analysis of the flanking sequences is that the ends of transposed *Tos17* copies are intact. In contrast, rearrangements of terminal regions of T-DNA are induced quite often (Castle et al. 1993, Krysan et al. 1996). Because there is no induction of rearrangements of the *Tos17* ends, it is easy to isolate flanking sequences and screen for mutants by PCR (see the following section).

4.2.2 Screening of mutants of a specific gene by using PCR

The principle of PCR screening of mutants is shown in Fig. 5. Four combinations of two gene-specific primers (G1, G2) and two *Tos17*-specific primers (T1, T2) were used in separate reactions to detect *Tos17* insertions irrespective of their position and orientation. Two primer combinations produce PCR products if mutants carry insertions between T1 and T2 primer sites. If mutants carry insertions outside this region, only one primer combination produces PCR products. To carry out PCR screening efficiently, DNAs used were extracted from pooled samples. A two- or three-dimensional pooling system can be used. For screening of a small population, a two-dimensional system was used (Fig. 6). Leaves of 30 regenerated rice lines from a row or a column were mixed to make a pool and DNA was extracted from each pool. Thus, 900 lines can be converted to 60 DNA pools, which are subjected to PCR reaction. If this population contains a mutant of a target gene, a PCR product is produced from one row and one column; thus, the mutant line can be addressed on the matrix. To further reduced the number of PCR reactions, superpools can be used.

To examine the feasibility of PCR screening, 1750 regenerated rice plants have been screened by using specific primers to different genes encoding receptor kinases and zinc finger proteins, disease resistance-related genes, and genes involved in development such as homeobox genes and phytochrome genes. After PCR reaction, amplified fragments were electrophoresed and visualized by staining with EtBr. To confirm that the amplified fragments were derived from the mutant genes, the fragments were sequenced. Mutants of the homeobox gene (S. Matsuoka et al., unpublished), genes with sequence similarity to disease resistance-related genes *MLO* of barley and *LSD1* of Arabidopsis (H. Hirochika et al., unpublished), the phytochrome A gene (H. Kanegae et al., unpublished), and MAP kinase genes (R. Hirochika et al., unpublished) have been isolated. Details of the analysis of the mutant of the homeobox gene are described by S. Matsuoka et al. in this volume. Degenerate primers for the conserved sequence of *myb*-like genes and serine/threonine protein kinase genes were also used for PCR screening (H. Hirochika et al., unpublished). One mutant of a *myb*-like gene and four mutants of serine/threonine kinase genes have been identified. Degenerate primers should be useful to screen for mutants of gene families.

All the results described here demonstrate the feasibility of the PCR screening

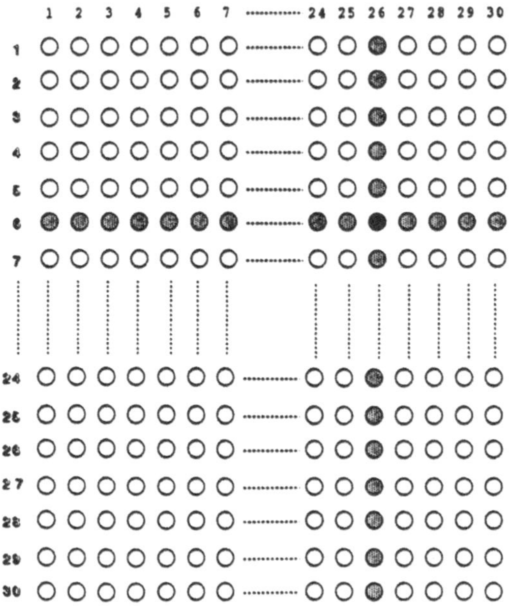

Fig. 6. Two-dimensional pooling of DNAs of regenerated rice lines. Each circle represents an independent line. Leaves of 30 lines from a row or a column are mixed to make a pool and DNA is extracted from each pool. These DNAs are subjected to PCR reaction. If an amplification product is produced from one row (No. 6) and one column (No. 26), the mutant line can be addressed on the matrix (shown by a *filled circle*).

of mutants. To make this system generally applicable to any gene of interest, mutations induced by *Tos17* insertion must be saturated. The number of regenerated rice plants required for saturation mutagenesis can be estimated based on the following assumptions. (1) *Tos17* preferentially transposes into genes, as has been suggested by the analysis of the transposition target sites (Hirochika et al. 1996). (2) The total number of rice genes is 30,000. (3) The number of transposed *Tos17* copies per regenerated plant is five. By using the equation of Clarke and Carbon (1976), a total of 20,000 regenerated rice lines are estimated to be required to give 95% saturation. We are now in the process of screening mutants and making 20,000 regenerated rice plants for saturation mutagenesis.

5 Conclusion

Insertional mutagenesis followed by cloning of causative genes is one of the most important methods for plant molecular biology. Insertion elements such as transposable elements of maize have been efficiently used in many heterologous species as well as in homologous plant (Sundaresan 1996). Another type of insertion element, T-DNA, has also been proved to be very useful in Arabidopsis (Azpiroz-Leehan and Feldmann 1997). The studies reviewed in this article demonstrated the feasibility of transposon-tagging in rice using an endogenous retrotransposon *Tos17*, which is silent under normal conditions and can be activated by tissue culture. Although the first transposon-tagging in eukaryotes used the *copia* retrotransposon (Bingham et al. 1981), most transposon-tagging experiments have been carried out using class II elements. One major reason for the limited use of retrotransposons is their low frequency of transposition. As discussed above, the transposition frequency of the rice retrotransposon *Tos17* is high enough for practical use. It has also been shown that *Tos17* can be used as a powerful tool for reverse genetic studies of rice. Because of an explosion in the number of sequenced genes of rice with unknown function (Sasaki et al. 1994, Yamamoto and Sasaki 1997, Sasaki 1997), development of a systematic method of discovering the biological functions of these genes is extremely important. Reverse genetic strategies using *Tos17* should enable this. Furthermore, it should be possible to develop new methods for genome analysis by using the unique features of retrotransposons, as has been nicely demonstrated in yeast (Smith et al. 1996). To establish the systems for forward and reverse genetic studies, about 20,000 regenerated rice lines are required for saturation mutagenesis with *Tos17*. Considering the ease of tissue culture and regeneration of rice, this goal should be accomplished in the near future.

Acknowledgments. This study was supported by a project grant from the Ministry of Agriculture, Forestry, and Fisheries of Japan, a grant-in-aid for Scientific Research on Priority Areas from the Ministry of Education, Science and Culture of Japan (No. 06278103), a grant for the enhancement of Center-of-Excellence, the special coordination funds for promoting science and technology in Japan, and the Program for Promotion of Basic Research Activities for Innovative Biosciences.

References

Azpiroz-Leehan R, Feldmann KA (1997) T-DNA insertion mutagenesis in *Arabidopsis*: going back and forth. Trends Genet 13:152-156

Bajaj YSP (1990) Somaclonal variation - origin, induction, cryopreservation, and implications in plant breeding. In: Bajaj YSP (ed) Somaclonal variation in crop. Springer-Verlag, Berlin, pp 4-48

Ballinger DG, Benzer S (1989) Targeted gene mutations in *Drosophila*. Proc Natl Acad Sci USA 86:9402-9406

Bancroft I, Dean C (1993) Transposition pattern of the maize element *Ds* in *Arabidopsis thaliana*. Genetics 134:1221-1229

Bancroft I, Jones JDG, Dean C (1993) Heterologous transposon tagging of the *DRL1* locus in *Arabidopsis*. Plant Cell 5:631-638

Bensen RJ, Johal GS, Crane VC, et al (1995) Cloning and characterization of the maize *An1* gene. Plant Cell 7:75-84

Bingham PM, Levis R, Rubin G (1981) Cloning of DNA sequences from the *white* locus of *D. melanogaster* by a novel and general method. Cell 25: 693-704

Brettell RIS, Dennis ES, Scowcroft WR, Peacock WJ (1986) Molecular analysis of a somaclonal mutant of maize alcohol dehydrogenase. Mol Gen Genet 202:235-239

Bouchard ML, Cote S (1993) The *Drosophila melanogaster* developmental gene *g1* encodes a variant zinc-finger-motif protein. Gene 125:205-209

Castle LA, Errampalli D, Atherton TL, et al (1993) Genetic and molecular characterization of embryonic mutants identified following seed transformation in *Arabidopsis*. Mol Gen Genet 241:504-514

Clarke L, Carbon J (1986) A colony bank containing synthetic Col El hybrid plasmids representative of the entire *E. coli* genome. Cell 9:91-99

Das L, Martienssen R (1995) Site-selected transposon mutagenesis at the *hfc109* locus in maize. Plant Cell 7:287-294

Dennis ES, Brettell RIS, Peacock WJ (1987) A tissue culture induced Adh null mutant of maize results from a single base change. Mol Gen Genet 210:181-183

Doolittle RF, Feng D-F, Johnson MS, McClure MA (1989) Origin and evolutionary relationships of retroviruses. Q Rev Biol 64:1-30

Gierl A (1995) The *En/Spm* transposable element of maize. In: Saedler H, Gierl A (eds) Transposable elements. Springer, Berlin Heidelberg New York, pp 132-160

Grandbastien M-A (1998) Activation of plant retrotransposons under stress conditions. Trends Plant Sci 3:181-187

Grandbastien M-A, Spielman A, Caboche M (1989) *Tnt1*, a mobile retroviral-like transposable element of tobacco isolated by plant cell genetics. Nature 337:376-380

Hiei Y, Ohta S, Komari T, Kumashiro T (1994) Efficient transformation of rice (*Oryza sativa*) mediated by *Agrobacterium* and sequence analysis of the boundaries of the T-DNA. Plant J 6:271-282

Hirochika H (1993) Activation of tobacco retrotransposons during tissue culture. EMBO J 12:2521-2528

Hirochika H (1997) Retrotransposons of rice: their regulation and use for genome analysis. Plant Mol Biol 35:231-240

Hirochika H, Fukuchi A (1992) Transposable elements in rice plants. Jpn Agric Res Quart 25:230-237

Hirochika H, Fukuchi A, Kikuchi F (1992) Retrotransposon families in rice. Mol Gen Genet 233:209-216

Hirochika H, Otsuki H, Yoshikawa M, et al (1996) Autonomous transposition of the tobacco retrotransposon *Tto1* in rice. Plant Cell 8:725-734

Hirochika H, Sugimoto K, Otsuki Y, et al (1996) Retrotransposons of rice involved in mutations induced by tissue culture. Proc Natl Acad Sci USA 93:7783-7788

Izawa T, Ohnishi T, Nakano T, et al (1997) Transposon tagging in rice. Plant Mol Biol 35:219-229

Kaiser K, Goodwin SF (1990) "Site-selected" transposon mutagenesis of *Drosophila*. Proc Natl Acad Sci USA 87:1686-1690

Koes R, Souer E, Van Houwelingen A, et al (1995) Targeted gene inactivation in petunia by PCR-based selection of transposon insertion mutants. Proc Natl Acad Sci USA 92:8149-8153

Krysan P, Young J, Tax F, Sussman M (1996) Identification of transferred DNA insertions within *Arabidopsis* genes involved in signal transduction and ion transport. Proc Natl Acad Sci USA 93:8145-8150

Larkin PJ, Scowcroft WR (1981) Somaclonal variation - a novel source of variability from cell culture for plant improvement. Theor Appl Genet 60:197-214
Liu Y-G, Whittier RF (1995) Thermal asymmetric interlaced PCR: automatable amplification and sequencing of insert end fragment from P1 and YAC clones for chromosomal walking. Genomics 25:674-681
Long D, Martin M, Sundberg E, et al (1993) The maize transposable element system *Ac/Ds* as a mutagen in *Arabidopsis*: identification of an *albino* mutation induced by *Ds* insertion. Proc Natl Acad Sci USA 90:10370-10374
Martin E, Nussaume L, Quesada A, et al (1996) Molecular identification of zeaxanthin epoxidase of *Nicotiana plumbaginifolia*, a gene involved in abscisic acid biosynthesis and corresponding to the *ABA* locus *of Arabidopsis thaliana*. EMBO J 15:2331-2342
McKinney EC, Ali N, Traut A, et al (1995) Sequence-based identification of T-DNA insertion mutations in *Arabidopsis*: actin mutants *act2-1* and *act4-1*. Plant J 8:613-622
Motohashi R, Ohtsubo E, Ohtsubo H (1996) Identification of *Tnr3*, a *Suppressor-Mutator/ Enhancer*-like transposable element from rice. Mol Gen Genet 250:148-152
Noma K, Nakajima R, Ohtsubo H, Ohtsubo E (1997) *RIRE1*, a retrotransposon from wild rice *Oryza australiensis*. Genes Genet Syst 72:131-140
Okagaki RJ, Wessler SR (1988) Comparison of non-mutant and mutant waxy genes in rice and maize. Genetics 120:1137-1143
Otsuki Y (1990) A visual manual for the protoplast culture system of rice. Food and Agriculture Research Development Association, Tokyo
Peschke VM, Phillips RL (1991) Activation of the maize transposable element *Suppressor-mutator* (*Spm*) in tissue culture. Theor Appl Genet 81:90-97
Peschke VM, Phillips RL, Gengenbach BG (1987) Discovery of transposable element activity among progeny of tissue culture-derived maize plants. Science 238:804-807
Pouteau S, Huttner E, Grandbastien MA, Caboche M (1991) Specific expression of the tobacco *Tnt1* retrotransposon in protoplasts. EMBO J 10:1911-1918
Ruegger M, Dewey E, Hobbie L, et al (1997) Reduced naphthylphthalamic acid binding in the *tir*3 mutant of *Arabidopsis* is associated with a reduction in polar auxin transport and diverse morphological defects. Plant Cell 9:745-757
Sasaki T (1997) The rice genome project in Japan. Proc Natl Acad Sci USA 95:2027-2028
Sasaki T, Song J, Koga-Ban Y, et al (1994) Towards cataloguing all rice genes: large scale sequencing of randomly chosen cDNAs from a callus cDNA library. Plant J 6:615-624
Siebert PD, Chenchik A, Kellogg DE, et al (1995) An improved PCR method for walking in uncloned genomic DNA. Nucleic Acids Res 23:1087-1088
Smith V, Chou KN, Lashkari D, et al (1996) Functional analysis of the genes of yeast chromosome V by genetic footprinting. Science 274:2069-2074
Sundaresan V (1996) Horizontal spread of transposon mutagenesis: new uses for old elements. Trends Plant Sci 1:184-190
Takeda S, Sugimoto K, Otsuki H, Hirochika H (1998) Transcriptional activation of the tobacco retrotransposon *Tto1* by wounding and methyl jasmonate. Plant Mol Biol 36:365-376
Tan BC, Schwartz SH, Zeevart JA, McCarty DR (1997) Genetic control of abscisic acid biosynthesis in maize. Proc Natl Acad Sci USA 94:12235-12240
Xiong Y, Eickbush TH (1990) Origin and evolution of retroelements based upon their reverse transcriptase sequences. EMBO J 9:3353-3362
Yamamoto K, Sasaki T (1997) Large-scale EST sequencing in rice. Plant Mol Biol 35:135-144

Zwaal RR, Broeks A, van Meurs J, et al (1993) Target-selected gene inactivation in *Caenorhabditis elegans* by using a frozen transposon mutant bank. Proc Natl Acad Sci USA 90:7431-7435

4
Ac/Ds as tools for rice molecular biology

Takeshi Izawa and Ko Shimamoto

Laboratory of Plant Molecular Genetics, Nara Institute of Science and Technology, 8916-5 Takayama, Ikoma, Nara 630-0101, Japan

1 Introduction

A model system in biology provides valuable concepts that can be shared by diverse biological systems. Rice, one of the most important crops in the world, is becoming an important model plant, mainly due to accumulation of fundamental information on its genome and development of efficient transformation system (Hiei et al. 1994; Izawa and Shimamoto 1996; Harushima et al. 1998). Especially, comparative genetics in cereal makes rice a practical model in monocotyledonous plant species (Devos and Gale 1997). The availability of an efficient transposon tagging system in rice would provide the use for insertional mutagenesis with rice. Although a rice retrotransposon that can be activated in culture has been recently isolated and its use as a insertional mutagen is suggested (Hirochika et al. 1996), no other active endogenous transposons have been reported in rice. Therefore, the maize transposable element *Ac/Ds* was introduced into rice and demonstrated to be highly active in rice (Izawa et al. 1991; Murai et al. 1991; Shimamoto et al. 1993; Sugimoto et al. 1994; Izawa et al. 1997).

The maize transposon *Ac/Ds* found by Barbara McClintock (Fedoroff 1989) is the first mobile genetic element that was recognized by human beings and has become a powerful tool for gene isolation, since the element can be employed to correlate biological phenomena with molecular interactions (Sundaresan 1996, Walbot 1992). Many plant genes have been isolated using transposons as molecular tags without information on biochemical properties of their gene products and their expression patterns.

Key words. transposon, *Ac/Ds*, insertional mutagenesis, transposon tagging, gene disruption, inactivation of transposon

The first cloning of a plant gene by transposon tagging, performed in 1984, was the maize *bronze* gene which encodes UDP-glucose: flavonoid 3-*O*-glucosyltransferase, a key enzyme of anthocyanin synthesis (Fedoroff et al. 1984). So far, a large number of genes have been isolated by endogenous transposons as tags (Walbot 1992). The use of transposons as a cloning tool, however, had been confined to plant species with active, well-characterized transposon systems, such as maize and snapdragon. Therefore, it was an epoch-making discovery when B. Baker and her colleagues (Baker et al. 1986; Baker et al. 1987) demonstrated active transposition of the maize *Ac*/*Ds* elements in transgenic tobacco. Afterwards, *Ac*/*Ds* has been introduced into various heterologous hosts and their activities were examined (Sundaresan 1996). In all species tested to date, the element was shown to exhibit some level of activity and to leave characteristic footprints, or empty donor sites (EDSs), with *Ac*/*Ds* upon transposition, suggesting that host factors involved in the *Ac*/*Ds* transposition process are widely conserved. The fact that *Ac* is highly active in tobacco and tomato (Jones et al. 1990; Yoder 1990; Osborne et al. 1991; Keller et al. 1993), but shows much reduced levels of activity in *Arabidopsis* (Schmidt and Willmitzer 1989; Keller et al. 1992; Dean et al. 1992), indicates a clear difference of transposition efficiency between species.

Seven years after Baker's report, a flower color gene was successfully cloned from petunia using *Ac* (Chuck et al. 1993). This is the first example for a gene to be cloned in heterologous host using an exogenous transposon. Concurrently, genes for leaf morphology and albino were isolated from *Arabidopsis* by the *Ac*/*Ds* system (Bancroft et al. 1993a; Long et al. 1993) and an *Arabidopsis* gene involved in male fertility was cloned by the use of another maize transposon system *En*/*Spm* (Aarts et al. 1993).

Directed tagging of target genes has been performed with *Ac*/*Ds* because they frequently transpose into linked loci from the original sites in several plant species. Three disease-resistant genes, the tomato *Cf-9* gene (Jones et al. 1994) for resistance to *Cladosporium fulvum*, the tobacco *N* gene (Whitham et al. 1994) for resistance to tobacco mosaic virus and the flax rust resistance gene, *L6* (Lawrence et al. 1995), and the *FAE1* gene (James et al. 1995) involved in fatty acid synthesis from *Arabidopsis* were cloned by directed transposon tagging. It is noteworthy that one *Ds* element located at 3 cM from the *Cf-9* gene was activated to tag the gene by the *Ac*TPase gene. Among approximately 160,000 progeny screened, at least 63 independent *Cf-9* mutants were obtained, of which 37 independent *Ds* insertions were detected. These results indicate effectiveness of transposon tagging from a linked transposon to a target locus. However, this powerful method of gene isolation has not been applied yet to major cereals that do not carry active transposable elements.

With development of PCR technology to efficiently amplify DNA fragments flanking over molecular tags such as T-DNA and transposons, targeted gene inactivation by PCR-based selection has been reported using endogenous transposons in petunia and maize (Koes et al. 1995; Mena et al. 1996). T-DNA has been also used for PCR-based mutant screening in *Arabidopsis* (McKinney et al. 1995). There is one example reported using *Ac*/*Ds* in a heterologous system, tomato (Cooley et al. 1996). Considering the tremendous accumulation of genome information such as expression sequence tags (ESTs) of rice, rice can be a good target to develop a gene disruption system.

In this chapter, we review behaviors of *Ac*/*Ds* in rice and discuss the possibility of the use of *Ac*/*Ds* as tools for molecular genetic studies in rice.

2 Structure of *Ac/Ds*

The autonomous *Ac* element was cloned from the *waxy* locus of maize (Fedoroff et al. 1983; Beherens et al. 1984). The *Ac* element is about 4.6 kb in length and has short, 11-base-pair imperfect inverted terminal repetitions (IRs). A single transcriptional unit has been identified and extends over most of the element's length and encodes a 3.5-kb mRNA (Figure 1). This mRNA has an ORF of 807 amino acids that is referred as *Ac*TPase, the product of which may function as a transposase in heterologous plant species. Although the *Ac* element can transpose by the action of *Ac*TPase encoded by itself, both the insertion and excision of the non-autonomous *Ds* elements basically depend on the presence of active *Ac*TPase expression. In the absence or inactivation of the autonomous element, the non-autonomous elements neither insert nor excise (Fedoroff 1989). The endogenous *Ds* of maize have proved surprisingly varied in structure, ranging from *Ac* elements with deletions to sequences that are virtually unrelated to that of the *Ac* element, except for the terminal IRs. Therefore, various synthetic *Ds* elements, which were *Ac* elements with synthetic deletions, have been introduced into several heterologous hosts and proved to function as *Ds* elements.

Ac/*Ds* elements undergo nonreplicative transposition and create the 8 bp of host sequence duplication at the site of element insertion (Fedoroff 1989). Upon excision, part or all of the duplication remains. The most common features of such sites, termed empty donor sites (EDSs), are deletions of several central nucleotides

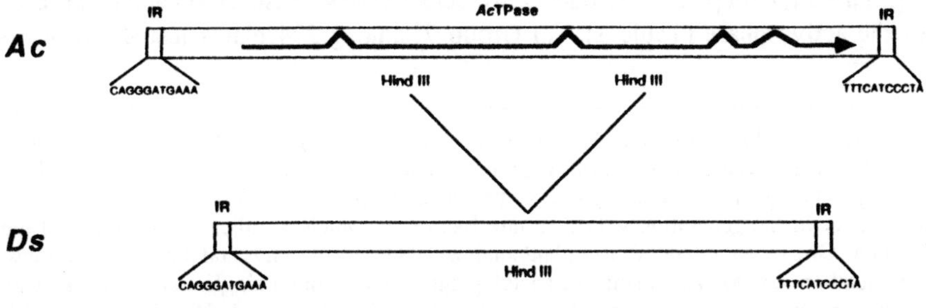

Fig. 1. Structure of *Ac*/*Ds* that was introduced into rice. The *Ac* was originally cloned from the *wx-m7* allele of maize (Beherens et al. 1984). The plasmid introduced into rice was described earlier (Izawa et al. 1991). A synthetic *Ds* that was an *Ac* derivative with a 1.6-kb deletion of a *Hin*dIII fragment was introduced into rice (Shimamoto et al. 1993). The *Ds* does not encode a functional *Ac*TPase.

and/or transversion of one or two central nucleotides. Once genes are tagged by *Ac/Ds*, this empty donor site allows stable or reverted phenotypes due to the excision of tagged *Ac/Ds*, both somatically and germinally. This characteristic is remarkable when compared with T-DNA and retrotransposon as insertional mutagens since T-DNA and retrotransposon never excise from the integration sites. Once reversion of phenotypes, which is correlated to the excision of *Ac/Ds*, is found, neither complementation tests nor correlation of independent mutations from several alleles with the phenotype is required for confirmation of tagging.

3 Transposition of *Ac/Ds* in rice

3.1 Behavior of the autonomous *Ac* element

First, we examined transposition of *Ac* in transgenic rice (Figure 1; Izawa et al. 1991). The 4.6-kb *Ac* element from the *wx-m7* allele of maize (Beherens et al. 1984) that was inserted between the CaMV35S promoter and the hygromycin-resistant (Hm^r) gene was introduced into rice protoplasts by electroporation (Shimamoto et al. 1989). When *Ac* is excised, the Hm^r gene is restored and rice cells acquire resistance to the antibiotic hygromycin. Hm^r-transformed calli appeared with frequencies of approximately 10^{-3} per transformed cells. Regenerated plants from *Ac*-containing calli showed distinct pattern of *Ac* insertions among different leaves of a single rice plant indicating that *Ac* transposes in leaves during development. Sequences of the footprints left between the 35S promoter and the Hm^r gene were similar to those that were detected in maize, transgenic tobacco, and transgenic *Arabidopsis* (Izawa et al. 1997).

We next analyzed 803 rice descendant plants derived from a single transformant carrying *Ac* through four successive generations (R4, R5, R6, and R7) by DNA gel blot analysis (our unpublished results). These plants can be divided into seven pedigree groups (our unpublished results). In four pedigree groups (Group 3, Group 4 [Figure 2], Group 6, and Group 7), we detected new transposition and inheritance of the active phase (Table 1). In Group 7, among 274 plants tested, 53 plants

Fig. 2a,b. Detailed analysis of *Ac* transposition in pedigree groups. Results of an inactive pedigree group (**a**) and an active one (**b**) are schematically presented. DNA gel blot analysis of 41 plants (**a**) and 141 plants (**b**) is summarized, respectively. One column represents a specifc *Ac* in the generation with a certain band size, which is indicated at the top of the column. One row corresponds to a plant analyzed. Naming of individual plants is as follows. The number next to "Ac" identifies an R1 plant derived from a single transformant. One hyphen indicates the proceeding of one generation. For instance, Ac26-1 indicates an R2 plant and Ac26-1--3 indicates an R4 plant, although the number of R3 plant is missing. When the band was present in the plant, a colored bar is presented. A *red bar* represents the presence of *Ac* according to the Mendelian transmission. A *blue bar* represents the presence of newly transposed *Ac*. A *green bar* indicates germinally transmitted *Ac*, which was identified as a new *Ac* band in the parental generation.

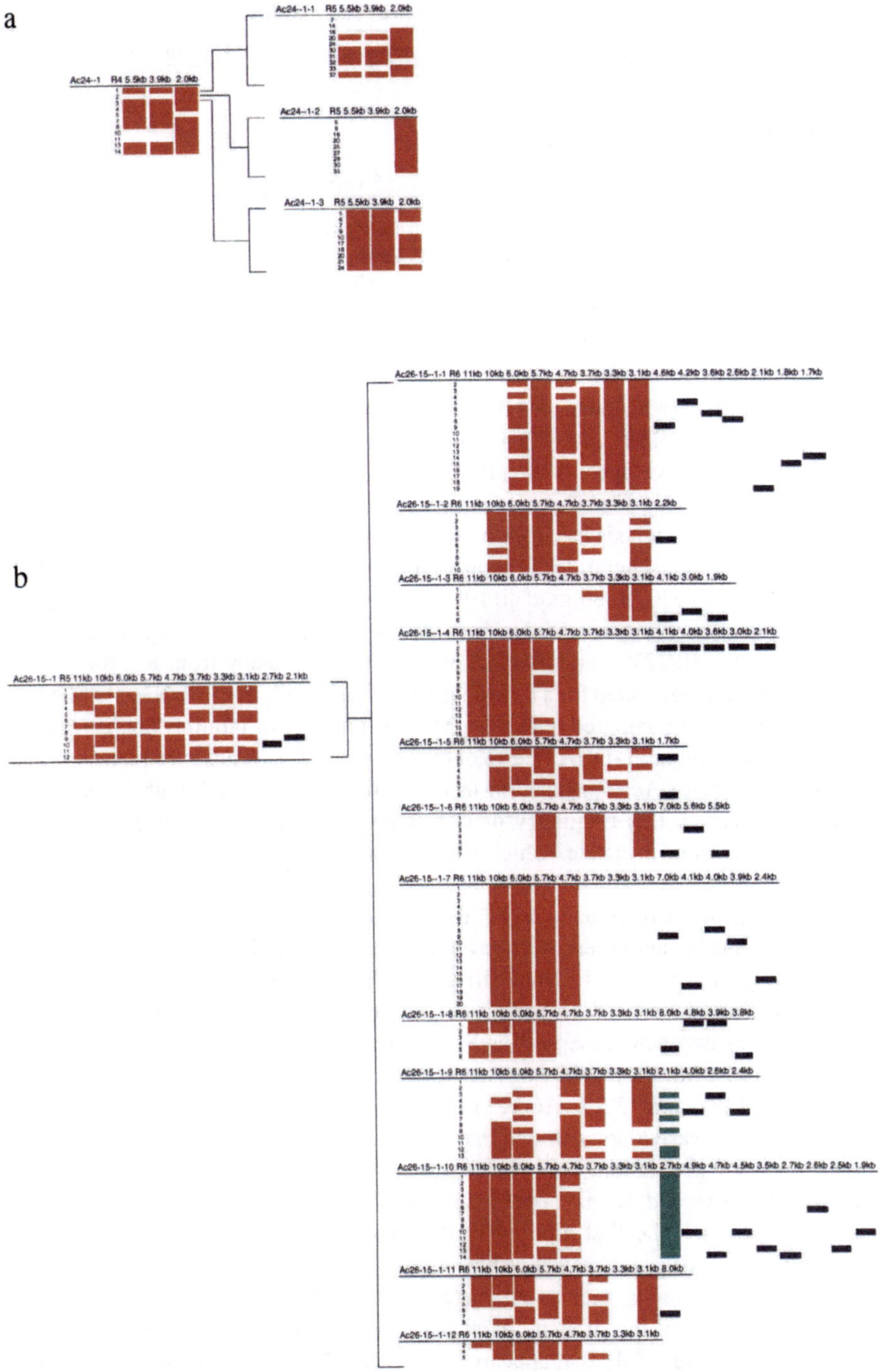
a
Ac24-1 R4 5.5kb 3.9kb 2.0kb
Ac24-1-1 R5 5.5kb 3.9kb 2.0kb
Ac24-1-2 R5 5.5kb 3.9kb 2.0kb
Ac24-1-3 R5 5.5kb 3.9kb 2.0kb
b
Ac26-15-1 R5 11kb 10kb 6.0kb 5.7kb 4.7kb 3.7kb 3.3kb 3.1kb 2.7kb 2.1kb
Ac26-15-1-1 R6 11kb 10kb 6.0kb 5.7kb 4.7kb 3.7kb 3.3kb 3.1kb 4.6kb 4.2kb 3.6kb 2.6kb 2.1kb 1.8kb 1.7kb
Ac26-15-1-2 R6 11kb 10kb 6.0kb 5.7kb 4.7kb 3.7kb 3.3kb 3.1kb 2.2kb
Ac26-15-1-3 R6 11kb 10kb 6.0kb 5.7kb 4.7kb 3.7kb 3.3kb 3.1kb 4.1kb 3.0kb 1.9kb
Ac26-15-1-4 R6 11kb 10kb 6.0kb 5.7kb 4.7kb 3.7kb 3.3kb 3.1kb 4.1kb 4.0kb 3.6kb 3.0kb 2.1kb
Ac26-15-1-5 R6 11kb 10kb 6.0kb 5.7kb 4.7kb 3.7kb 3.3kb 3.1kb 1.7kb
Ac26-15-1-6 R6 11kb 10kb 6.0kb 5.7kb 4.7kb 3.7kb 3.3kb 3.1kb 7.0kb 5.6kb 5.5kb
Ac26-15-1-7 R6 11kb 10kb 6.0kb 5.7kb 4.7kb 3.7kb 3.3kb 3.1kb 7.0kb 4.1kb 4.0kb 3.9kb 2.4kb
Ac26-15-1-8 R6 11kb 10kb 6.0kb 5.7kb 4.7kb 3.7kb 3.3kb 3.1kb 8.0kb 4.8kb 3.9kb 3.8kb
Ac26-15-1-9 R6 11kb 10kb 6.0kb 5.7kb 4.7kb 3.7kb 3.3kb 3.1kb 2.1kb 4.0kb 2.6kb 2.4kb
Ac26-15-1-10 R6 11kb 10kb 6.0kb 5.7kb 4.7kb 3.7kb 3.3kb 3.1kb 2.7kb 4.9kb 4.7kb 4.5kb 3.5kb 2.7kb 2.6kb 2.5kb 1.9kb
Ac26-15-1-11 R6 11kb 10kb 6.0kb 5.7kb 4.7kb 3.7kb 3.3kb 3.1kb 8.0kb
Ac26-15-1-12 R6 11kb 10kb 6.0kb 5.7kb 4.7kb 3.7kb 3.3kb 3.1kb

Table 1. Frequency of *Ac/Ds* into coding sequences and retrotransposons in rice

Line	*Ac/Ds* Insertions[a]	Into genes[b]	Into retrotransposons[c]	Transmission to progeny[d]
Ac26-15–1(*Ac*)	44	8	3	4 (/6)
Ac26-6 (*Ac*)	55	13	1	6 (/11)
B10-1 (*Ds*)	18	3	0	Not tested
Total	117	24 (21%)	4 (3.4%)	

[a] 117 independent clones flanking over *Ac/Ds* were cloned and partially sequenced.
[b] 24 clones exhibited significant homology with known sequences in databases.
[c] Four clones exhibited significant homology with retrotransposons.
[d] Among insertions into coding sequences, 17 clones were examined by PCR and DNA gel blot analyses. 10 insertions into the genes were confirmed to be transmitted to the progeny.

(19.3%) carried newly transposed *Ac*. The frequency of *Ac* transposition was 40.9% (9/22), 22.5% (32/142), and 8.3% (10/121) in Group 3, Group 4, and Group 6, respectively. On average, 18.6% (104/559) was the frequency of *Ac* transposition in these active lines.

DNA gel blot analysis has been particularly useful in studies of *Ac* transposition through successive generations in tomato, tobacco, and petunia (Belzile et al. 1989; Hehl and Baker 1990; Robbins et al. 1994). Hehl and Baker analyzed 123 transgenic tobacco plants carrying *Ac* and demonstrated *Ac* transposition at frequencies of 0% (0/29) to 83% (10/12). In petunia, 24.6% (15/61) of progeny from R3, R4 and R5 families, which segregated for a transposed *Ac* from the original T-DNA site, showed new transposition compared with 0% (0/94) of progeny in families that carry only an *Ac* at the original T-DNA site (Robbins et al. 1994). These frequencies were similar to those of *Ac* transposition in rice shown above. Although it cannot be simply compared, this frequency of rice is also comparable to the frequency of germinal reversion in maize, which was estimated to be 1% to 17% (Greenblatt 1968).

The frequencies mentioned do not necessarily indicate the number of new transposition events because there were several plants carrying the same transposition in the progeny of a plant and plants carrying multiple newly transposed *Ac*. Therefore, based on the size of *Ac* band, new *Ac* transposition events were counted in rice. Twenty two new transposition events were observed in Group 7, indicating the frequency of 8.0% (22/274). The frequency in other lines was 36.3% (8/22), 30.3% (43/142), and 13.2% (16/121) in Group 3, Group 4, and Group 6, respectively. These clearly demonstrated that *Ac* transposition activity was inherited stably even in R7 generation.

Newly transposed *Ac* that appeared in these pedigrees were roughly categorized into two groups. The first group included new *Ac* that appeared uniquely in the progeny. *Ac* in this group may have been transposed in a cell that became a germline cell of the parent plants, such as a gamete in a floret, and transmitted into a progeny plant, or in a cell of the progeny plant that became a major part of the leaf. The second group included *Ac* that appeared in several plants of the progeny. *Ac* in this group may represent relatively early transpositions in cells of the parent plants, which became significant parts of the germline, such as gametes in a panicle.

There is one exception, in which the newly transposed *Ac* appeared in all 14 R7 progeny plants of *Ac* 26-1--3-14-4 (our unpublished results). This suggests that germline cells of the R6 parent plant carried this transposition homozygously. Since only genetic transmission can make transposed *Ac* homozygous, this transposition may have occurred in the *Ac* 26-1--3-14 (R5) plant, in which we were not able to detect the band, and it may be due to the chimeric nature of the R5 and R6 plants.

A burst of *Ac* transposition was observed in the progeny of *Ac* 26-15--1-4 (R5) plant (data not shown). Sixteen progeny plants of the R5 plant were subjected to the DNA gel blot analysis, and the result indicated that the R5 plant carried four homozygous *Ac* and one heterozygous *Ac*. All R6 progeny except one did not carry any newly transposed *Ac*, whereas the R6 plant of *Ac* 26-15--1-4-2 carried five newly transposed *Ac* in addition to five *Ac* transmitted from the R5 plant. This demonstrated that activity of *Ac* transposition at certain insertion sites can be affected epigenetically. A similar rapid proliferation of *Ac* has been reported in a line of transgenic tomato (Yoder 1990).

3.2 Behavior of the non-autonomous *Ds* element in rice

A non-autonomous synthetic *Ds* element that was an *Ac* derivative carrying a 1.6-kb internal deletion has also been examined (Figure 1; Shimamoto et al. 1993). Hm^r calli were obtained only when co-transformed with the *Ac*TPase gene driven by the CaMV35S promoter, indicating that there is no endogenous *Ac*TPase activity in rice and that *Ds* can excise only in the presence of the introduced *Ac*TPase gene. EDS sequences of *Ds* were similar to those found in the case of *Ac* in rice (Izawa et al. 1997).

Interestingly, a callus line, named No. 17, did not carry the *Ac*TPase gene and still carried several transposed *Ds* elements, suggesting that *Ds* was excised from the introduced plasmid by the action of the transiently expressing *Ac*TPase gene and that *Ds* was integrated into the rice genome during early stages of protoplast culture. A regenerated plant derived from this callus line, plant 17-1, carried four copies of stable *Ds* that were transmitted to the progeny. Cloning of these *Ds* elements revealed that two of them were structurally intact while the others had a 200-bp deletion and contained complicated DNA rearrangements including a 2.7-kbp insertion, respectively (Izawa et al. 1997). Since all the four *Ds* have the intact 5'- and 3'-region, they are potentially capable of transposition when *Ac*TPase is provided. Segregation analysis of 21 progeny from the plant 17-1 demonstrated that the four *Ds* elements were segregated in the next generation in a Mendelian fashion. Further molecular analysis revealed that plant 17-1 carried four functional *Ds* elements at three different chromosomes and did not have the *Ac*TPase gene. Therefore, this plant may be suitable as a parent plant for transposon-induced mutagenesis.

To induce transposition of *Ds*, the plant 17-1 was crossed with the transgenic plant carrying a single active 35S-*Ac*TPase gene. In total, 69 F1 plants were obtained carrying both *Ds* elements and the 35S-*Ac*TPase gene. DNA gel blot analyses of F_2 progeny revealed that germinal transposition of newly transposed *Ds* in rice occurred with high frequency (Izawa et al. 1997). While some of the bands were sharp and unique, due to late germinal transposition, others were sharp and common to siblings, due to early germinal transposition. We also observed unique and faint bands probably caused by somatic transposition.

It has been reported that the frequency of *Ds* transposition varies from species to species. Elevated level of *Ac*TPase activity was shown to confer high frequencies of *Ds* excision in *Arabidopsis* (Bancroft and Dean 1993b, 1993c; Swinburne et al. 1992) and tobacco (Scofield et al. 1992). When the 35S promoter was used to drive the *Ac*TPase gene, the frequency of germinal transposition was 10%-74% in *Arabidopsis* and 6.8% in tobacco. A previous report indicated that only approximately 20% of *Arabidopsis* F_2 plants carrying the excision maker inherited a transposed *Ds* after activation by the 35S-*Ac*TPase (Swinburne et al. 1992). Therefore, germinal reinsertion frequencies of *Ds* in *Arabidopsis* might be in the range of 2%-15% when the *Ac*TPase gene was driven by the CaMV35S promoter. This estimated frequency of germinal transposition of *Ds* by 35S-*Ac*TPase was apparently lower than those in rice. However, to precisely estimate the frequency of *Ds* transposition in rice, further analysis is needed.

4 Inactivation of *Ac/Ds* in rice

Despite the highly active transposition of *Ds* in the F_1 plants, we were unable to detect efficient transposition of *Ds* in later generations even in the presence of the 35S-*Ac*TPase gene. Only a few lines still show active transposition of *Ds* in the F_5 generation. For example, in the presence of the *Ac*TPase gene, only one line carries actively transposing *Ds* elements among 13 F_4 lines originated from a F_2 plant, B10-1 (Izawa et al. 1997). This indicates that the *Ds* frequently becomes inactive in rice during successive generations after activation by the 35S-*Ac*TPase in F_1 plants.

To overcome the inactivation of *Ds* transposition in rice, we have tried to reactivate inactive elements. Since it is known that the tissue culture process activates silent transposable elements in maize (Bretell and Dennis 1991), we applied plant regeneration via protoplasts derived from four lines carrying *Ds* elements and the *Ac*TPase gene to see whether the culture process activates transposition of the inactive *Ds* elements. DNA gel blot analysis demonstrated that *Ds* was highly reactivated in two lines, while reactivation of *Ds* was observed only in some progeny in another line (Izawa et al. 1997). Since no *Ds* transposition was found in a line in which the *Ac*TPase gene was absent, the observed reactivation of *Ds* via protoplast culture was likely to require the function of the *Ac*TPase. These results clearly demonstrated that inactive *Ds* can be reactivated in protoplast culture procedure in the presence of the *Ac*TPase gene. Therefore, it may be possible to increase the frequency of reversion events from mutant lines even with inactive *Ds* elements in this way.

On the other hand, *Ac* basically remains active even in R7 generations as described above. However, our results also demonstrated occurrence of *Ac* inactivation and inheritance of inactive state in rice (our unpublished results). Among 803 plants comprising seven pedigree groups examined, we did not detect any transposition in Group 1, Group 2, and Group 5 although Group 1, Group 2, and Group 3 contained 2 copy (5.0 kb, 2.8 kb by *Hin*dIII), 3 copy (Figure 2; 5.5 kb, 3.9 kb, 2.0 kb), and 2 copy (5.2 kb, 4.0 kb) of *Ac*, respectively (data not shown). All the *Ac* elements in Group 1 and Group 2 were inactive while inactivation was initiated at the R4 generation in Group 5 (data not shown). Here, the change from inactive to active states has never been observed. McClintock first reported that *Ac* undergoes cyclical changes in phase of activity in maize (McClintock 1963). No such changes were observed in rice with this population size.

Epigenetic inactivation of *Ac/Ds* has been first described by McClintock (McClintock 1963) and involvement of DNA methylation has been suggested in maize (Fedoroff and Chandler 1994; Brutnell and Dellaporta 1994). Interestingly, inactivation of *Ac/Ds* has not yet been reported in any heterologous systems. These facts may suggest that there are different mechanisms in plants between transposition and inactivation of *Ac/Ds* and that no mechanisms to inactivate *Ac/Ds* exist in dicotyledonous plants such as tobacco, tomato, petunia, and *Arabidopsis*.

Chomet et al. (1987) demonstrated DNA methylation at the *Pvu* II site inside *Ac* in an inactive line of *wx-m7*, not in its active line. Furthermore, complete DNA methylation of *Hpa*II sites in *Ac* in a spontaneously arisen inactive derivative *wx-m9 Ds-cy* was observed while the active *Ac* was methylated at one end of the element (Schwartz and Dennis 1987). We have tried to find the change of methylation state of the inactive *Ac*, but have not observed any significant change so far (data not shown). It has been shown that changes in the methylation profile of *Ac* correspond to changes in *Ac* activity through the differential accumulation of the *Ac* transcripts (Brutnell and Dellaporta 1994). Therefore, more extensive analysis needs to be performed to determine whether DNA methylation and *Ac*TPase expression are involved in inactivation of *Ac* in rice.

5 Frequent gene disruption by *Ac/Ds*

There might be a preference for insertion sites of transposons, called hot spots. For instance, the yeast retrotransposon *Ty1* integrates specifically into tRNA genes (Ji et al. 1993), and there might be hot spots in the *P* gene of maize for *Ac* (Grotewold et al. 1991). Therefore, for insertional mutagenesis by transposons, this frequency should be assessed for practicality. In addition to *Arabidopsis*, rice is a suitable material to address this question because of accumulated information on its genome and cDNA. Although this type of approach has not been much reported yet, there is an example of a rice retrotransposon termed *Tos17*. Eight flanking sequences of it were analyzed, and four showed significant homology with known clones in databases (Hirochika et al. 1996).

We examined 99 flanking sequences over the 5'-region of *Ac*, and 21 flanking sequences showed significant homologies. (our unpublished results). Furthermore, 18 flanking sequences were examined over the 5'-region of *Ds*, and three clones exhibited some homologies (our unpublished results). The frequency of *Ac*, and *Ds* transposition into known coding regions, were 21.2% and 16.7%, respectively (Table 1). Homology with nontranscribing regions such as promoters and nontranslating region such as introns of rice genes could not be searched at the time. In addition, only partial information on coding regions such as of EST clones of rice is available in databases at present. Therefore, the frequency of *Ac/Ds* transposition into rice genes is likely to be underestimated.

Interestingly, ten clones showed significant homology with the *Xa21* gene of rice (Song et al. 1995). *Xa21* gene is a disease resistance gene conferring resistance to a rice pathogen, *Xanthomonas oryzae* pv. *oryzae*, and encodes a receptor-like kinase consisting of leucine-rich repeats and a serine/threonine protein kinase. Considering the overlapped sequences, at most eight genes containing domains similar to those of *Xa21* had *Ac* insertions. It has been shown that *Ac* transposes into closely linked loci (Jones et al. 1990) and that the *Xa21* locus comprises a cluster of tandemly aligned *Xa21*-related genes (Song et al. 1997). Therefore, the frequent transposition into *Xa21*-related genes may reflect closely linked transposition in a cluster of *Xa21*-related genes in the rice genome. Seven other clones contained homologies with various other genes, whose functions have been well characterized, such as those of chloroplastic aldolase (Tsutumi et al. 1994), alcohol dehydrogenase, cytochrome P450, *RAD2* endonuclease (Habraken et al. 1993), and pectin esterase (Figure 3). The other four clones showed significant homology with EST clones in databases whose functions are unknown. Two of them matched almost completely with rice EST clones at the nucleotide sequence level. Furthermore, only four clones show homologies with retrotransposons in databases. On the other hand, genomic clones flanking over *5'Ds* contained homology with β-tublin, the *Xa-21* gene, and an EST clone of rice.

These results from random sequencing may give an insight on the rice genome composition. Recently, a 1.9-Mb genomic sequence from chromosome 4 of *Arabidopsis* revealed that the average size of a gene is 4.8 kb in *Arabidopsis* and that about 19% of the sequence consists of five classes of repeats (Bevan et al. 1998). However, only ten retrotransposons at six places were dispersed in this region. This contrasts with the genome structure of the *Adh1* region of maize, where multiple retrotransposons of different types comprise about 50% of the region (SanMiguel et al. 1996). This expansion of retroelement numbers is proposed to be the principal mechanism for increasing the genome size in higher plants (Wessler et al. 1995). Among 99 flanking clones of rice, only four clones show significant homologies with Pol proteins of the maize retrotransposon. These flanking sequences may represent a partial genome composition of rice due to frequent transposition of *Ac* into linked loci. Therefore, this result suggests that retroelements may not comprise major parts of the rice genome, unlike that of maize. The total copy number of retrotransposons in the rice genome was estimated to be about 1000, which may correspond to around one-hundredth of the rice genome (Hirochika

a

```
15-46            :    1 bp AADAVVAADGSGDYTTIAAAVAAAPSKSTKRYVIYIKXGTYNELITIGQNTWNLTLIGDGMDVTIITGNQS-VGGGVSLMSKTGTVSKWHF 270 bp
                           ++DA+VA DG+G++T+++ AV AAP+ S +RYVIYIK GTY+E + I ++ WNL +IGDGMD+TII+GN+S V G +++ S T +VS   F
pectin esterase  :  206 aa SVDAIVAQDGTGNFTNVTDAVLAAPDYSMRRYVIYIKRGTYKENVEIKKKKWNLMMIGDGMDATIISGNRSFVDGWTTFRSATFAVSGRGF 296 aa
```

b

```
6-47             :   51 bp ELLEYLGMPVLRAKGEGEALCAQLNNQGHVDACITSDSDAFLFGAKTVIK 200 bp
                           ELL   G+P + A  E EA CA+L      VD  IT DSD FLFG   + K
RAD2             :  777 aa ELLSRFGIPYITAPMEAEAQCAELLQLNLVDGIITDDSDVFLFGGTKIYK 826 aa

6-47             :  331 bp ECYNMADIESGLGLKRKQMVAMALLVGSDH 420 bp
                           E Y+    I    LGL RK M+ +A L+GSD+
RAD2             :  836 aa EFYDAESILKLLGLDRKNMIELAQLLGSDY 865 aa
```

Fig. 3a,b. Examples of flanking sequences having homology with genes in databases. On flanking sequences, the nucleotide number is presented. The nucleotide number 1 of all flanking sequences corresponds to the insertion site adjacent to the 5' inverted repeat of *Ac*. On the other hand, the amino acid number is presented in database clones. The amino acid number counted from the first methionine of homologous clones is shown. **a** The 15-46 flanking clone shows homology with a pectin esterase. **b** The 6-47 flanking clone shows homology with an endonuclease of *Saccharomyces cerevisiae*, *RAD2* (Habraken et. al. 1993). *RAD2* is involved in nuclear excision repair and related to the *XP-G* gene, which causes a human genetic disease, *Xeroderma pigmentosum* (O'Donovan et al. 1994). Discontinuous homologies are likely to be due to introns in the flanking sequence.

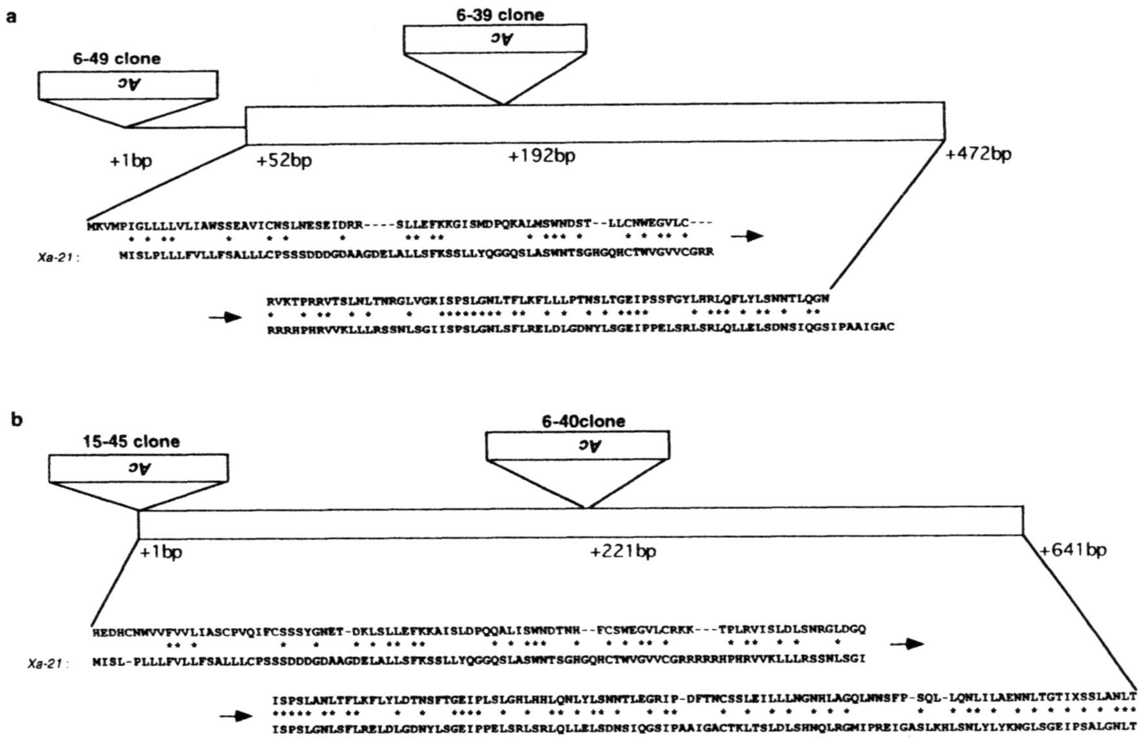

Fig. 4a,b. Schematic presentation of *Ac* insertion into closely linked loci. **a** Both the 6-49 and 6-39 clones were inserted with a 192-bp distance into the same gene of rice. The putative gene encodes a *Xa-21*-related protein. **b** Both the 15-45 and 6-40 clones were inserted with a 221-bp distance into the same gene. This gene also may encode another *Xa-21*-related protein.

1997). Therefore, this estimation is consistent with our results. Retroposons such as long interspersed nuclear elements (LINEs), which encode Gag and Pol polyproteins, and short interspersed nuclear elements (SINEs) are also significant components of plant genome but copy numbers of these elements are not yet estimated in rice. Alternatively, it is possible that *Ac* may preferentially transpose into coding sequences in rice.

Seventeen *Ac* inserted into genes were examined by DNA gel blot analysis and PCR for germinal transmission to the progeny (our unpublished results). Ten independent *Ac* insertions were confirmed to be transmitted to the progeny; seven *Ac* insertions tested were not transmitted. The failure of transmission suggests that the insertion was somatic or the mutation was dominant lethal. It is also possible that the number of progeny examined was not sufficient to confirm the transmission.

A series of sequences of flanking region over the 5'-region of *Ac* revealed that some sequences had overlapping regions (Figure 4). Two flanking clones, 6-49 and 6-39, from a pedigree group contained overlapping regions indicating *Ac* transposition into closely linked loci (192 bp). In addition, another flanking clone, 15-40, in a pedigree group also contained the insertion site of *Ac* at 221 bp from that with clone 6-40 in another group. These *Ac* insertions of clones 15-45 and 6-40 were derived from different pedigree groups that originated from the same R1 plant. This suggests that the original *Ac* was located around the inserted gene and transposed independently into the same gene in both groups. Regarding the three *Xa21* homologs, transmission of *Ac* in the genes was confirmed; however, insertion sites detected in the progeny were different from the original sites. These results would reflect that *Ac* preferentially transpose into closely linked loci.

6 Rice mutants generated by *Ac/Ds*

Various mutants in F_2 plants originated from 69 F_1 plants (*Ds* x 35S-*Ac*TPase) were screened (Izawa et al. 1997). First of all, visible mutants were scored, including stripe, albino, virescent, chlorina, dwarf, twin, and chlorosis. Generally, the number of plants displaying mutant phenotypes was not in good agreement with the Mendelian law. For example, 7 of 97 F_2 plants (7.2%) from a single F_1 plant exhibited a dwarf phenotype. Most mutations found in this experiment might be not caused by somaclonal mutation in culture, but may be generated by the action of *Ds* transposition. In addition, it is noteworthy that multiple mutant phenotypes were often observed in F_2 seedlings, and frequencies of visible mutations in the F_2 generation varied among F_1 plants (0%–32%).

It has been reported that 559 families of *Arabidopsis* were subjected to a broad screening for mutants exhibiting visible phenotypes, each family representing at least one germinal *Ds* excision event (Altmann et al. 1995) . Furthermore, 43 families representing at least one germinal *Ac* excision were also subjected to mutant screening in this experiment. Plants with abnormal phenotypes were observed in the progeny of 218 of these 602 families. When 134 families of these primary mutants were subjected to a second screening, 48 families turned out to contain plants with heritable defects. Similarly, 1000 families of *Arabidopsis* carrying approximately 500 independent germinal transposition events of *Ac* were screened for mutants showing visible phenotypes. Fifty different families segregated mutations in their progeny (Bhatt et al. 1996). In our experiment, 48 of 69 families originated from different F_1 plants exhibited visible mutant phenotypes (Izawa et al. 1997). Since the frequencies of visible mutants were much higher than those found in *Arabidopsis*, the frequency of germinal *Ds* transposition in rice is likely to be higher than that of *Arabidopsis*.

In F_3 and F_4 generations, new visible mutations were rarely found. This is consistent with the observation that *Ds* become inactive in F_3 and F_4 generations. Visible mutants were often found in progeny of regenerated plants from protoplasts of four inactive lines carrying both *Ds* and the *Ac*TPase gene (Izawa et al. 1997), whereas none was found among about 500 progeny of regenerated plants from a line carrying only *Ds* (data not shown). DNA gel blot analysis confirmed reactivation of *Ds* transposition in the four lines. These results indicate that activity of *Ds* transposition was well correlated to the generation of the visible mutations. Therefore, these mutations are likely to be caused by insertion of the *Ds* elements and/or EDSs.

6.1 Analysis of blast disease-sensitive mutants

The mutants which become sensitive against the infection of an incompatible race of blast fungus were further analyzed in detail. A japonica rice cultivar, Toride 1, which carries *Pi-z*^{t}, a gene conferring resistance to blast disease, was used for introduction of *Ac*/*Ds* (Izawa et al. 1991; Shimamoto et al. 1993). The *Pi-z*^{t} locus is one of the 14 resistance loci identified in a differential set of Japanese cultivars and is located on chromosome 6 (Kiyosawa 1984). Toride 1 was generated by introgression of the *Pi-z*^{t} gene from an indica cultivar, TKM 1, into Norin 8, which is susceptible to most races of the blast fungus. To screen mutations of the blast resistance genes caused by *Ds* insertions, about 8000 F_2 seedlings were examined and 97 susceptible plants from 29 F_1 plants were selected (Shimamoto et al. 1996).

Further analysis of F_3 progeny demonstrated that eight F_2 lines exhibited the stable S phenotype and other two lines exhibited the stable R phenotype. These phenotypes were apparently due to mutations, and they seemed to be in the homozygous state for mutations. The other 36 F_2 lines gave rise to plants segregating for resistant and susceptible phenotypes in the F_3 generation. About 50% of susceptible F_2 lines exhibited only the resistant phenotype in the F_3 generation.

To examine whether mutations found in these *Ds* lines were mapped at the *Pi-z*^{t} locus, mutant lines were crossed with a Japonica cultivar, Norin 8. Norin 8 is a recurrent parent used to generate Toride 1 and does not carry *Pi-z*^{t} genes. This genetic analyses of the mutants revealed that at least seven mutant lines were shown to have mutations at the *Pi-z*^{t} locus (Izawa et al. 1997).

To examine if these mutations were linked with specific *Ds* insertions, genomic DNAs of several F_3 plants were isolated and subjected to DNA gel blot analysis. So far, we have genetically analyzed 11 stable mutants and did not find any *Ds* linked to the phenotypes. A possible interpretation of these results is that *Ds* is frequently excised from the *Pi-z^t* locus, giving rise to new mutations due to EDSs. Similar phenomena were often observed in both *Ac*-induced *N* mutants of tobacco and *L6* mutants of flax (Lawrence et al. 1993; Whitham et al. 1994). In addition, deletions of the *L6* locus detected among the flax plants carrying *Ac* might be induced by chromosome breakage by *Ac* (Lawrence et al. 1993). Therefore, it would be interesting to examine the chromosomal structures at the *Pi-z^t* locus of our mutants. Recently, we identified that a *Ds*, termed *Ds80*, is located at about 25 cM from the *Piz^t* locus on the rice chromosome 6. Therefore, we are now attempting to screen new mutants from the line carrying *Ds80* in order to clone the *Pi-z^t* gene.

7 Perspectives

The genome of rice has been extensively analyzed because of its small genome size and its importance as a major crop (Izawa and Shimamoto 1996; Sasaki and Moore 1997). For further advancement of the genomics of rice, methods to systemically analyze functions of genes whose sequences are available in databases need to be developed. For this purpose, the use of a retrotransposon that is activated in culture has been suggested (Hirochika et al. 1996; Hirochika 1997). In this review, we propose the possibility of the maize transposon *Ac/Ds* as a tool for functional genomics of rice so as to isolate novel genes by means of changes of phenotypes and to inactivate genes whose sequences are available in databases using PCR-based screening of disruptants.

References

Aarts MGM, Dirkse WG, Stiekema WJ, et al (1993) Transposon tagging of a male sterility gene in *Arabidopsis*. Nature 363: 715-717

Altmann T , Felix G, Jessop A, et al (1995) *Ac / Ds* transposon mutagenesis in *Arabidopsis thaliana*: mutant spectrum and frequency of *Ds* insertion mutants. Mol Gen Genet 247: 646-652

Baker B, Schell J, Lorz H, et al (1986) Transposition of the maize controlling element *Activator* in tobacco. Proc Natl Acad Sci USA 83: 4844-4848

Baker B, Coupland G, Fedoroff NV, et al (1987) Phenotypic assay for excision of the maize controlling element *Ac* in tobacco. EMBO J 6: 1547-1554

Bancroft I, Jones JDG, Dean C (1993a) Heterologous transposon tagging of the DRL locus in Arabidopsis. Plant Cell 5: 631-638

Bancroft I, Dean C (1993b) Transposition pattern of the maize element *Ds* in *Arabidopsis thaliana*. Genetics 134: 1221-1229

Bancroft I, Dean C (1993c) Factors affecting the excision frequency of the maize transosable element *Ds* in *Arabidopsis thaliana*. Mol Gen Genet 240: 65-72

Beherens U, Fedoroff N, Laird A, et al (1984) Cloning of *Zea mays* controlling element *Ac* from the *wx-m7* allele. Mol Gen Genet 194: 346-347

Belzile F, Lassner MW, Tong Y, et al (1989) Sexual transmission of transposed *Activator* elements in transgenic tomatoes. Genetics 123: 181-189

Bevan M, Bancroft I, Bent E, et al (1998) Analysis of 1.9 Mb of contiguous sequence from chromosome 4 of *Arabidopsis thaliana*. Nature 391: 485-488

Bhatt AM, Page T, Lawson EJ, et al (1996) Use of *Ac* as an insertional mutagen in *Arabidopsis*. Plant J 9: 935-945

Bretell RIS, Dennis ES (1991) Reactivation of a silent *Ac* following tissue culture is associated with heritable alterations in its methylation. Mol Gen Genet 229: 365-372

Brutnell TP, Dellaporta SL (1994) Somatic inactivation and reactivation of *Ac* associated with changes in cytosine methylation and transposase expression. Genetics 138: 213-225

Chomet PS, Wessler S, Dellaporta, SL (1987) Inactivation of the maize transposable element *Activator* (*Ac*) is associated with its DNA modification. EMBO J 6: 295-302

Chuck G, Robbins T, Nijjar C, et al (1993) Tagging and cloning of a petunia flower color gene with the maize transposable element *Activator*. Plant Cell 5: 371-378

Cooley, MB, Goldsbrough AP, Still DW, et al (1996) Site-selected insertional mutagenesis of tomato with maize *Ac* and *Ds* elements. Mol Gen Genet 252: 184-194

Dean C, Sjodin C, Page T, et al (1992) Behaviour of the maize transposable element *Ac* in *Arabidopsis thaliana*. Plant J 2: 69-81

Devos KM, Gale MD (1997) Comparative genetics in the grasses. Plant Mol Biol 35: 3-15

Fedoroff NV, Wessler S, Shure M (1983) Isolation of the transposable maize controlling elements *Ac* and *Ds*. Cell 35: 235-242

Fedoroff NV, Furtek DB, Nelson OE (1984) Cloning of the *bronze* locus in maize by a simple and generalizable procedure using the transposable element *Activator* (*Ac*). Proc Natl Acad Sci USA 81: 3825-3829

Fedoroff N (1989) Maize transposable elements. In: Berg D, Howe M (eds) Mobile DNA. American Society of Microbiology, Washington, DC, pp 375-411

Fedroff NV, Chandler V (1994) Inactivation of maize transposable elements. In: Paszkowski J (ed) Homologous recombination and gene silencing in plants. Kluwer Academic Publishers, Dordrecht, The Netherlands, pp 349-385

Greenblatt IM (1968) The mechanism of *Modulator* transposition in maize. Genetics 58: 585-597

Grotewold E, Athma P, Peterson, T (1991) A possible hot spot for *Ac* insertion in the maize *P* gene. Mol Gen Genet 230: 329-331

Habraken Y, Sung P, Prakash L, et al (1993) Yeast excision repair gene *RAD2* encodes a single-stranded DNA endonuclease. Nature 366: 365-368

Harushima Y, Yano M, Shomura A, et al (1998) A high-density rice genetic linkage map with 2275 markers using a single F_2 population. Genetics 148: 479-494

Hehl R, Baker B (1990) Properties of the maize transposable element *Activator* in transgenic tobacco plants: a versatile inter-species genetic tool. Plant Cell 2: 709-721

Hiei Y, Ohta S, Komari T, et al (1994) Efficient transformation of rice (*Oryza sativa* L.) mediated by *Arobacterium* and sequence analysis of the boundries of the T-DNA. Plant J 6: 271-282

Hirochika H, Sugimoto K, Otsuki Y, et al (1996) Retrotransposon of rice involved in mutations induced by tissue culture. Proc Natl Acad Sci USA 93: 7783-7788

Hirochika H (1997) Retrotransposons of rice: their regulation and use for genome analysis. Plant Mol Biol 35: 231-240

Izawa T, Miyazaki C, Yamamoto M, et al (1991) Introduction and transposition of the maize transposable element *Ac* in rice (*Oryza sativa* L.) Mol Gen Genet 227: 391-396

Izawa T, Shimamoto K (1996) Becoming a model plant: the importance of rice to plant science. Trends Plant Sci 1: 95-99

Izawa T, Ohnishi T, Nakano T, et al (1997) Transposon tagging in rice. Plant Mol Biol 35: 219-229

James DW Jr, Lim K, Keller J, et al (1995) Directed tagging of the *Arabidopsis FATTY ACID ELONGATION 1 (FAE1)* gene with the maize tranposon Activator. Plant Cell 7: 309-319

Ji H, Moore DP, Blomberg MA, et al (1993) Hotspots for unselected *Ty1* transposition events on yeast chromosome III are near tRNA genes and LTR sequences. Cell 73: 1007-1018

Jones JD, Carland F, Lim E, et al (1990) Preferential transposition of the maize element *Activator* to linked chromosomal locations in tobacco. Plant Cell 2: 701-707

Jones DA, Thomas CM, Hammond-Kosack KE, et al (1994) Isolation of the tomato *Cf-9* gene for resistance to *Cladosporium fulvum* by transposon tagging. Science 266: 789-793

Keller J, Lim E, James DW, et al (1992) Germinal and somatic activity of the maize element *Activator* (*Ac*) in *Arabidopsis*. Genetics 131: 449-459

Keller J, Jones JD, Harper E, et al (1993) Effects of gene dosage and sequence modification on the frequency and timing of transposition of the maize element *Activator* (*Ac*) in tobacco. Plant Mol Biol 21: 157-170

Kiyosawa S (1984) Establishment of differential varieties for pathogenicity test of rice blast fungus. Rice Genet Newsl 1: 95-97

Koes R, Souer E, van Houwelingen A, et al (1995) Targeted gene inactivation in petunia by PCR-based selection of transposon insertion mutants. Proc Natl Acad Sci USA 92: 8149-8153

Lawrence G, Finnegan J, Ellis J (1993) Instability of the *L6* gene for rust resistance in flax is correlated with the presence of a linked *Ac* element. Plant J 4: 659-669

Lawrence GJ, Finnegan EJ, Ayliffe MA, et al (1995) The *L6* gene for flax rust resistance is related to the *Arabidopsis* bacterial resistance gene *RPS2* and the tobacco viral resistance gene *N*. Plant Cell 7: 1195-1206

Long D, Martin M, Sundberg E, et al (1993) The maize transposable element system *Ac / Ds* as a mutagen in *Arabidopsis*: Identification of an *albino* mutation induced by *Ds* insertion. Proc Natl Acad Sci USA 90: 10370-10374

McClintock B. (1963) Further studies of gene-control systems in maize. Carnegie Inst Washington Year Book 62: 486-493

McKinney EC, Ali N, Traut A, et al (1995) Sequence-based identification of T-DNA insertion mutations in *Arabidopsis*: actin mutants *act2-1* and *act4-1*. Plant J. 8: 613-622

Mena M, Ambrose BA, Meeley RB, et al (1996) Diversification of C-function activity in maize flower development. Science 274: 1537-1540

Murai N, Li ZJ, Kawagoe Y, et al (1991) Transposition of the maize *Activator* element in transgenic rice plants. Nucleic Acids Res 19: 617-622

O'Donovan A, Davies AA, Moggs J, et al (1994) XPG endonuclease makes the 3' incision in human DNA nucleotide excision repair. Nature 371: 432-435

Osborne BI, Corr CA, Prince JP, et al (1991) *Ac* transposition from a T-DNA can generate linked and unlinked clusters of insertions in the tomato genome. Genetics 129: 833-844

Robbins TP, Jenkin M, Courtney-Gutterson N (1994) Enhanced frequency of transposition of the maize transposable element *Activator* following excision from T-DNA in *Petunia hybrida*. Mol Gen Genet 244: 491-500

SanMiguel P, Tikhonov A, Jin YK, et al (1996) Nested retrotransposons in the intergenic regions of the maize genome. Science 274: 765-768

Sasaki T, Moore G. (Eds) (1997) *Oryza*: from molecule to plant. Kluwer Acdemic Publishers, Dordrecht, The Netherlands, pp 1-251

Schmidt R, Willmitzer L (1989) The maize autonomous element *Activator* (*Ac*) shows a minimal germinal excision frequency of 0.2%-0.5% in transgenic *Arabidopsis thaliana* plants. Mol Gen Genet 220: 17-24

Schwartz D, Dennis E (1986) Transposase activity of the *Ac* controlling element in maize is regulated by its degree of methylation. Mol Gen Genet 205: 476-482

Scofield SR, Harrison K, Nurrish SJ, et al (1992) Promoter fusions to the Activator transposase gene cause distinct patterns of Dissociation excision in tobacco cotyledons. Plant Cell 4: 573-582

Shimamoto K, Terada R, Izawa T, et al (1989) Fertile transgenic rice plants regenerated from transformed protoplasts. Nature 337: 274-276

Shimamoto K, Miyazaki C, Hashimoto H, et al (1993) Trans-activation and stable integration of the maize transposable element *Ds* cotransfected with the *Ac* transposase gene in transgenic rice plants. Mol Gen Genet 239: 354-360

Shimamoto K, Ohnishi T, Kume Y, et al (1996) Toward molecular isolation of the blast resistance gene in rice by the maize *Ac*/*Ds* elements. In: Mills D, Kunoh H, Keen N, et al (eds) Molecular aspects of pathogenicity and resistance: requirement for signal transduction. APS Press, St. Paul, MN, pp 177-183

Song WY, Wang GL, Chen L, et al (1995) A receptor kinase-like protein encoded by the rice disease resistance gene *Xa21*. Science 270: 1804-1806

Song WY, Pi LY, Wang GL, et al (1997) Evolution of the rice *Xa21* disease resistance gene family. Plant Cell 9: 1279-1287

Sugimoto K, Otsuki Y, Saji S, et al (1994) Transposition of the maize *Ds* element from a viral vector to the rice genome. Plant J 5: 863-871

Sundaresan V (1996) Horizontal spread of transposon mutagenesis: new uses for old elements. Trends Plant Sci 1: 184-190

Swinburne J, Balcells L, Scofield SR, et al (1992) Elevated levels of *Activator* transposase mRNA are associated with high frequencies of *Dissociation* excision in *Arabidopsis.* Plant Cell 4 : 583-595

Tsutsumi K, Kagaya Y, Hidaka S, et al (1994) Structural analysis of the chloroplastic and cytolasmic aldolase-encoding genes implicated the occurrence of mulitiple loci in rice. Gene 141: 215-220

Walbot V (1992) Strategies for mutagenesis and gene cloning using transposon tagging and T-DNA insertional mutagenesis. Annu Rev Plant Physiol Plant Mol Biol 43: 49-82

Wessler SR, Bureau TE, White SE (1995) LTR-retrotransposons and MITEs: important players in the evolution of plant genomes. Curr Opin Genet Dev 5: 814-894

Whitham S, Dinesh-Kumar SP, Choi D, et al (1994) The product of the tobacco mosaic virus resistance gene *N*: similarity to *Toll* and the interleukin-1 receptor. Cell 78: 1-20

Yoder JI (1990) Rapid proliferation of the maize transposable element *Activator* in transgenic tomato. Plant Cell 2: 723-730

Part 2
Development

5
Morphogenetic mutants of rice

Yasuo Nagato[1], *Jun-Ichi Itoh*[1], *and Hidemi Kitano*[2]

[1] Graduate School of Agricultural and Life Sciences, University of Tokyo, Tokyo 113-8657, Japan
[2] Faculty of Agriculture, Nagoya University, Nagoya 464-8601, Japan

1 Introduction

Morphogenesis in plants has long interested a large number of plant scientists. The most successful example of genetic approach is the ABC model for floral organ identity (Bowman et al. 1991, Coen and Meyerowitz 1991). It is noted that the isolation of various flower mutants have enabled the establishment of the ABC model. Thus, for elucidating genetic program involved in morphogenesis, we need to identify and characterize morphogenetic mutants.

To date, a large number of morphogenetic mutants have been identified in rice, which are interesting from the aspect of plant development. They include embryo mutants, those associated with vegetative characters (leaves and stem), and those of inflorescence and flowers. In this chapter, we mainly describe major morphogenetic mutants except those related to inflorescence and flowers which will be cited in the next chapter by J. Kyozuka.

2 Embryo mutants

Embryogenesis is a critical phase in plant development, during which fundamental body plan and two apical meristems are established. Thus, embryological study is of primary importance for understanding the regulatory mechanism of plant development. A large number of mutants have been identified in rice (Nagato et al. 1989, Kitano et al. 1993, Hong et al. 1995a), in maize (Clark and Sheridan 1991, Sheridan and Clark 1993) and in *Arabidopsis* (Errampalli et al. 1991, Jurgens et al. 1991, Castle and

Key words. rice, mutation, embryo, shoot apical meristem, radicle, vegetative phase, juvenile phase, adult phase, scutellum, coleoptile, phyllotaxy, endosperm, plastochron, leaf, dwarf

Meinke 1993). These embryo mutants will be quite useful for dissecting various morphogenetic events such as pattern formation, organ/tissue differentiation and morphogenesis of organs. To date, nearly 200 embryo mutants have been reported (Nagato et al. 1989, Kitano et al. 1993, Hong et al. 1995a). Based on their phenotypic characteristics, they can be categorized into six groups: (1) embryoless, (2) deletion of embryonic organs, (3) altered embryo size, (4) modified organ position, (5) aberrant morphology of organs, and (6) variable phenotypes. In addition, endospermless mutants are also known which cause the early degradation of endosperm and the enlargement of embryo. Before describing embryo mutants, it is useful to briefly summarize the developmental course of wild-type embryo in rice (Suetsugu 1953, Jones and Rost 1989). The embryo remains globular in shape for the first 3 days after pollination (Fig. 1a). At 4 days after pollination (4 DAP), differentiation of the coleoptile is recognized as a protrusion of cells on the ventral side, and soon shoot and radicle meristems become apparent (Fig. 1b). At 5 DAP, coleoptile and scutellum grow rapidly and the first leaf primordium is differentiated (Fig. 1c). Organogenesis is almost completed by 10 DAP, much earlier than in maize embryo (Fig. 1d). As shown in Fig.1, like other grasses, the rice embryo is highly complicated with three foliar leaves and various embryo-specific organs such as scutellum, coleoptile, and epiblast. In the scutellum, the epidermal cells (scutellar epithelium) take on a specialized morphology like a palisade cell.

2.1 Embryoless mutants

A mutation giving rise to embryoless seeds, *eml1* (*embryoless 1*), is known (Fig. 2b) (Hong et al. 1995a, b). In the *eml1* seeds, the embryo is degenerated at the early stage and is not observed in the mature seeds, but the endosperm formation is apparently normal. This mutant was first obtained by the presence of embryoless seeds with normal endosperm at a frequency of 25% among seeds of the panicle. Developmental analysis reveals that the embryo is degraded at the early developmental stage (around 3 DAP). Interestingly, homozygous *eml1* plants are recovered at a very low frequency, suggesting that this mutation is temperature sensitive and produces viable embryos depending on the temperature conditions. In field conditions (at around 30°C), almost all *eml1* seeds contain no embryo, comprising only endosperm. However, if the temperature becomes lower, the frequency of developing embryos becomes higher. At a low temperature, seeds are frequently observed that have normal or enlarged embryos but are defective in endosperm development. These results suggest that *eml1* primarily affects endosperm development: overdevelopment at high temperature and underdevelopment at low temperature. Thus, the extent of endosperm development will seriously affect the embryo development. This conclusion is reconfirmed by the analysis of embryo size mutants described in Section 2.3 of this chapter.

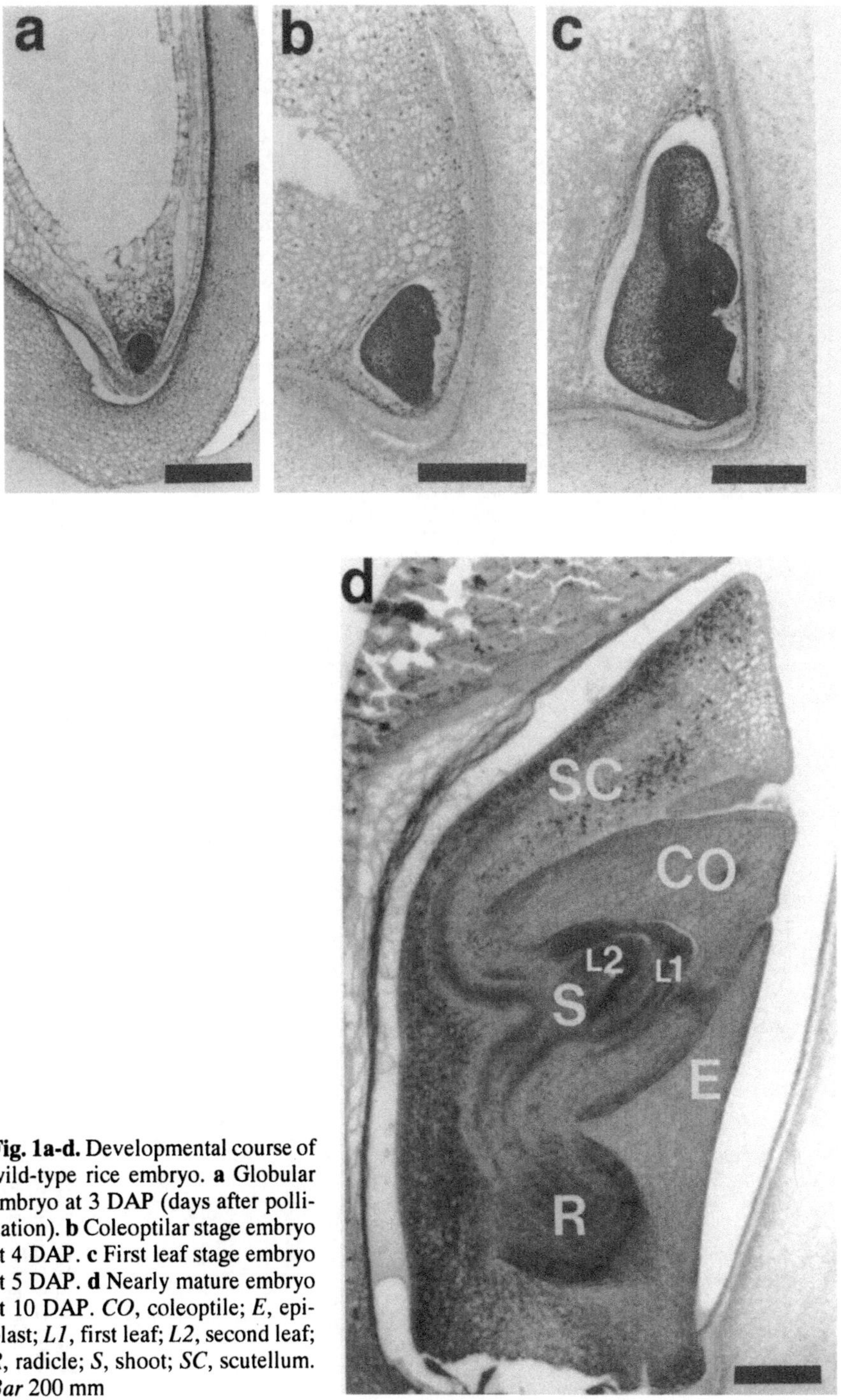

Fig. 1a-d. Developmental course of wild-type rice embryo. **a** Globular embryo at 3 DAP (days after pollination). **b** Coleoptilar stage embryo at 4 DAP. **c** First leaf stage embryo at 5 DAP. **d** Nearly mature embryo at 10 DAP. *CO*, coleoptile; *E*, epiblast; *L1*, first leaf; *L2*, second leaf; *R*, radicle; *S*, shoot; *SC*, scutellum. *Bar* 200 mm

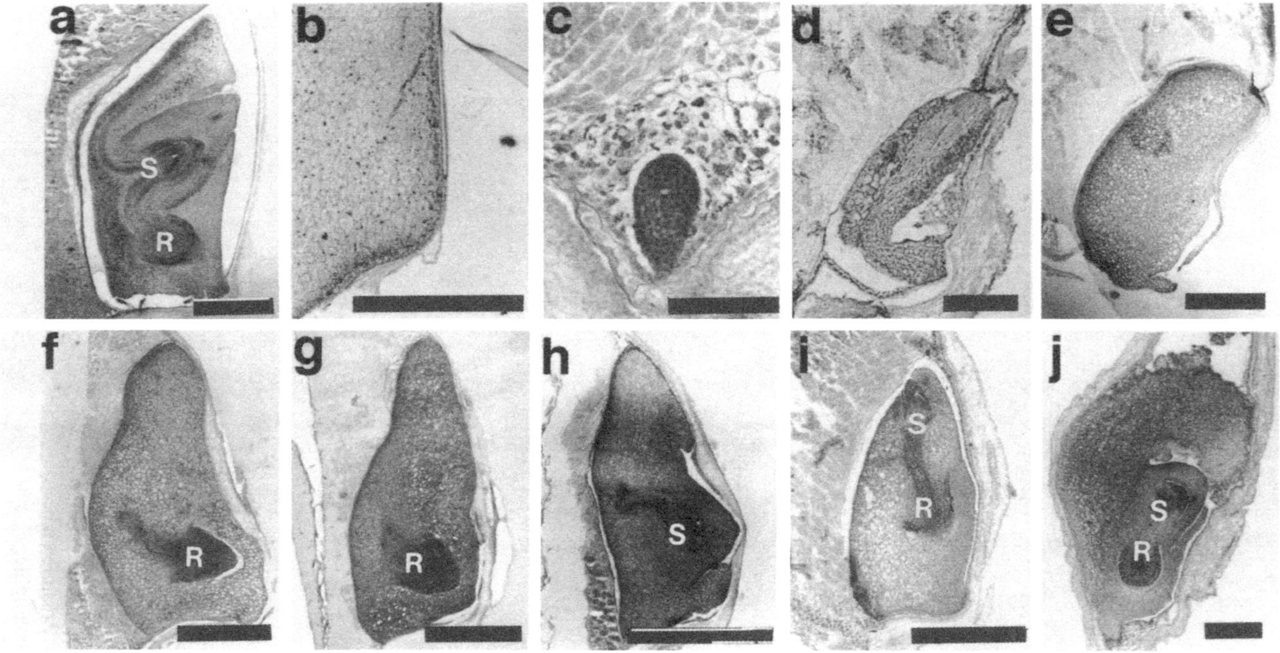

Fig. 2a-j. Phenotypes of mutant embryos at median longitudinal section. **a** Wild-type embryo. **b** *eml1* seed showing no embryo. **c** *gle1* embryo without organs. **d** *cle1* embryo without shoot and radicle. **e** *orl1* embryo without shoot and radicle. **f** *shl1* embryo lacking shoot but with normal radicle. **g** *shl2* embryo lacking shoot but with normal radicle. **h** *ral1* embryo lacking radicle. **i** *apd1* embryo with apically displaced shoot and radicle. **j** *enl1* large embryo lacking endosperm. *S*, shoot; *r*, radicle. *Bar* 200 mm in **b**,**c**; 500 mm in **a**, **d-j**

2.2 Mutations causing organ deletion

Four types of mutations are recognized in this group. In two types, both shoot and radicle are deleted, whereas the other two types delete either shoot or radicle.

In *gle* (*globular embryo*) mutants, embryos remain globular in shape and no apparent organ differentiation is observed (Fig. 2c). At least three loci (*gle1–gle3*) are associated with this phenotype. These phenotypes are not due to the developmental arrest at the normal globular stage, because *gle* embryos become larger than the wild-type globular embryo and continue to grow until the late stage of seed development. The *GLE* genes are involved in the overall differentiation of embryo, not in the specific organ differentiation.

Two types of mutations, *cle* (*club-shaped embryo*) and *orl* (*organless*), do not differentiate both shoot and radicle, similarly with *gle* mutations (Fig. 2d,e). However, detailed anatomical examination indicates that palisade-shaped cells characteristic to scutellar epithelium exist in the whole epidermis of *cle1* embryo and in the epidermis facing the endosperm in *orl1* embryo. Further, an in situ hybridization experiment probed with the α-amylase gene, which is expressed specifically in the scutellar epithelium of the embryo, shows strong hybridization signals in the epidermis of *orl1* and *cle1* embryos. Thus, it is considered that the scutellum is normally differentiated and functional in both *cle1* and *orl1* embryos. However, the mature *cle1* embryo comprises highly vacuolated cells and is not viable. In addition, normal dorsoventrality is not recognized in *cle1*, as evidenced by the palisade cells surrounding the whole embryo. On the other hand, the *orl1* embryo is composed of normal cells and exhibits normal embryonic polarity, as the position of scutellar epithelium and the tissue specificity of *OSH1* (one of rice homeobox genes) expression are conserved (Satoh et al. 1996). Then, it is considered that *CLE1* functions in the establishment of embryonic polarity and/or basic regionalization and that *ORL1* is associated with the differentiation of shoot and radicle after the fundamental embryonic pattern is established. These two genes exemplify complicated processes that are operating during early embryogenesis.

Since shoot apical meristem (SAM) and root apical meristem control the plant development after germination, the elucidation of regulatory mechanisms leading to the formation of these two meristems is highly important. Nine shootless mutants have been isolated, derived from four independent loci, *SHL1* (*SHOOTLESS1*), with two alleles, *SHL2* with five alleles, *SHL3* with one allele, and *SHL4* with one allele. Eight mutations at *SHL1*, *SHL2*, and *SHL4* cause indistinguishable phenotypes: complete loss of shoot accompanying the loss of coleoptile and epiblast, and the normal radicle formation (Fig. 2f,g). Their radicles are normally germinated after imbibition. By the anatomical observation of epithelium and in situ hybridization experiment using the α-amylase gene as a probe, it is shown that the scutellum is normally formed (Sugimoto et al. 1998). In another mutant, *shl3*, in addition to deletion of the shoot, the radicle is exogenously produced and cells in the remaining tissues are highly vacuolated. Studies by in situ hybridization probed with *OSH1* reveal that *SHL1* and *SHL2* function in the upstream of *OSH1* and *SHL3* and *SHL4* in the downstream or independently of *OSH1* (Satoh et al., unpublished data).

The phenotypes of shootless mutants give important suggestions on the controversial problem in grass embryos: which organ(s) are homologous to cotyledon(s) (Brown 1960). The most popular interpretation is that coleoptile and scutellum constitute the grass cotyledon (Eames 1961). However, the two organs do not share developmental origin, since the differentiation of coleoptile and epiblast depend on the presence of SAM but the scutellum is produced independently of SAM. Thus, the above interpretation is inadequate. The grass cotyledon will be either the scutellum or coleoptile (epiblast), depending on the definition of cotyledon to be a first lateral organ of SAM or a first leaf-like organ independent of SAM.

In contrast to a large number of shootless mutants, only three radicleless mutants derived from two loci, *RAL1* (*RADICLELESS 1*) and *RAL2*, have been recovered (Fig. 2h). In the *ral* embryos, the shoot is apparently normal and can germinate. The *ral* embryos seem to be truncated in the apical-basal direction and normally produce adventitious roots after germination. Thus, the radicleless mutants are not defective in root-producing ability, but may be defective because of the deletion of the basal embryonic region from which the radicle develops.

As described above, numerous types of organ deletions have been detected, suggesting that a complicated cascade of genes exists in the early development of the embryo. Further detailed analyses of the mutants will provide fruitful results for understanding the pattern formation and the determination of organ differentiation.

2.3 Mutations associated with embryo size

Two types of mutations affecting embryo size are expected, reduction and enlargement. In rice, both types have been recovered (Hong et al. 1996). Since embryo size mutants have not been reported in maize and *Arabidopsis*, these mutations are unique to rice and valuable for elucidating the mechanism of embryo size regulation. Four giant embryo mutations are derived from a single locus, *GE* (*GIANT EMBRYO*). At least three loci, *RE1*(*REDUCED EMBRYO1*), *RE2*, and *RE3*, reduce the embryo size by recessive mutations. The giant embryo mutants show 1.2–1.5 times longer embryos and have a reduced endosperm (Fig. 3b,c). In the embryos, only the scutellum is enlarged but the sizes of shoot and radicle are not affected. Interestingly, the number of cells constituting the scutellum is comparable to that of the wild type. Thus, the giant embryos are mostly the result of the enlargement of scutellum cells. On the other hand, the embryos of *re1*, *re2*, and *re3* are less than half of the wild-type embryo in length, and show a reduction of all the embryonic organs including apical meristems and enlargement of the endosperm (Fig. 3d,e). The size reduction of organs is due to the small number of cells, not the small cell size. Neither the reduced nor the giant embryo mutations modify postembryonic development, indicating that *RE* and *GE* genes function seed specifically.

As embryo size and endosperm size are negatively correlated in both *re* and *ge* mutants, it is not certain whether *GE* and *RE* function in embryo or in endosperm. Developmental observation shows that a large number of endosperm cells located near the embryo degenerate in *ge*. In contrast, the basal cells of *re* endosperm over-

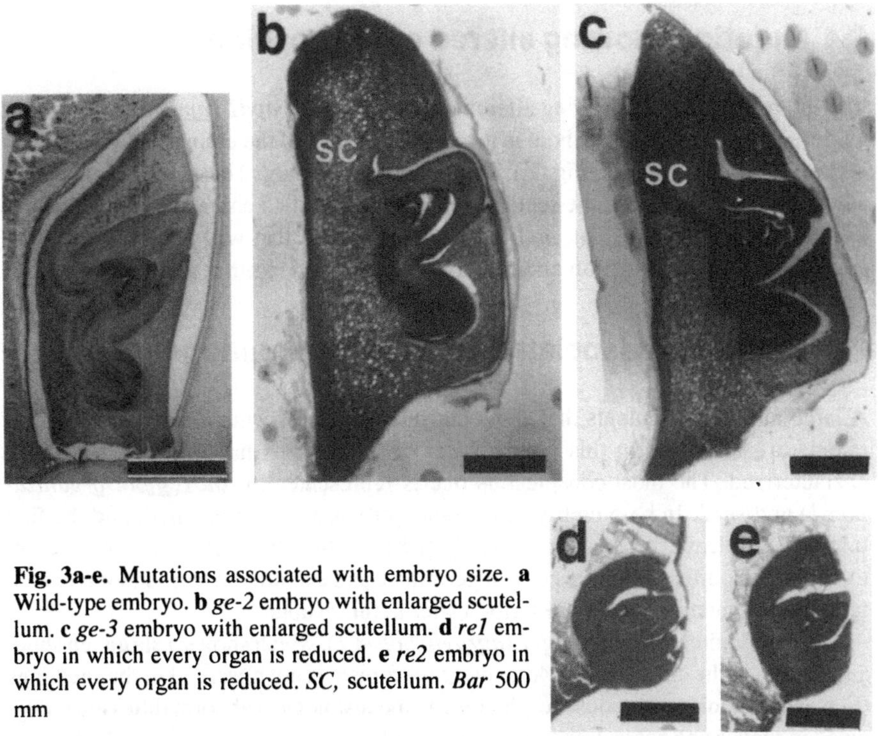

Fig. 3a-e. Mutations associated with embryo size. **a** Wild-type embryo. **b** *ge-2* embryo with enlarged scutellum. **c** *ge-3* embryo with enlarged scutellum. **d** *rel* embryo in which every organ is reduced. **e** *re2* embryo in which every organ is reduced. *SC*, scutellum. *Bar* 500 mm

develop, resulting in direct contact with the embryo. These facts suggest that *GE* and *RE* regulate the endosperm development. To confirm this, double mutants between embryo size mutants and embryo-specific mutants were produced. In the double mutants, *ge-2 cle1* and *ge-2 gle1*, the embryo is identical to *cle1* or *gle1* whereas the endosperm is reduced as in *ge-2*. That is, the embryo shows a *cle1* or *gle1* phenotype but the endosperm shows the *ge-2* phenotype. Accordingly, it is concluded that *GE* regulates endosperm development.

The double mutant, *re2 cle1*, shows *cle1* embryo and enlarged endosperm of *re2*. In addition, the *ge-2 re1* double mutant have *re1* embryo and endosperm, not the *ge-2* endosperm. Thus, *RE1* and *RE2* are also involved in the regulation of endosperm development, and the embryo will self-regulate its organ size according to the available space.

Apart from the embryological standpoint, *GE* and *RE* are the first important genes identified that are associated with the endosperm development. Because the abnormalities of *ge* and *re* endosperms are recognized in the basal region as early as 2 DAP, these genes will regulate the early proliferation of the basal endosperm cells.

2.4 Mutation showing altered organ position

The interesting mutant showing altered organ position is *apd1* (*apical displacement 1*). In the *apd1* embryo, the shoot is formed at the apex of the embryo and the radicle at the center of the embryo (Fig. 2i). This results in an enlarged basal region of embryo and underdevelopment of the scutellum. Although detailed characterization of *apd1* has not yet been carried out, *apd1* is a unique mutant that will be associated with embryonic pattern formation and has not been reported in other plant species.

2.5 Mutations associated with organ morphogenesis

A large number of mutants that show morphological abnormalities of shoot and/or radicle are classified in this group. Most of the mutants have not yet been well characterized. The most conspicuous one is represented by *sho1*(*shoot organization 1*) and *sho2*. In both embryos, the coleoptile is not differentiated, and the first through third leaves are malformed (Fig. 4a,b). On the other hand, radicle development is not modified. These seeds can germinate and exhibit a strange shoot architecture. These mutations will be described later in this chapter.

Another unique mutant is *agr1* (*agravitropism 1*), which shows a shrunken shoot and radicle in the embryo (Fig. 4c), and gravitropism is not observed in the seedlings (Fig. 4d). In addition, various morphological and histological abnormalities are recognized in leaves and roots. Other mutants include those showing severe defects in all the organs and those with abnormalities in either shoot or radicle.

2.6 Mutations causing variable embryonic phenotypes

A considerable number of mutants show a wide range of phenotypes in spite of being a single mutation, judged from the segregation rate. For example, the *vep1* (*variable embryo phenotypes 1*) embryo exhibits at least four kinds of phenotypes similar to *gle1*, *orl1*, *shl* and morphological abnormalities (Fig. 5). In the *vep2* embryo, a different phenotypic spectrum is observed: large organless, shootless, radicleless, and aberrant organ morphology. These variable phenotypes suggest that each *vep* gene regulates the expression of genes functioning in diverse developmental pathways.

2.7 Endospermless mutants

As an interesting mutant involved in endosperm development, Kageyama et al. (1991) identified a recessive endospermless mutant (*enl1*) (Fig. 2j) in which the endosperm starts to develops but degenerates at an early stage (ca. 3 DAP). As a result of endosperm degradation, a very large embryo is produced. In the *enl1* embryo, only the scutellum is enlarged, and the sizes of other organs are not affected, as in *giant embryo* (Hong et al. 1996). It is considered that the *enl1* gene regulates the cell

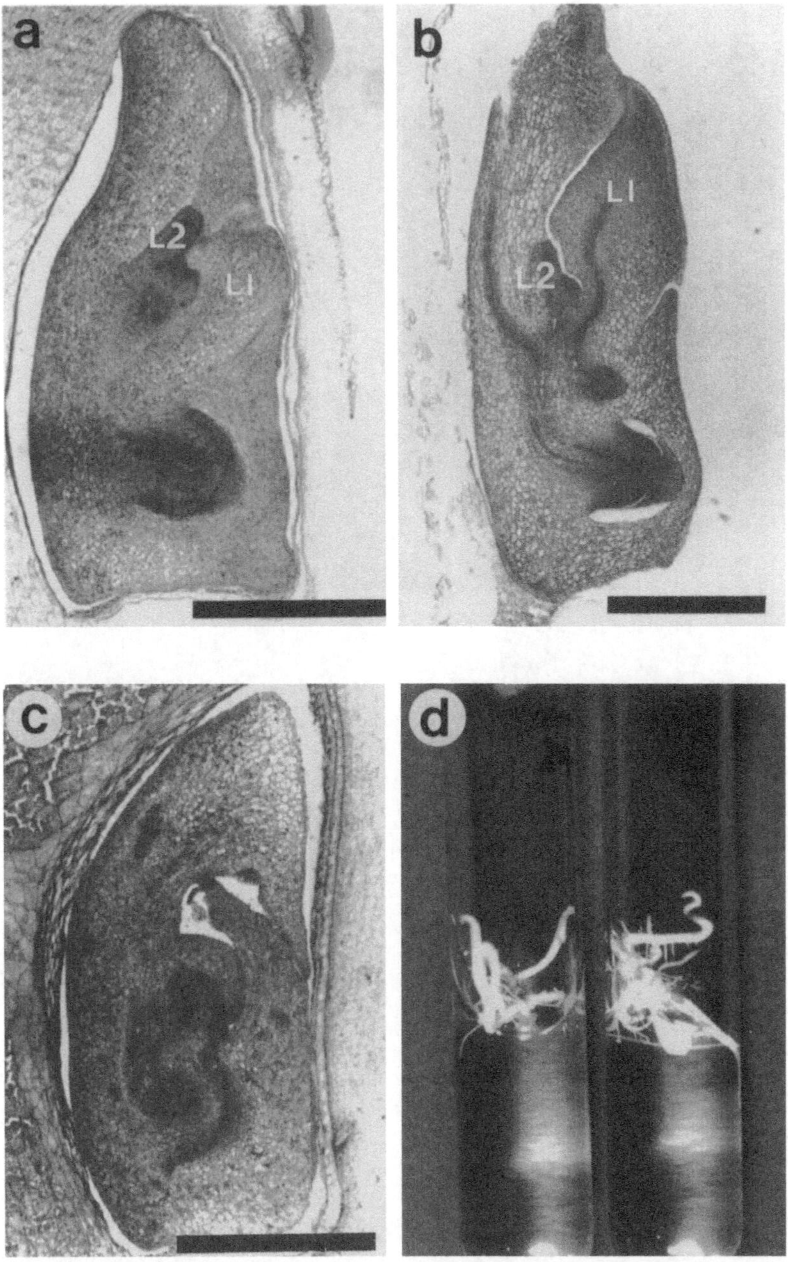

Fig. 4a-d. Embryo mutants showing abnormal morphology. **a** *sho1* embryo in which coleoptile is deleted and leaves are malformed but radicle is normal. **b** *sho2* embryo in which coleoptile is deleted and leaves are malformed, but radicle is normal. **c** *agr1* embryo with shrunken shoot and radicle. **d** Seedlings of *agr1* showing loss of gravitropism. *L1*, first leaf; *L2*, second leaf. *Bar* 500 mm

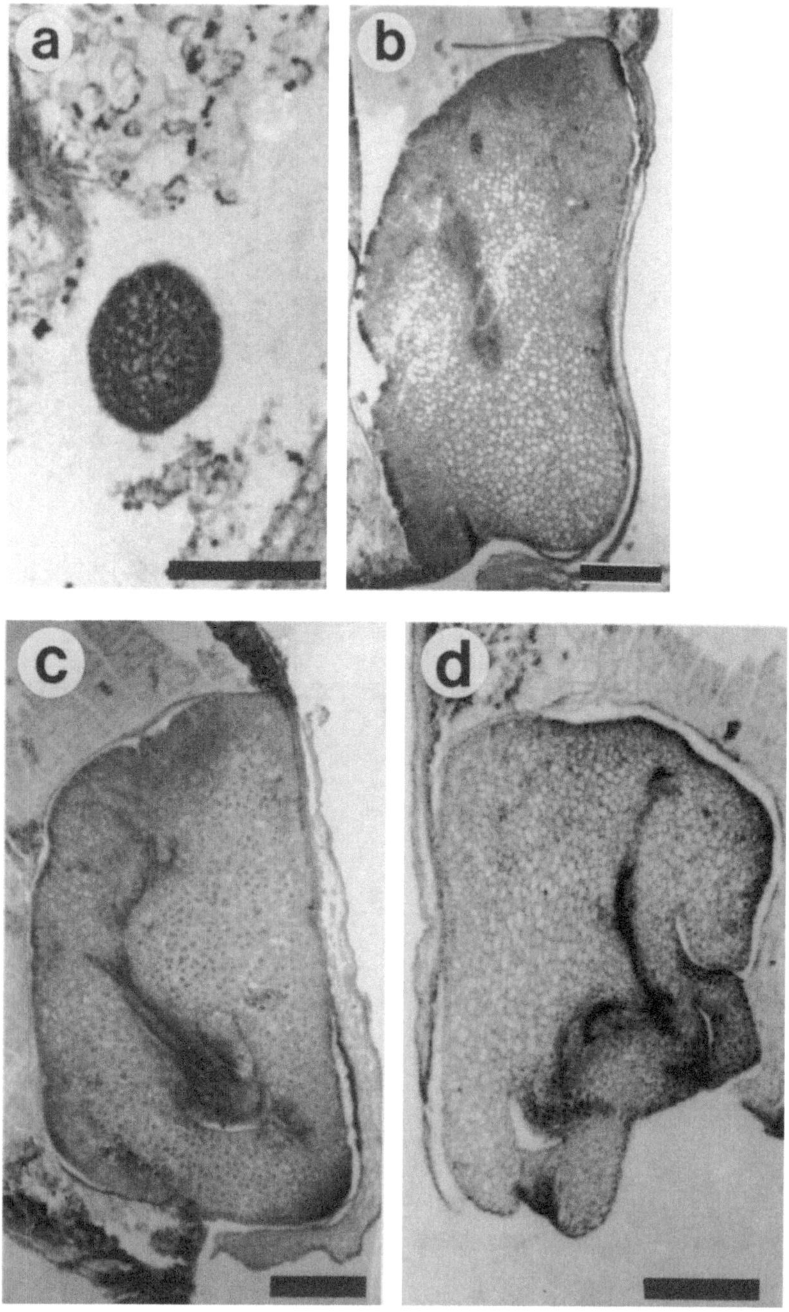

Fig. 5a-d. Variable phenotypes observed in *vep1* embryos. **a** Small organless globular embryo. **b** Large organless embryo. **c** Shootless embryo. **d** Embryo with malformed shoot and radicle. *Bar* 100 mm in **a**, 300 mm in **b-d**

proliferation at the early stage of endosperm development.

Because cereal endosperms are the most important staple diet in many countries, genes regulating endosperm development such as *ENL1*, *GE* and *RE1* should be more intensively studied. As described, a large number of embryo mutants have been identified in rice. Although most of them have not been well characterized, these mutants will be useful materials for elucidating developmental programs of the embryo such as pattern formation, regionalization, regulation of organ positions, size regulation, organ determination, and morphogenesis.

3 Mutants on vegetative morphological traits

The postgermination period of plants is called the vegetative phase. In the vegetative phase, plants regularly produce leaves, lateral shoots, and stem. The plant body thus produced is composed of repetitive units called phytomers comprising leaves, axial buds, and internodes. The production of leaves shows a high regularity in many plants including rice. The spatial distribution of leaf primordia around the shoot apex is called phyllotaxy, and the time span of two successive primordia is called the plastochron. In the vegetative phase of wild-type rice, the phyllotaxy is 1/2 alternate, and the plastochron is 3–5 days. The similar shoot architectures observed among wild-type plants are mainly due to the stability of phyllotaxy and plastochron.

Leaf shape and structure also affect plant architecture. Because leaves are agronomically important as a photosynthetic organ and the variation in leaf shape is conspicuous, various mutants have been identified. In rice, internodes start to elongate around the onset of the reproductive phase; the culm length is also an important factor regulating final plant form. Because plant height is an important agronomical character, many dwarf mutants have been isolated in rice as well as in other grain crops.

3.1 Shoot architecture

Both phyllotaxy and plastochron affect plant architecture. In rice, *sho1*, which was previously described in Section 2.5 as an embryo mutant, affects both plastochron and phyllotaxy (Tamura et al. 1992). In *sho1*, phyllotaxy is random, and the plastochron is very short, to be shorter than 1 day, resulting in a strange plant architecture (Fig. 6). In addition, leaves are malformed, small, narrow like a tendril, and lacking bulliform cells and bundle sheath. It is considered that defects in SAM will cause these overall abnormalities of shoot. The SAM of *sho1* is flat and wide in diameter, markedly different from the normal domelike structure. This structural modification of SAM seems to affect the position, production rate, and development of leaf primordia. Recently a similar mutation on another locus, *sho2*, was reported (Hong et al. 1995a). Interestingly, mutant phenotypes of both *sho1* and *sho2* are limited to the early phase (juvenile phase), and almost normal phenotypes are recovered in the adult phase.

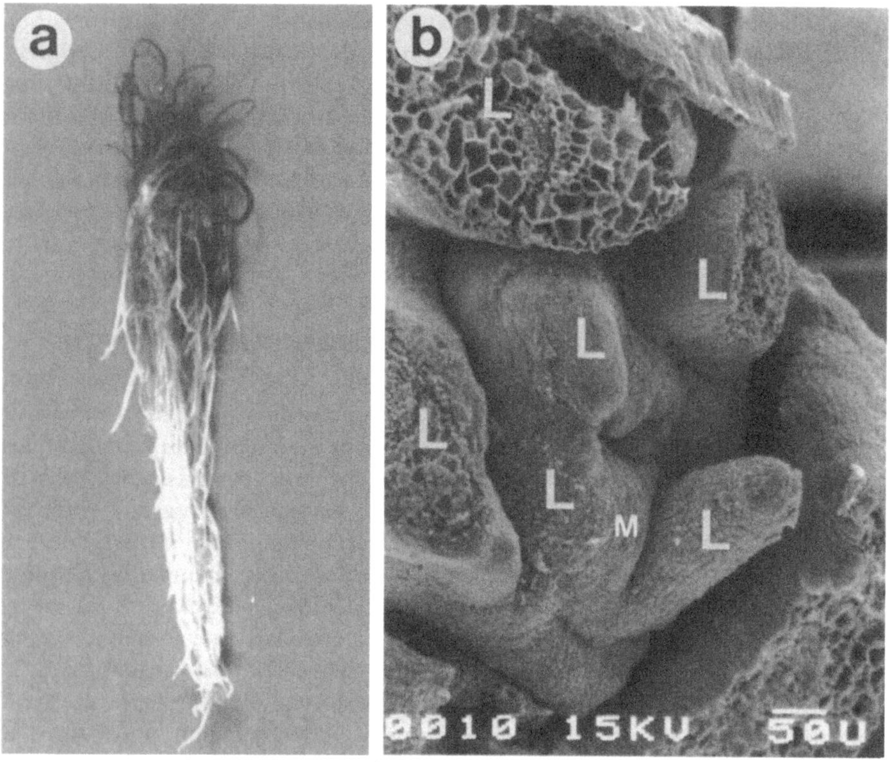

Fig. 6a,b. Seedlings of *sho1* showing abnormal shoot architecture. **a** A seedling about 3 weeks after germination. A large number of small and fine leaves are seen. Roots are apparently normal. **b** Scanning electron microscopy of *sho1* shoot apex. Leaves (*L*) are randomly positioned, and apical meristem (*M*) is flat.

Accordingly, *SHO1* and *SHO2* would regulate the SAM structure in the juvenile phase. Mutations similar to *sho1* and *sho2* have not been reported in other species.

A mutant affecting plastochron, not phyllotaxy, was ever reported (Hu 1961). In the *lhd* (*leafy head*) mutant, plastochron was reduced to about 1/2 of the wild type in the vegetative phase but the phyllotaxy was normal 1/2 alternate, indicating that phyllotaxy and plastochron are genetically separated. That is, *lhd* produced leaf primordia twice as fast as the wild type. This mutant also showed abnormalities in the reproductive phase: panicles were degenerated and shoots were ectopically produced. Although *lhd* would be a very interesting mutant, unfortunately *lhd* was lost, and we cannot analyze it further. Quite recently, however, two mutants phenotypically very similar to *lhd* have been identified. They were allelic and designated as *pla1-1* (*plastochron 1*) and *pla1-2* (Itoh et al. 1998). In *pla1-1* and *pla1-2*, plastochron in the vegetative phase is reduced to nearly half of the wild type. In the reproductive phase, primordia of primary rachis branches are transformed into vegetative shoots in a varying degree depending on the strength of alleles. These mutants will be very useful for dissecting temporal regulation of SAM development. It is noted that no mutations showing modified plastochron have been identified in other species.

Phyllotaxy, which is another interesting trait spatially characterizing the SAM, has interested many botanists according to its regularity and occasional drastic changes during ontogeny. In maize, *abphyl* is known to show various phyllotaxy in vegetative phase: decussate, spiral, and alternate (Greyson and Walden 1972, Greyson et al. 1978). It is revealed that *abphyl* has larger SAM than the wild type. In rice, no mutant has been reported which specifically modifies the phyllotaxis. Recently, however, we have identified a mutant showing decussate phyllotaxy instead of 1/2 alternate in the early vegetative phase (Itoh et al., unpublished data). Preliminary examination indicates that this mutant has large SAM. Together with the maize *abphyl* mutant, it is suggested that the phyllotaxy is regulated by the size (or shape) of SAM.

3.2 Leaves

A considerable number of mutations associated with leaf development have been identified in rice. However, since they have been utilized as chromosome linkage markers, they have not been developmentally well characterized. Some interesting mutants on leaf traits are described next.

drooping leaf (*dl*): This mutant is characterized by the drooping leaf phenotype due to the lack of midrib (Fig. 7a) (Iwata and Omura 1971). An interesting phenotype is also observed in the flower of the strong allele in which the pistil is homeotically transformed into stamens (Fig. 7b) (Nagasawa et al., unpublished data). Thus, *DL* gene is associated with midrib development in the vegetative phase and with pistil development in the reproductive phase. Similar mutants in which both midrib and pistil development are affected have been reported in *Panicum* (Fladung et al. 1991), pearl millet (Rao et al. 1988), and barley (Tsuchiya 1962). Therefore, *DL* and homologous genes will be widely conserved in grasses.

narrow leaf (*nal*): At least three loci, *NAL1*, *NAL2*, and *NAL3* have been reported whose double or triple recessive mutations cause narrowness of the leaf blade (Fig. 8) (Mori et al. 1973, Iwata and Omura 1975, Yoshimura et al. 1982). Narrow-leaf mutants have been well characterized in maize (Scanlon et al. 1996, Scanlon and Freeling 1997). A double recessive plant of two unlinked mutations of maize, *narrow sheath 1* (*ns1*) and *ns2*, shows remarkable narrowness of leaf sheath and blade, and this phenotype is suggested to be caused by the deletion of the leaf domain including the margin. To date, rice *nal* mutants have not been developmentally analyzed.

rolled leaf (*rl*): The leaves of *rl* are narrow, fail to extend the blade, and roll adaxially. At least three loci, *RL1–RL3*, are associated with the *rl* phenotype (Jones 1952, Mori et al. 1973, Nagao et al. 1964). This phenotype is similar to the leaf blade when a plant is under water-deficient stress. *rl* plants tend to be dwarf and to set a smaller panicle. In addition, dehiscence of anthers is not complete. The developmental mechanism causing the rolling of the leaf blade is not known.

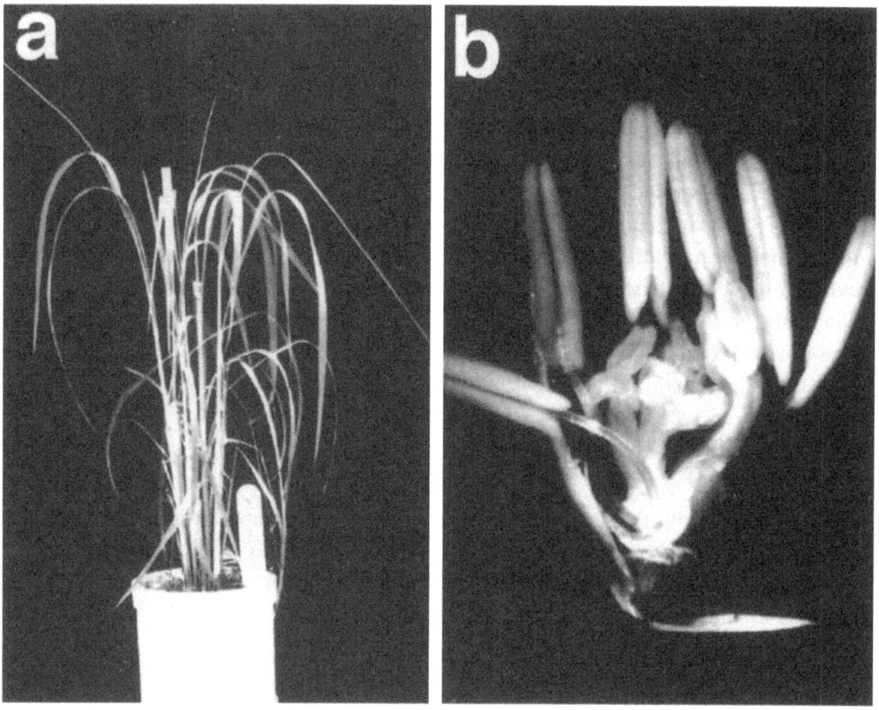

Fig. 7a,b. Phenotypes of *dl* mutant. **a** Phenotype in vegetative phase. Leaves are drooping due to the lack of midrib. **b** Flower phenotype of strong *dl* allele. In the central region, pistil is homeotically transformed into stamens.

glabrous leaf and hull (*gl*) and *Hairy leaf* (*Hl*): *gl* is identified as a mutant that has no trichomes on the leaf surface (Nagao et al. 1960, Nagao and Takahashi 1963, Nakane et al. 1977). Trichomes are also not observed on the hull. Thus, the development of trichomes on the leaves and hulls is regulated by the same gene. Two dominant complementary genes, *Hla* and *Hlb*, for long pubescence of leaves and hulls are known. When *Hla* and *Hlb* coexist with *gl*, the hairs on the leaves are markedly shortened (Nagao and Takahashi 1963, Kinoshita and Takahashi 1968).

dripping-wet leaf (*drp*): To date, eight recessive mutants are known, *drp1–drp8* (Nagao et al. 1964). In *drp* mutants, the leaf surface is highly hydrophilic. Scanning electron microscopy reveals that in *drp* leaves the epicuticular wax is reduced (Nakamura and Hattori 1996).

liguleless (*lg*): Ligule and auricle are small organs which develop on the adaxial side of the boundary between leaf sheath and blade. These organs are commonly observed in grasses, but rice has a well-developed ligule compared with the other grasses.

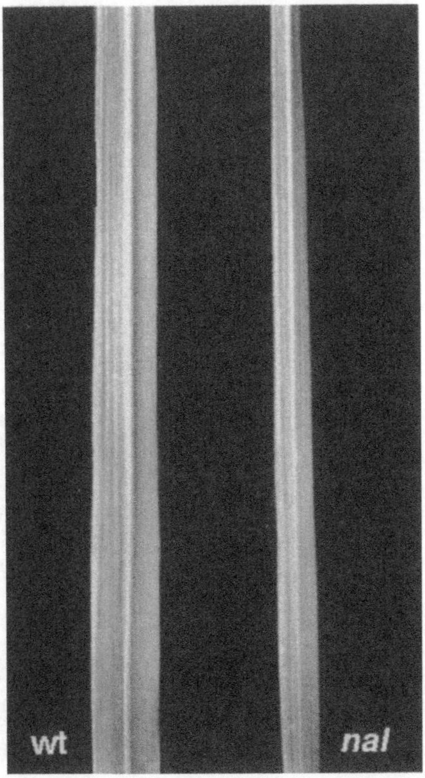

Fig. 8. Leaf phenotype of *nal* mutant. In *nal* leaf, the width of leaf blade is nearly half of that of the wild type.

Two genes, *lg* (*liguleless*) and *aul* (*auricleless*), are reported which affect ligule and auricle development. In the *lg* mutant, both ligule and auricle are completely deleted (Fig. 9) (Nagao and Takahashi 1963). In *aul*, the auricle is absent and the ligule is rudimentary (Librojo and Khush 1986).

In maize, two genes, *LG1* (*LIGULELESS 1*) and *LG2*, are known to be required for ligule/auricle development (Harper and Freeling 1996). Recently, *LG1* and *LG2* have been cloned and estimated to encode transcription factors (Moreno et al. 1997, Walsh et al. 1997). Since maize and rice genomes are generally colinear and the chromosomal locations of rice *LG* and maize *LG1* are syntenous, *LG* and *AUL* of rice may be counterpart genes of maize *LG1* and *LG2*, respectively.

3.3 Dwarfism

Because culm length is an important agronomical trait, dwarfism has been intensively studied, compared with other morphological traits. Dwarfism is generally defined as the mutation causing the reduction of plant height. To date, about 60 dwarf genes have been identified, including semidwarf ones such as *sd1* which is utilized in prac-

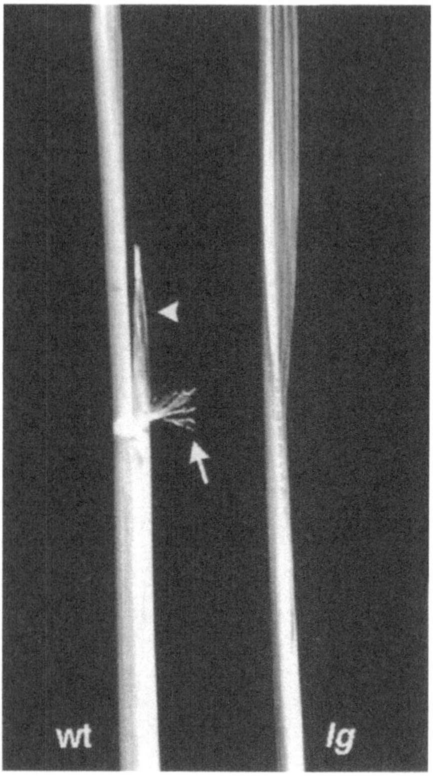

Fig. 9. Leaf of *lg* mutant. Leaf of *lg* lacks ligule (*arrowhead*) and auricle (*arrow*), which are observed in the boundary between blade and sheath of the wild type

tical breeding (Futsuhara et al. 1986, Kamijima et al. 1996). Among them, more than 30 genes were mapped on the chromosomes. In spite of a large number of dwarf mutations having been identified, their functions have not been well examined. In addition, several mutants were lost.

The rice stem consists of nodes and internodes. When the shoot enters into the reproductive phase, the upper four to six internodes start to elongate. In the wild type, internode length is largest in the uppermost one and then gradually reduced basipetally (Fig. 10). It is noted that the internodes are usually numbered basipetally from the uppermost one as first, second, third, and so on.

Dwarf mutants exhibit a variety of internode elongation patterns. Takeda (1977) classified the dwarf mutants into five different types, i.e., **dn**, **dm**, **d6**, **sh**, and **nl**, based on the relative length of each internode against culm length (Fig. 10), although it is noted that there exist mutants whose internode elongation patterns do not fit this classification.

The **dn** type similarly affects the elongation of all the internodes, resulting in a pattern of internode elongation similar to that of the wild type. In this type, *d18* and *d12* are involved. The *d18-h* mutant shows a typical miniature phenotype, its final plant height being only 10–15 cm (Fig. 11). This locus has multiple alleles showing

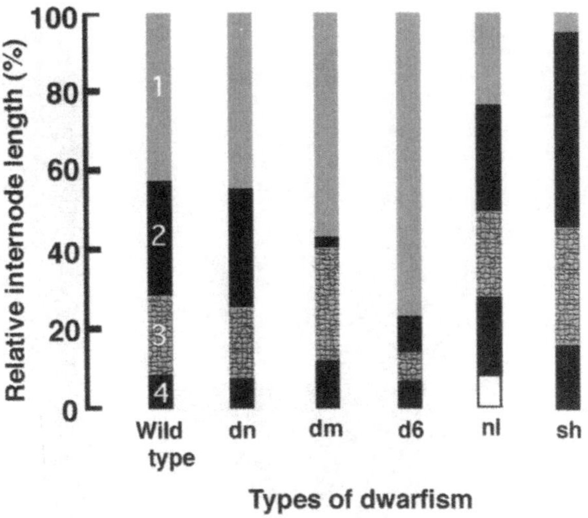

Fig. 10. Schematic representation of five types of dwarf phenotype. Each internode length is shown as the percent of the total culm length. (Revised after Takeda 1977)

different dwarf phenotypes, all of which can recover the wild-type phenotype by the application of gibberellins (Shinbashi et al. 1976). Thus, *D18* is involved in the gibberellin biosynthesis. **dm** type is characterized by the specific shortening of the second internode (Fig. 11). This specific pattern is caused independently by *d1*, *d2*, and *d11* genes, which also affect the growth of the other organs such as leaf and panicle. Besides these recessive dwarf genes, a dominant gene showing **dm** type, *Ssi1*, is also identified. *Ssi1* shows semidwarfness and less pleiotropic effects on the other organs (Xiong et al. 1997). Another unique dwarf type is **d6**, in which all the internodes except the first one are shortened severely. Recently, it has been revealed that *d6* is caused by a loss-of-function mutation in the rice homeobox gene, *OSH15* (see Chapter 7). **sh** and **nl** types are represented by *shp* (*sheathed panicle*) and *nl* (*neck leaf*), respectively. In the former type, only the first internode is specifically shortened, and in the latter type, all the internodes show a similar length as a result of the reduction of the first internode and the elongation of the basal internodes.

Internode elongation depends on the cell division and cell elongation activity of the intercalary meristem (IM) differentiated at the basal region of each internode. Many of the **dn**-type genes suppress internode elongation by decreasing the number of cells rather than reducing cell size. In the **dm** type (Fig.10), IM is defective specifically in the second internode, although it is normally developed in the other internodes. In the second and the lower internodes of **d6** type, both cell division and cell elongation in IM are restricted intensively, and epidermal, hypodermal, and vascular tissues also exhibit abnormal morphologies. On the other hand, *d12* and its allelic mutation *d50* affect both division and elongation of parenchymatous cells derived from the inner region of IM.

Wide variation in the internode elongation patterns and morphological abnormalities observed among dwarf mutants mean that dwarf genes have multiple functions at various developmental stages and at various tissues. Thus, the dwarf mu-

Fig. 11a,b. Phenotypes of dwarf mutants. **a** Nearly mature plant of *d18-h*. **b** Internodes of wild-type (*left*) and dm-type mutant (*right*). Numerals in **b** show internode number counted from the top. In the dm-type mutant, the second internode is severely reduced. *Arrows* indicate nodes. For explanation of each type, see text.

tants are good materials to elucidate the regulatory mechanisms associated with growth and development.

4 Concluding remarks

As described here, many morphogenetic mutants have been identified in rice. It is widely accepted that morphogenetic mutants are indispensable for dissecting developmental events, which attract many plant scientists. In rice, many of the morphogenetic mutants have been thus far utilized as genetic linkage markers and have not been developmentally analyzed. Rice (grasses) is specialized in many embryonic, vegetative, and reproductive traits. Accordingly, studies of the developmental program of rice are important from the aspect of plant evolution as well as from the ontogenic point of view. It is expected that the existing mutants will be efficiently utilized for understanding rice plants as an important staple diet and also as model monocotyledonous plants.

(For information on the other morphogenetic mutants, see homepage, http://www.shigen.nig.ac.jp/NIG_rice/rice.html)

References

Bowman JL, Smyth DR, Meyerowitz EM (1991) Genes directing flower development in *Arabidopsis*. Plant Cell 1:37-52

Brown WV (1960) The morphology of the grass embryo. Phytomorphology 10:215-223

Castle LA, Meinke DW (1993) Embryo-defective mutants as tools to study essential functions and regulatory processes in plant embryo development. Semin Dev Biol 4:31-39

Clark JK, Sheridan WF (1991) Isolation and characterization of 51 embryo-specific mutations of maize. Plant Cell 3:935-951

Coen ES, Meyerowitz EM (1991) The war of the whorls: genetic interactions controlling flower development. Nature 353:31-37

Eames AJ (1961) Morphology of the angiosperms. McGraw-Hill, New York, pp 309-368

Errampalli D, Patton D, Castle L, et al (1991) Embryonic lethals and T-DNA insertional mutagenesis in *Arabidopsis*. Plant Cell 3:149-157

Fladung M, Bossinger G, Rosb GW, et al (1991) Correlated alterations in leaf and flower morphology and rate of leaf photosynthesis in a *midribless* (*mbl*) mutant of *Panicum maximum* Jacq. Planta 184:356-361

Futsuhara Y, Kikuchi F, Rutger JN (1986) Gene symbols for dwarfness. Rice Genet Newslett 3:8-10

Greyson RI, Walden DB (1972) The *ABPHYL* syndrome in *Zea mays*. I. Arrangement, number and size of leaves. Am J Bot 59:466-472

Greyson RI, Walden DB, Hume AJ (1978) The *ABPHYL* syndrome in *Zea mays*. II. Pattern of leaf initiation and the shape of the shoot apical meristem. Can J Bot 56:1545-1550

Harper LC, Freeling M (1996) Interactions of *liguleless1* and *liguleless2* function during ligule induction in maize. Genetics 144:1871-1882

Hong S-K, Aoki T, Kitano H, et al (1995a) Phenotypic diversity of 188 rice embryo mutants. Dev Genet 16:298-310

Hong S-K, Aoki T, Kitano H, et al (1995b) Temperature-sensitive mutation, *eml1*, affects both embryo and endosperm development in rice. Plant Sci 108:165-172

Hong S-K, Kitano H, Satoh H, et al (1996) How is embryo size genetically regulated in rice? Development 122:2051-2058

Hu CH (1961) An X-ray induced panicle-degenerating mutant in rice. Jpn J Breed 11:19-23

Itoh JI, Hasegawa A, Kitano H, et al (1998) A recessive heterochronic mutation, *plastochron 1*, shortens the plastochron and elongates the vegetative phase in rice. Plant Cell 10:1511-1521

Iwata N, Omura T (1971) Linkage analysis by reciprocal translocation method in rice plants (*Oryza sativa* L.). II. Linkage groups corresponding to the chromosomes 5,6,8,9,10 and 11. Sci Bull Fac Agr Kyushu Univ 25:137-153 (in Japanese with English summary)

Iwata N, Omura T (1975) Studies on the trisomics in rice plants (*Oryza sativa* L.). III. Relation between trisomics and genetic linkage groups. Jpn J Breed 25:363-383

Jones JW (1952) Inheritance of natural and induced mutations in Carloro rice and observations on sterile Caloro types. J Hered 43:81-85

Jones TJ, Rost TL (1989) Histochemistry and ultrastructure of rice (*Oryza sativa*) zygotic embryogenesis. Am J Bot 76:504-520

Jurgens G, Mayer U, Ruiz, RAT, et al (1991) Genetic analysis of pattern formation in the *Arabidopsis* embryo. Development (Suppl) 1:27-38.

Kageyama Y, Fukuoka H, Yamamoto K, et al (1991) The rice plant bearing endospermless grains: a novel mutant induced by gamma-irradiation of tetraploid rice (*Oryza sativa* L.). Jpn J Breed 41:341-345

Kamijima O, Tanisaka T, Kinoshita T (1996) Gene symbols for dwarfness. Rice Genet Newslett 13:19-25

Kinoshita T, Takahashi M (1968) Supplemental report on genes responsible for pubescence of glumes and leaves in rice plant. Genetical studies on rice plant. XXXII. Mem Fac Agric Hokkaido Univ 6:364-370 (in Japanese with English summary)

Kitano H, Tamura Y, Satoh H, et al (1993) Hierarchical regulation of organ differentiation during embryogenesis in rice. Plant J 3:607-610

Librojo AL, Khush GS (1986) Chromosomal location of some mutant genes through the use of primary trisomics in rice. In: Rice genetics. IRRI, Manila, Philippines, pp 249-255

Moreno MA, Harper LC, Krueger RW, et al (1997) *liguleless1* encodes a nuclear-localized protein required for induction of ligules and auricles during maize leaf organogenesis. Genes Dev 11:616-628

Mori K, Kinoshita T, Takahashi M (1973) Linkage relationships of genes for some mutant characters of rice kept in Kyushu University. Genetical studies on rice plant. LV. Mem Fac Agric Hokkaido Univ 8:377-385 (in Japanese with English summary)

Nagao S, Takahashi M (1963) Trial construction of twelve linkage groups in Japanese rice. Genetical studies on rice plant. XXVII. J Fac Agric Hokkaido Univ 53:72-130

Nagao S, Takahashi M, Kinoshita T (1960) Genetical studies on rice plant. XXV. Inheritance of three morphological characters, pubescence of leaves and floral glumes, and deformation of empty glumes. J Fac Agric Hokkaido Univ 51:299-314

Nagao S, Takahashi M, Morimura K (1964) Genetical studies on rice plant. XXVIII. Causal genes and their linkage relationships of some morphological characters, introduced from foreign rice varieties. Mem Fac Agric Hokkaido Univ 5:89-96 (in Japanese with English summary)

Nagato Y, Kitano H, Kamijima O, et al (1989) Developmental mutants showing abnormal differentiation in rice embryo. Theor Appl Genet 78:11-15

Nakamura K, Hattori K (1996) Genetic analysis of epicuticular structure involving a dripping-wet leaf mutant of rice. Int Rice Res Note 21:39-40

Nakane A, Kikuchi M, Yokoo M, et al (1977) Inheritance of several characters in a cross between Japanese and USA rice varieties. Annu Rep Div Genet Natl Inst Agric Sci Japan 1977:10-11

Rao SA, Mengesha MH, Reddy CR (1988) Characteristics, inheritance, and allelic relationships of midribless mutants in pearl millet. J Hered 79:18-20

Satoh Y, Hong SK, Tagiri A, et al (1996) A rice homeobox gene, *OSH1*, is expressed prior to organ differentiation in a specific region during early embryogenesis. Proc Natl Acad Sci USA 93:8117-8122

Scanlon MJ, Freeling M (1997) Clonal sectors revealed that a specific meristematic domain is not utilized in the maize mutant *narrow sheath*. Dev Biol 182:52-66

Scanlon MJ, Schneeberger RG, Freeling M (1996) The maize mutant *narrow sheath* fails to establish leaf margin identity in a meristematic domain. Development 122:1683-1691

Sheridan WF, Clark JK (1993) Mutational analysis of morphogenesis of the maize embryo. Plant J 3:347-358

Shinbashi N, Kinosita T, Takahasi M, et al (1976) Genetical studies on rice plants. LXVi. Genetic behaviors of the two dwarfs, Hosetsu dwarf and Kotake-Tamanishiki, and their character expressions. Mem Fac Agric Hokkaido Univ 10:69-75

Suetsugu I (1953) Developmental morphology of the embryo in rice varieties. Bull Natl Inst Agric Ser D 4:23-52

Sugimoto N, Takeda G, Nagato Y, et al (1998) Temporal and spatial expression of the α-amylase gene during seed germination in rice and barley. Plant Cell Physiol 39:323-333

Takeda K (1977) Internode elongation and dwarfism in some gramineous plants. Gamma Field Symp 16:1-20.

Tamura Y, Kitano H, Satoh H, et al (1992) A gene profoundly affecting shoot organization in the early phase of rice development. Plant Sci 82:91-99

Tsuchiya T (1962) Radiation breeding in two-rowed barley. Seiken Ziho 14:21-34

Walsh J, Waters CA, Freeling M (1997) The maize gene *liguleless2* encodes a leucine-zipper protein involved in the establishment of the leaf blade-sheath boundary. Genes Dev 11:208-218

Xiong W, Uozu S, Hattori K, et al (1997) Dominant gene, *Ssi1*(t) located on the chromosome 1 shows the dm-type internode elongation pattern accompanied by a semidwarfness. Rice Genet Newslett 14:45-47

Yoshimura A, Iwata N, Omura T (1982) Linkage analysis by reciprocal translocation method in rice plants (*Oryza sativa*). III. Marker genes located on chromosome 2, 3, 4 and 7. Jpn J Breed 32:323-332

6
Flower development of rice

Junko Kyozuka

Laboratory of Plant Molecular Genetics, Nara Institute of Science and Technology, 8916-5 Takayama, Ikoma, Nara 630-0101, Japan

1 Introduction

Our knowledge of the genetic network governing floral development has increased significantly in the last several years. Genetic and molecular studies with two dicot plants, *Antirrhinum* and *Arabidopsis*, have shown that the genetic network controlling flower development is conserved at least in the two dicot species (for review, see Coen and Meyerowitz 1991, Weigel and Meyerowitz 1994, Weigel 1995, Yanofsky 1995, Ma 1998). However, it is not answered yet whether the findings obtained from studies on *Arabidopsis* and *Antirrhinum* can be applied as a general model to other distantly related species such as grasses.

Monocots diverged from dicots 180 million years ago. Many features of floral and inflorescence development of monocot species are distinct from those of dicot species. Among the monocot species, the flowers in the grasses are the most highly diverged. Understanding the genetic and molecular basis for morphological evolution is one of the fundamental issues of modern biology. How did grass species acquire unique floral morphologies during the course of evolution? How many genes were involved in the process? What was the key factor for creating a novel morphology? The remarkable diversity of grass inflorescence and floral structures make them good candidates for a model system to study morphological evolution among angiosperms.

Rice (*Oryza sativa* L.), which produces a flower typical of grass species, has several advantages as a model species for studying molecular biology (Izawa and Shimamoto 1996). Although molecular studies on rice flower development have just begun, a growing number of genes are now being cloned. In this chapter, first,

Key words. rice, panicle development, flower development, FLORICAULA/LEAFY, MADS-box genes, homologs, transgenic rice

morphology and development of rice panicle and flowers are summarized. Then, rice mutants in which normal development of panicle or flowers are perturbed are described in Section 3. Finally, recent progress toward identification and characterization of genes which may possess importance in rice flower development will be discussed in Sections 4 and 5.

2 Morphology of rice flowers and inflorescences

A rice shoot apex produces leaves distichously during the vegetative developmental stage. The first sign of the transition to reproductive growth is an increase of the apex size. Then, a young panicle apex produces several bracts of the panicle branches. Primary branches grow out from the axil of each bract. Successive branch primordia develop from the base of the apical meristem upward with 2/5 phylotaxy. At this stage, bract hairs develop on the bracts and finally cover the developing panicle completely. The bract hairs are obstacles for observation of early stages of rice panicle development. The primary branches initiated later grow faster than the branches that arose earlier. Therefore, the developmental stage of later arose branches soon precedes that of earlier branches. Secondary branches or sometimes a higher order of branches are produced on the primary branches (Fig. 1A). Differentiation of the spikelet primordia starts after the initiation of secondary branch primordia. The rice spikelet is pedicellate and borne on a rachilla of primary or higher order of branches. Usually, standard japonica type rice cultivars take around 30 days to complete the whole process from the transition to opening the first floret (Hoshikawa 1989).

There have been long debates regarding the interpretation of structures of grass spikelets. Here I describe the generally accepted explanation for morphology and development of rice flowers and panicles. An individual rice flower contains a gynoecium, six stamen, and two lodicules (Fig. 1B). The gynoecium, the innermost organ, consists of a single carpel and bears a short, bifurcated stigma. According to Bossinger et al. (1992), lodicules represent the inner whorl of the perianth, while the outer whorl of perianth is missing in flowers of grass species although some discussion regarding the identity of lodicules still exists. Each flower is in the axil of a bract, termed the lemma, and the stalk of each flower bears a bracteole called the palea. The whole structure including lemma, palea, and flower is referred to as a floret. Two extra bracts, which are sterile and termed the lower and upper glumes, envelop a floret. Glumes, lemma, palea are typically membranous catapylles, scale leaves. The glumes, lemma, and palea are first differentiated in this order, successively in alternate phyllotaxis. After that, organ differentiation of the rice spikelet proceeds inward. The whole unit, glumes plus florets, is termed the spikelet. The number of florets present in a single spikelet varies depending upon species. In rice, a spikelet contains only a single floret.

In grass species, the arrangement of spikelets rather than individual flowers is used to describe the morphology of the inflorescences. Rice inflorescence is termed

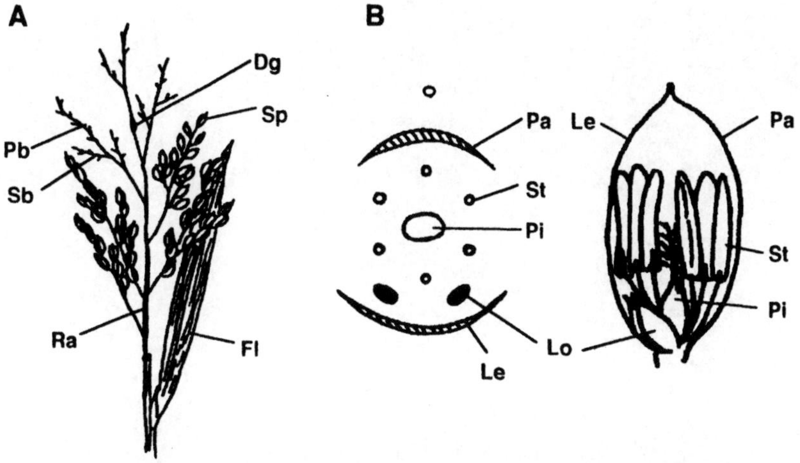

Fig. 1A,B. Rice panicle and flower structure. **A** A rice panicle (inflorescence). The panicle contains primary branches (*Pb*), secondary (*Sb*), and sometimes higher order of branches and spikelets. Primary branches are arranged along the main axis (*Ra*, rachis). Two extra bracts, which are sterile and termed as lower and upper glumes, envelop a floret. *Dg*, degenerated growth point. **B** Diagram of a rice spikelet (*right*) and schematic diagram of a transverse section through a flower (*left*). A flower is composed of a pistil (*Pi*), six stamens (*St*), and two lodicules (*Lo*). A rice flower is enclosed by a pair of bract-like structures called the lemma (*Le*) and palea (*Pa*). The structure including lemma, palea, and a flower is referred to as a floret. The whole unit, glumes plus florets, is termed the spikelet. A spikelet is enclosed by two sterile glumes called rudimentary glumes.

a panicle because it is highly branched. Each branch of the panicle terminates with a spikelet containing one fertile floret (Fig 1A). Therefore, the rice panicle is classified as a determinate inflorescence. One interesting feature of rice panicle development is that the apical meristem is not converted to floral meristem as is the case in typical determinate inflorescences. Instead, the apical meristem of rice panicle axis loses its activity and degenerates after the production of several primary branch primordia, leaving a scar called a degenerated growth point on the main rachis (Dg in Fig. 1A).

3 Mutants showing abnormal development of panicle and/or florets

A number of loci have been identified as genes involved in the regulation of rice panicle and flower development (Kinoshita and Takahashi 1991, Murai and Izawa 1994). More than 20 genes which control morphology of the panicle have been reported. In spite of a large collection of mutants, locations in a rice linkage map have been determined only for a small number of them.

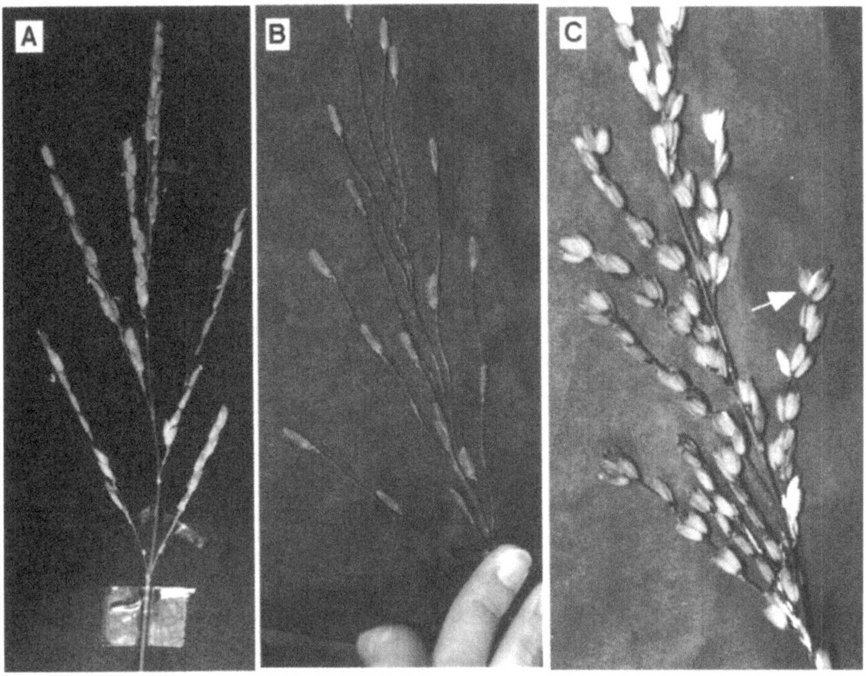

Fig. 2A-C. Mutants exhibiting abnormal panicle morphology. **A** Wild-type panicle. **B** A panicle of *lax* mutant. **C** A panicle of *Cl* mutant. *Arrow* indicates that three spikelets are produced in a cluster.

Basic architecture of the panicle is determined by branching pattern and the positions of flowers. Other factors such as elongation of rachis and rachilla also affect the panicle morphology. There are several mutants in which the branching pattern of panicle is altered. For example, a verticillate rachis, controlled by the *ri* gene, extends out a certain number of primary branches radially at the base of the panicle rachis. The lower part of the panicle axis is either missing or degenerated in *short panicle (sp1, sp2, sp3)* mutants. The number of secondary branches is increased in *Ur-1* mutant, which is incompletely dominant (Murai and Izawa 1994). Positions of flowers are affected in lax panicle *(lx)* and *Clustered-spikelets (Cl)* mutant plants (Fig. 2). *lx* plants produce panicles that differentiate florets only at the top of primary and secondary branches. In*Cl* mutant plants, spikelets are born in clusters consisting 3 to 4 spikelets (Murai and Izawa 1994). *Cl* is also an incompletely dominant mutation. Several defects are observed in the panicle of *pap1* mutant plants (Takahashi et al. 1996). The panicle of the *pap1* mutant plant produce increased number of primary panicle branches, whereas the length of panicle is not affected. Furthermore, length of each branch is reduced, and secondary and tertiary branches are absent. In addition, glumes of the *pap1* mutants also exhibit various morphological abnormalities. These abnormalities suggest that *PAP1* plays

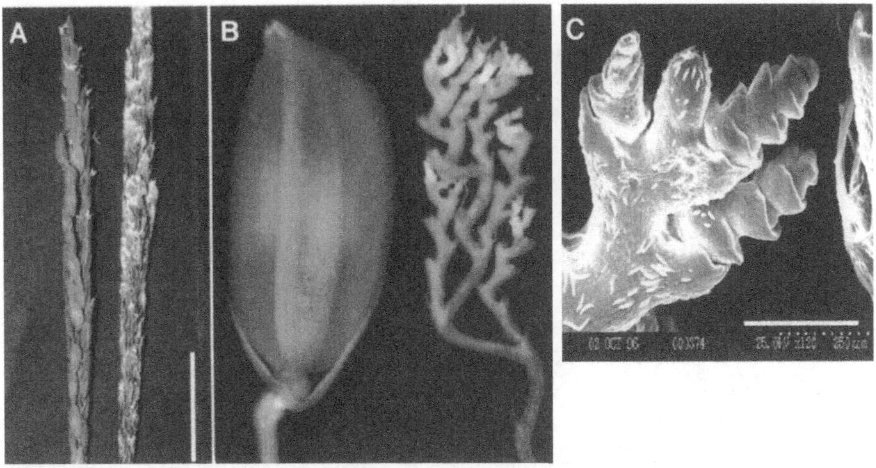

Fig. 3A-C. Panicles of *frizzy panicle (fzp)* and wild type. **A** Wild-type panicle. **B** Panicle (**A**) and floret (**B**) of wild type (*left*) and *fzp* mutant (*right*). *Bar* 2 cm. **C** SEM view of *fzp* mutant panicle. *Bar* 300 μm

important roles for the regulation of panicle phytomer development. A *dense panicle (ds)* mutant develops panicles with a dense seed setting, probably because of suppression of internode elongation in a panicle.

frizzy panicle (fzp) mutant (Mackill et al. 1991) is interesting with respect to determinacy of branch meristems. Initiation of the primary branches is not affected; thus, a normal number of primary branches are produced on a panicle axis (Fig. 3). Degeneration of the shoot apical meristem occurs after the initiation of primary branch primordia, as in the wild-type plants. However, florets are not initiated on the branches and a higher order of branches continue to form instead of spikelets (Fig. 3 B,C). As a result, an extraordinary order of branches are produced in the panicle of *fzp* mutant plants. This phenotype can be interpreted as a transformation of determinate floral meristems into indeterminate inflorescence meristems. The phenotype of *fzp* mutant indicates that the *FZP* gene is one of the major regulators in floral meristem initiation in rice. In *Arabidopsis LEAFY(LFY)* (Weigel et al. 1992) and *Antirrhinum FLORICAULA(FLO)* (Coen et al. 1990) mutants, similar transformation of floral meristems into shoots occurs. However, our analysis of the *RFL* gene (Kyozuka et al. 1998), a rice *LFY/FLO* homolog, suggested that it is unlikely that *RFL* is the causal gene for *fzp* mutant (Kyozuka, unpublished results).

A number of mutants affecting floret morphology have been also reported. Lemma and palea are the organs mostly affected. On the other hand, fewer mutants showing morphological alterations in lodicules, stamens, or pistils are reported, probably because it is difficult to observe the inner organs enclosed by a palea and a lemma.

Numbers of stigma and pistil are increased in *dl* (*Drooping Leaf*) mutants

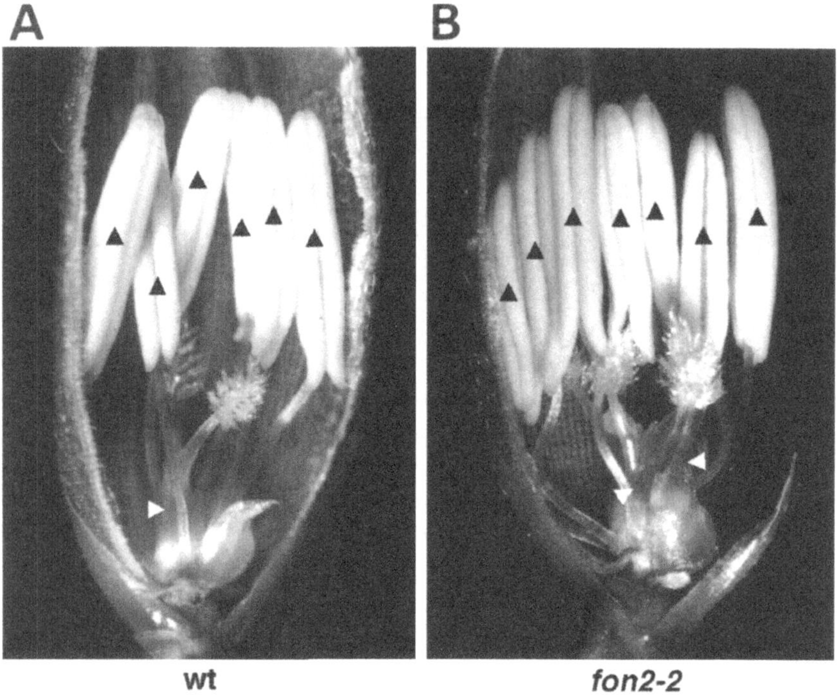

Fig. 4A,B. Florets of wild type and *fon2* mutant. **A** Wild type. **B** *fon2-2*. A wild-type floret contain six stamens (indicated by the *black arrowhead*) and a single pistil with two stigmas (indicated by the *white arrowhead*). The number of stamens and pistils increased to seven and two, respectively, in the floret shown here.

(Nagasawa et al. 1996a). Homeotic transformation of pistils to stamens is also observed in severe *dl* mutant plants. In addition to the floral abnormalities, midrib formation is affected in the *dl* plants. In *floral organ number 1 (fon1)* and *floral organ number 2 (fon2)* mutants, floral meristems are enlarged and the number of lodicules, stamens, and carpels is increased (Nagasawa et al. 1996b; Fig. 4). In addition, homeotic alterations of lodicules and anthers were observed. The phenotypes are similar to *Arabidopsis clavata1 (clv1)* (Clark et al. 1993) and *clavata3(clv3)* mutants (Alvarez and Smyth 1994; Clark et al. 1995), although there are clear differences. Both vegetative and reproductive shoot apical meristems are affected in *clv1* and *clv3* mutants. In contrast, the size of the shoot apical meristem at the vegetative developmental stage was not affected in *fon1* and *fon2* mutants. *CLV1* encodes a receptor-type protein kinase (Clark et al. 1997), suggesting the involvement of a kinase-dependent signal transduction pathway in the regulation of meristem size. It will be quite interesting to reveal whether *FON1* and *FON2* are homologous to *CLV1* and *CLV3* and a similar signal transduction pathway regulates meristem size in similar but slightly different ways in rice.

4 Genes involved in early steps of rice floral development

The genetic and molecular network controlling floral initiation is highly conserved, at least in *Arabidopsis* and *Antirrhinum*. Several genes have been identified as meristem identity genes, which regulate transition from inflorescence to flower. Among them, *FLO/LFY* and *Antirrhinum SQUAMOSA (SQUA*, Huijser 1992) and its *Arabidopsis* counterpart, *APETALA1 (AP1*, Mandel et al. 1992), play major roles in floral meristem initiation. Inflorescence-like shoots instead of flowers were formed in the mutants with *FLO/LFY* null alleles. Conversion of flowers to shoots also occurs in *squa* mutant plants. In contrast, *ap1* mutants exhibit less severe phenotypes compared to *flo, squ,* or *lfy* mutants, and specification of floral meristem identity is only partially affected. *FLO/LFY* and *SQU/AP1* orthologues have been isolated from other plant species including grass. This suggests that common molecular mechanisms control floral initiation even in distantly related species such as grass species. Our group has isolated *FLO/LFY* and *SQU/AP1* homologs from rice, and they are named *RFL(**Rice** FLO/**LFY**)* (Kyozuka et al. 1998) and *RAP1A(**Rice AP1A**)* and *RAP1B (**Rice AP1B**)* (Kyozuka and Shimamoto, unpublished results), respectively.

FLO/LFY-like genes isolated from dicot species so far show high sequence conservation (Fig. 5). Especially, the amino acid sequences in the C-terminal half exhibit extensive homology. *RFL* encodes a protein of 379 amino acids. Although the deduced amino acid sequence of *RFL* possesses high similarity with *FLO/LFY* proteins in the C-terminal half, its overall homology is lower than that of other dicot *FLO/LFY* orthologs. *FLO/LFY* genes are supposed to encode transcription factors. *RFL* shares the characteristics of transcription factors commonly seen in *FLO/LFY* and their dicot orthologs, such as prolin stretch in the N-terminal region and the presence of an acidic region. A putative NLS sequence is present in *RFL* at the same location as the *FLO/LFY* genes. Localization of RFL protein in nuclei was confirmed by using the *GFP* gene as a reporter (Abe et al., unpublished results). The sequence of *RFL* demonstrated that it is a *FLO/LFY* homolog of rice that arose from a common ancestor gene; however, diversification of the amino acid sequence occurred. High conservation of their sequences in the C-terminal half suggests its importance for their ancestral function. Ectopic expression of *LFY* but not *RFL* by CaMV35S promoter caused transformation of inflorescence meristems into floral meristems in *Arabidopsis* (Weigel and Nilsson 1995, Kyozuka et al. 1998). This suggests that the protein encoded by *RFL* does not work as LFY protein does in *Arabidopsis*.

The expression pattern of *RFL* also diverged from that of *FLO/LFY*. Both *FLO/LFY* are expressed in early floral meristems, consistent with their roles in specifying floral meristem identity. Northern blot analysis showed that *RFL* is expressed predominantly in young panicle but not in mature florets, leaves, or roots. *RFL* expression is activated in the shoot apical meristem immediately after the transition from vegetative to reproductive growth. *RFL* RNA was detected in all the

```
RFL      MDPNDAFSAA HPFRWD---- -LGPPAP--- APV--PPPPP PPPPPPPANV ---------- PREL---EEL VAGYGVRMST VARISELGFT ASTLLAMTER 7 7
FLO      ***-D**--- L-*K**HRT- A*PQ*NRLLD *-*--A**-- **--***QA- -PSYSM---R ****GGL*** FQA**I*YY* A*K*A***** VN***D*RDE 8 2
LFY      ***-EG*TSG L-***NPTR- A*VQ-**PP- --V--P**LQ QQPV-TPQT- -AAFGM---R ---*GGL*G* FGP**I*FY* A*K*A***** ****VG*KDE 8 2
BOFH     ***-EG*TSG L-***NPTR- VMVQ-**TP- --I--P**QQ QSPA-TPQT- -AAFGM---R ---*GGL*G* FGP****FY* A*K*A***** ****VG*KDE 8 2
NFL      ***-E****S L-*K**PRG- AMPPPTRLLE *A*--A**** **AL***QPL SAAYSI---K T***GGL*** FQA**I*YY* A*K*A***** VN***D*KDE 9 2
UNI      ***-***T*S L-*K**PRT- V*ST-**SPR PQLL-DYAVT *TT--APMTY HPA-RL---- ****GGL*** FQA**I*YY* A*K*A***** V***VD*KDD 8 8
populus  ***-E**T*S L-*K**TRA- MVPH*NRLLE M-*--***QQ **A------- -AAFAV---R ****CGL*** FQA**I*YY* A*K*A***** VN***D*KDE 8 3
pine     ***-ESFS*A F-*K**QRPP A*AP*QMQRS *GLEAQRIFH DFGV*NA*AM AASNNSSSC- RK**NCL*** FRN**V*YI* LTKMVDM*** VN**VN***Q 9 7

RFL      ELDDMMAALA GLFRWDLLLG ERFGLRAALR AERGRLMSLG G----RHHGH QSGSTVDGAS Q---EVLSDEH DMAGSGGMGD DDNGR-RMVT GKKQ-----  164
FLO      ***E**NS*C QI******V* **Y*IK**V* ***R*IDEEE ----VRRRHL LL---G*T-T H-ALDA**QE ----GLSEEP VQQEKEAMGS G-GGGVGGVW 1 68
LFY      **EE**NS*S HI***E**V* **Y*IK**V* ***R**QEEE EEESSRRRHL LLSAAG*SGT HHALDA**QE DDWTGLSEEP VQQQDQTDAA GNNGG----- 177
BOFH     **E***NS*S HI***E**V* **Y*IK**V* ***R**QEEE EEESSRRRHL LLSAAG*SGT HLALDA**QE DDWTGLSQEP VQHQDQTDAA GINGG----- 1 77
NFL      ******NS*S QI***E**V* **Y*IK**I* ***R**EEEE ----LRRRGH LL---S**GT N-ALDA**QE ----GLSEEP VQQQ-EREAV G-SGGGGTTW 1 78
UNI      ******NS*S QI******V* **Y*IK**I* ***R**DEEE ----IKRRGL LS---G*T-T N-ALDA**QE ----GLSEEP VVQ-REKEAM G-SGG-GSTW 1 72
populu   ***E**NS*S QI******V* **Y*IK**V* ***R**DEED ----PRRRQL LS---G*NNT N-TLDA**QE ----GFSEEP VQQ--DKEAA G-SGGRGT-W 167
pine     ****LVRT*V EIY*VE**V* *KY*IKS*I* **KR**EEAE RKRMEQLFVD VDGRKI*EN  --MLDT**QE ---GLSVEEP QGDNAIILSQ NNTSANFPLN 193

RFL      AKKGSAA-RK GKKA       RRKKVDDLRL DMQEDEMDCC DEDGGGGSES TESSAGGGGG -ERQREHPFV VTEPGEVARA KKNGLDYLFH LYEQCRLFLL 2 56
FLO      EMM*A-GG** APQRRRKNY- --KGRSRM-- ASMEEDDDDD DDETEG---- --AEDDENIV S********I *********G ********** ********** 2 56
LFY      GGS*YWDAGQ GKMKKQQQQR RRKK-PMLTS VETDEDVNEG ED*D*MDN-- --GNG*S*L* T********I *********G ********** ********** 2 72
BOFH     GRG*YWEAGQ TTIKKQQQ-R RRKK-RLYVS -ETDDDGNEG ED*D*MDI-- --VNG-S*V* M********I *********G ********** ******E*** 2 69
NFL      EVVAAAGGGR M*QRRRKKV- VAAGREKRGG ASAEEDEETE EGQEDDWN-- --INDAS**I S********I *********G ********** ******D**I 2 73
UNI      E-VAVVEERR KRQQIRRR-- ----RMKMKG NDHGENEEGE E*EED--N-- --ISG**V** G********I *********G ********** ******E**I 2 59
populus  EAVAA-GER- K*QSGRKK-- ---GQRK--- -VVDLDGDDE HG-------- -------*AI C********I *********G ********** ******D**I 2 41
pine     LNAGMDPVLI LQNSGHLGTT VSGLIGMPDT NYGSEQT----KACKKQKRRR SKD*--*ED* E********I ******L**G *********D *****GK*** 2 87

RFL      QVQSMAKLHG HKSPTKVTNQ VFRYAKKVGA SYINKPKMRH YVHCYALHCL DEEASDALRR AYKARGENVG AWRQACYAPL VDISARHGFD IDAVFAAHPR 3 56
FLO      ***TI**ER* E*C******* *******A** N********* ********** DEA**N**** *F*E****** *******K** *A*A**Q*W* **TI*N**** 3 56
LFY      ***TI**DR* E*C******* *******S** ********** ********** *****N**** *F*E****** S******K** VN*AC***W* *****N**** 3 72
BOFH     ***TI**DR* E*C******* *******S** N********* ********** *****N***S *F*V****** S******K** ***AC***W* *****N**** 3 69
NFL      ***NI**ER* E*C******* *******A** ********** ********** *****N**** *F*E****** *******K** *A*A**Q*W* **TI*N**** 3 73
UNI      ***AI**ER* E*C******* *******A** ********** ********** ***V*NE*** GF*E****** *******K** *A*A**Q*W* ***I*N**** 3 59
populus  ****I**ER* E*C******* *******A** ********** ********** **DA*N**** *F*E****** *******K** *A*AS*Q*W* **SI*N**** 3 41
pine     D**HI**ER* E*C******* ***H**HS** G********* ********** *I*Q*NR*** ***E****** *******Y** *AMAKDN*W* *EG**NKHEK 3 87

RFL      LAIWYVPTRL RQLCHQARSS HAAAAAALPP PLF------- ---------- --       389
FLO      *S******** *****AE*** A*V**TSSIT G--------- ---GGPADHL PF       396
LFY      *S******** *****LE*NN AV******VG GISCTGSSTS GRGGCGGDDL RF       424
BOFH     *S******** *****LE*NN AE****T*VG GISCRDRL-- -RLDALG--F N-       415
NFL      ********** *****SE**N A****SSSVS G--------- ---GVGDHLP HF       413
UNI      *S******K* *****AE*NG A*--*SSSVS V--------- ---GT-THLP F-       395
populus  *A******K* ****YAE*NS ATS--SSSVS G--------- ---T-GGHLP F-       377
pine     *R******K* *****LEK*K Q--------- ---SHL---- ---------- --       411
```

Fig. 5. Comparison of amino acid sequences encoded by *FLO*/*LFY* orthologs. *RFL* (Kyozuka et al. 1998), *FLO* (Coen et al. 1990), *LFY* (Weigel et al. 1992), *UNI* (Hofer et al. 1997), *NFL* (Kelley et al. 1995), and *BOFH* (Anthony et al. 1993) were isolated from rice, *Antirrhinum*, *Arabidopsis*, pea, tobacco, cauliflower, populus (U93196), and pine (U76757), respectively.

layers of very young panicle producing primary branch primordia. But, interestingly, it was absent in the primary branch differentiation sites and branch primordia. As development of branches proceeds, *RFL* RNA accumulation localized in the developing branches except for the apical meristem of the branches. Downregulation of the *RFL* expression was observed again in incipient or developing secondary branches.

During rice panicle development, the apical meristem of the panicle axis loses its activity and degenerates after the production of several primary branch primordia. *RFL* RNA accumulation in the shoot apical meristem seemed to correlate with the activity of the meristem. The *RFL* RNA expression started to diminish in the corpus of the panicle axis at the middle stage of primary branch differentiation. After all the primordia of primary branches had initiated and the shoot apex had lost its meristimatic activity, *RFL* RNA disappeared entirely from the main axis of the young panicle. After the branch formation stage, no detectable level of *RFL* RNA expression was found in the developing panicle, including floral meristems. Accumulation of *RFL* RNA is kept under detectable level in the apex of branch where a terminal floret is formed. This implies that floral meristems are produced without a detectable level of *RFL* RNA and therefore *RFL* RNA is not required for the initiation of floret meristems. Taken together, the expression pattern of *RFL* suggests that it plays a role which is distinct from that of *LFY* in *Arabidopsis* or *FLO* in *Antirrhinum.*

Our analysis of *RFL* showed that both regulatory and functional changes occurred in the *RFL* gene in the course of evolution. *RFL* is a good example showing how the same genes are used in different ways in distantly related species such as rice and *Arabidopsis.* To further elucidate the function of *RFL* in rice panicle development, loss-of-function mutants with the defective *RFL* gene should be obtained and analyzed. Also, isolation and analysis of *FLO/LFY* genes from other grass species will reveal whether rice is the only exception or whether the regulatory and functional changes found in *RFL* are generally seen in *FLO/LFY* genes in the Gramineae family.

Homologs of *AP1/SQUA*, another meristem identity gene, may play the major role for floral meristem initiation in rice. *SQUA/AP1* are members of the MADS-box gene family. The MADS domain is a DNA-binding/dimerization region present in transcription factors from many organisms. Plant MADS-domain proteins contain another conserved region called the K-domain, which is involved in protein—protein interactions. The deduced amino acid sequences for *RAP1A* and *RAP1B,* putative rice *SQUA/AP1* homologs, share 89% and 88% amino acid identity with *AP1* through the MADS-domain and 70% and 73% identity through the first 157 amino acids which include MADS-domain, I-region, and K-domain, whereas no significant homology is found in the C-terminal region between *RAP1A, RAP1B,* and *AP1*. The *RAP1A* and *RAP1B* show the highest sequence similarity to *ZAP1,* a maize *AP1* ortholog (Mena et al. 1995). Their deduced protein sequences share about 70% identity. Interestingly, their homology is extended to the C-terminal region.

The C-terminal region is the most variable part of the MADS-box proteins, both in sequence and in length. Krizek and Meyerowitz (1996) showed that the C-terminal region can be swapped among AP1, AP3, PI, and AG without affecting the specific functions of the proteins. On the other hand, ectopic expression of *AG* without the C-terminal region caused the *ag* mutant phenotype, suggesting that the truncated AG protein worked in a dominant-negative manner (Mizukami et al. 1996). A mutant allele of *Arabidopsis CAL* gene which causes severe cauliflower-like phenotype when combined with *ap1* mutation has amino acid exchanges in the C-region (Kempin et al. 1995). Furthermore, small stretches of conserved amino acids are found in the C-terminal region in homologous MADS-box proteins in different species. Taken together, function of the C-terminal region could be attributed to transcriptional activation but not to the determination of the functional specificity (Riechman and Meyerowiz 1997). It remained to be answered whether the conservation in the C-terminal region of ZAP1, RAP1A, and RAP1B contribute to Gramineae-specific function.

Overexpression of *RAP1A* by CaMV35S promoter caused very early flowering in rice (Kyozuka et al., unpublished results). This is similar to the results obtained by Mandel and Yanofsky (1995) which showed that 35S-*AP1* caused early flowering in *Arabidopsis*. In contrast to 35S-*RAP1A*, no abnormality was observed in transgenic rice plants carrying 35S-*RAP1B* with respect to the flowering time and the floral and panicle morphology. Thus, early flowering is a 35S-*RAP1A*-specific phenomenon although *RAP1A* and *RAP1B* share high sequence conservation. Early flowering was observed when another rice MADS-box gene, *OsMADS1*, was expressed ectopically in transgenic tobacco (Chung et al. 1994). It was shown that *OsMADS1* but not *OsMADS2* or *OsMADS4* caused early flowering in tobacco (Chung et al. 1994). *OsMADS1* encodes a protein exhibiting the highest homology to *Arabidopsis AGL2*, whose function is unknown. Effect of the gain and loss of function effect of *OsMADS1* should be examined in rice to reveal its role in rice floral development.

5 Genes involved in floral organ development

As described in Section 2, floral structures of grass species are distinct from those of typical dicot species. In particular, the perianth is highly modified. Lodicules have been the most puzzling organs in grass flowers. Generally, they are thought to represent the perianth in grass flowers (Bossinger et al. 1992); however, there is no molecular evidence supporting the homology between the two organs, lodicules and dicot perianth. Recent progress of our understanding on genes controlling floral development is expected to provide clues to interpreting the structures of grass flowers. Genetic analysis in *Arabidopsis* and *Antirrhinum* has demonstrated that several genes called floral organ identity genes act alone or in combination to specify floral organs (for review, see Yanofsky 1995, Riechmann and Meyerowitz 1997). Pattern formation of floral organs is explained by a simple model called the ABC model. According to the ABC model, *AP1, APETALA2(AP2), APETALA3(AP3),* and *PISTILATA(PI)* are involved in the differentiation of sepals and petals in *Arabidopsis*. Carpels and stamen are formed in whorl 1 and 2 in the strong *ap2* mutant. Petals are missing or converted to leaf-like structures in *ap1* mutant plants. Petals are transformed into sepals in *ap3* or *pi* mutants. These floral organ identity genes, except for *AP2,* encode proteins with MADS-domain. By analogy to the ABC model, changes in the MADS-box genes might be responsible for the morphological diversification in the perianth in grass species.

MADS-box genes represent a multigene family in plants and are divided in defined gene subfamilies according to their amino acid sequences. Most subfamily members share highly related functions in different plant species. This suggests that the emergence of the subfamilies may have been involved in the evolution of complex structures of seed plants. There are at least 28 MADS-box genes in *Arabidopsis* (Riechmann and Meyerowitz 1997). So far, 15 MADS-box genes have been isolated from rice. In addition, more than 13 rice expressed sequence tag (EST) clones with sequence similarity to MADS-box genes are present in the database (C74564, C72923, C75937, D22380, D46933, D15657, C73896, D45975, C74765, C26796, D47632, C73971, D47001). Phylogenic relationships of the rice and *Arabidopsis* MADS box genes are shown in Fig. 6. Rice MADS-box genes are found in all the subfamilies, supporting the notion that the major gene duplication of MADS-box genes took place in common ancestor of monocots and dicots (Purugganan et al. 1995). Therefore, it is likely that the basic features of floral development in monocot and dicot species are under the control of a conserved pathway. Then, an interesting question is how did molecular evolution in MADS-box genes, in their coding sequence and in their expression patterns, contribute to the morphological diversification and functional specialization of grass flowers?

Rice MADS-box genes *OsMADS2* and *OsMADS4* have high sequence similarity to *GLO/PI* (Chung et al. 1995). A sequence in K-domain and a segment of 16 amino acids at the C-terminal region, which are specifically conserved in *GLO/PI* orthologs in different plant species, are found in *OsMADS2* and *OsMADS4* at precisely same location, suggesting that they are rice *GLO/PI* orthologs. Since *GLO/PI* family genes are expressed in whorl 2 and 3 in later stages of flower development, analysis of *OsMADS2* and *OsMADS4* expression patterns might give a clue to understanding the interpretation for floral organs of rice, especially specification of the perianth. If they are expressed in lodicules, it is suggested that lodicules correspond to dicot petals. *OsMADS2* and *OsMADS4* show similar but slightly different patterns of RNA distribution. *OsMADS2* is expressed predominantly in anthers and carpels, whereas *OsMADS4* RNA was abundant in anthers but not in carpels. Since *GLO/PI* family genes are expressed in whorl 2 and 3 in later stages of flower development, expression of *OsMADS2* and *OsMADS4* in anthers is consistent with the notion that they are members of *GLO/PI* family. Unfortunately, expression of *OsMADS2* or *OsMAS4* in lodicules was not examined in the study of Chung et al. (1995). More detailed analyses on spatial and temporal expression pattern of *OsMADS2* and *OsMADS4*, especially expression in lodicules will provide a clue to

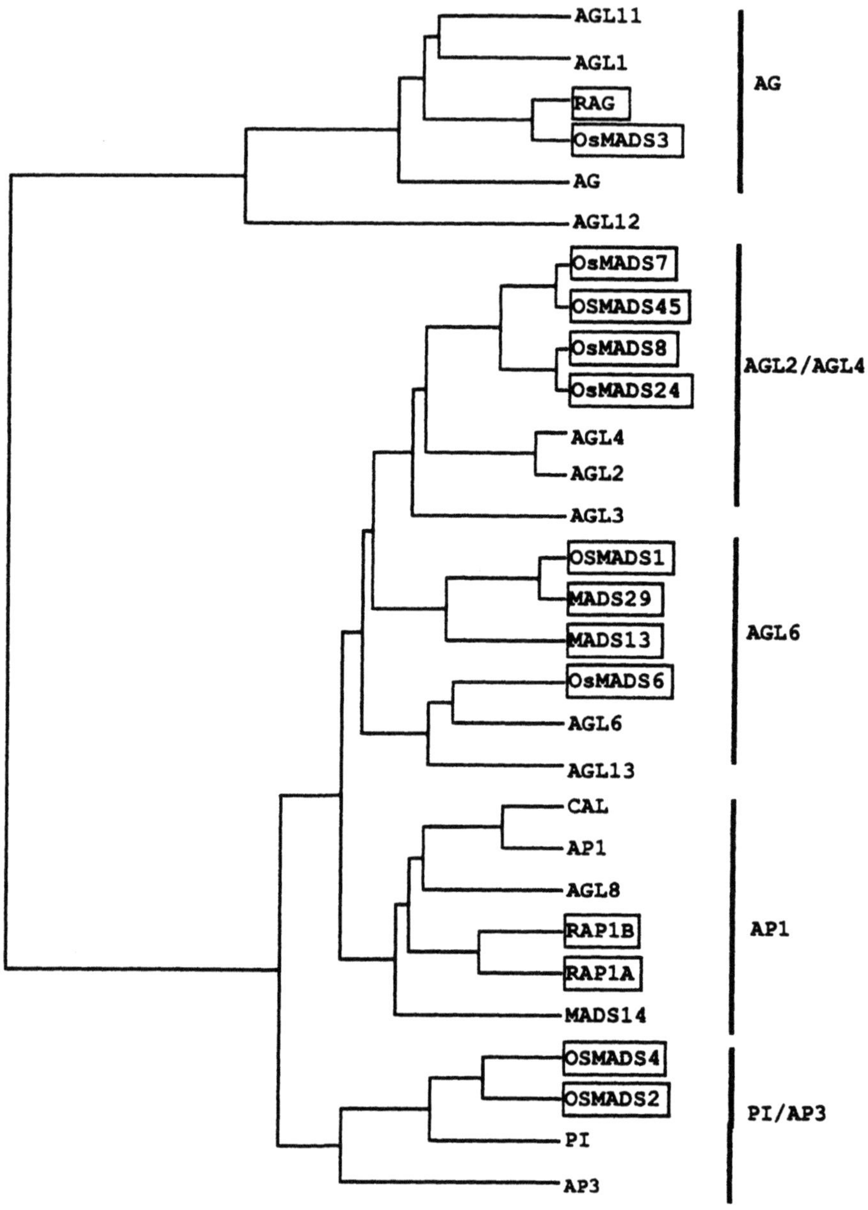

Fig. 6. Phylogenetic tree showing the relationship among rice and *Arabidopsis* MADS-box genes. The amino acid sequence of the entire coding region was used to make the tree. Rice MADS-box genes are presented in the *box*.

reveal the identity of lodicules.

OsMADS3 was reported as a rice *AGAMOUS(AG)* homolog (Kang et al. 1995). *AG* is a class C gene controlling the identity of reproductive organs and determinacy of floral meristems. The deduced protein sequence of *OsMADS3* shares 100% amino acid identity with *AG* in MADS domain and 64% in entire ORF. The MADS-domain of all the *AG* orthologs isolated from many flowering plants, including maize, is preceded by an extension of 14 to 53 amino acids; however, *OsMADS3* lacks the extension in the N-terminal region. We isolated *AG* -like cDNA (*RAG*) from rice (Kobayashi et al., unpublished results). The deduced amino acid sequence of *RAG* differs only in 4 amino acids except that RAG has 39 amino acids in the N-terminal. When compared with nucleotide sequences, *RAG* is different from *OsMADS3* in 14 nucleotides in 1.4 kb-cDNA, including 5' and 3' untranslated leader sequences. Since *OsMADS3* and *RAG* were isolated from different cultivars (*OsMADS3* was isolated from rice cv. M201, whereas *RAG* was isolated from cv. Toride 1), it is unclear whether *OsMADS3* and *RAG* represent distinct MADS-box genes in rice or whether they are the same gene. The differences in their nucleotide sequence might reflect the polymorphism between the two cultivars. Organ-specific expression of *OsMADS3* and *RAG* was examined by in situ hybridization and RT-PCR, respectively. Both *OsMADS3* and *RAG* are expressed predominantly in anthers and carpels but not in palea, lemma, or leaves. Ectopic expression of *OsMADS3* in transgenic tobacco under the control of CaMV35S promoter caused early flowering and homeotic alteration of sepals and petals to carpels and anthers, respectively. These results are consistent with the phenotype observed in transgenic *Arabidopsis* carrying CaMV35S-*AG* (Mizukami and Ma 1992). Transgenic *Arabidopsis* plants carrying 35S-*RAG* exhibited a less dramatic phenotype, the reduction in length of sepals and petals (Kobayashi et al., unpublished results). These phenotypes resembled weak phenotype of 35S-*AG Arabidopsis* plants. These imply that both OsMADS3 and RAG proteins share conserved function with AG.

To reveal the function of *RAG* in the context of rice floral development, we produced transgenic rice plants carrying 35S-*RAG*. Among 10 independent transgenic lines, similar abnormalities were found in 4 lines. These were increase of carpel and stigma numbers and malformation of stamens and pollen. Unfused carpels were observed. Since organs affected were well correlated with organ specificity of *RAG* expression, we suspected that the abnormalities were caused by co-suppression of *RAG* expression. To test this idea, the expression level of *RAG* in each floral organ was examined by RT-PCR analysis. In wild-type plants, *RAG* RNA is present in stamens and carpel but not in palea or lemma. In contrast, ectopic expression of the *RAG* RNA was detected in lemma, palea, and young leaves in the transgenic plants, indicating that the introduced 35S-*RAG* gene is transcribed. However, the level of *RAG* RNA accumulation in stamens and carpels in the transgenic plants was much lower than that in untransformed plants. The amount of *RAG* RNA expression was reduced to 8% and 24% in stamens and carpels, respectively. These results indicate that the expression of both endogenous *RAG* and introduced 35S-*RAG* gene was reduced in stamens and carpel, probably by cosuppression. As a results of the reduction of *RAG* transcript, normal development of stamens and carpel was affected in 35S-*RAG* plants.

Roles of MADS-box genes in floral development of monocot species were studied more extensively in maize (for review, see Veit et al. 1993, Schmidt and Ambrose 1998). Although maize flowers differentiate to staminate flowers in tassel and pistillate flowers in the ear, the basic structure of maize florets is quite similar to that of rice. A maize mutant plant in which function of *ZAG1,* a maize ortholog of *AG* (Schmidt et al. 1993), was disrupted by an insertion of a transposon has been obtained and analyzed (Mena et al. 1996). Surprisingly, development of reproductive organs was unaffected by the mutation. However, additional carpels were formed in florets. Close observation revealed that the floral meristem enlarged after stamen initiation and supernumerary carpels, which are unfused, arose repeatedly. This *zag1* mutant phenotype was interpreted as a loss of floral determinacy .

There are similarities between maize *zag1* and 35S-*RAG* phenotypes. Identity of reproductive organs was largely unaffected but the number of carpels that are unfused were increased in both *zag1* and 35S-*RAG* plants. Homeotic alteration of floral organs was not observed in either *zag1* or 35S-*RAG*. In the case of the 35S-*RAG*, incomplete suppression of the *RAG* expression might be an explanation for the lack of homeotic changes. For the *zag1* mutant, although the amount of *ZAG1* transcripts was highly reduced in reproductive organs of the mutant, it still remained to be determined whether the mutant has the null allele for *ZAG1*. Since the transposon insertion is found at the end of the K-box in the *zag1* mutant, it is possible that the protein encoded by the mutant *ZAG1* gene is partially functional and that the lack of homeotic transformation is due to incomplete knockout of *ZAG1* function. Alternatively, the differences in mutant phenotypes between *zag1* and *ag* were attributed to genetic redundancy of maize. Maize is allotetraploid and thus most genes including *ZAG1* are duplicated (Gaut and Doebley 1997). Another putative *AG* ortholog, *ZMM2* (Theissen et al. 1995), may control specification of reproductive organs in maize. Phylogenetic analysis showed that the duplication of *ZAG1* and *ZMM2* occurred before the divergence of common cereal species. Therefore, rice might have the duplicated gene of the *RAG* and this determines organ identity of reproductive organs in rice. Another possibility is that the mechanism controlling floral organ identity is different between grass and dicot plants and that organ identity for carpels and stamens is not determined by *AG* homologs in grass species. Analysis of more mutant alleles for *ZAG1* gene is required to make this point clear. Also, obtaining the loss-of-function mutant for the *RAG* gene is crucial to understanding the function of the *AG* orthologs in grass species.

Expression patterns of rice MADS-box genes, *OsMADS1* (Chung et al. 1995), *OsAMDS24*, and *OsMADS45* (Greco et al. 1997) were analyzed by northern blot analysis and in situ hybridization. *OsMADS1*, *OsMAD24,* and *OsMADS45* are all members of the *AGL2* subfamily. *AGL2* is expressed throughout the floral meristem after the early stages of floral development. It is also expressed in developing ovules, and after fertilization *AGL2* RNA was detected in embryos and seed coats (Ma et al. 1991, Flanagan and Ma 1994). *OsMADS1*, *OsAMDS24*, and *OsMADS45* are expressed predominantly in floral organs but not in leaves or roots. *OsMADS1* is expressed uniformly in young flower primordia, and in the mature flower its transcripts are localized in palea, lemma, and ovary. Ectopic expression of the *OsMADS1* driven by the 35S promoter resulted in early flowering and dwarf phenotype in tobacco; however, no homeotic alteration of floral organ was observed. *OsMADS24* and *OsMADS45* exhibited similar spatial expression patterns to each other. They are expressed in lodicules, stamens, and carpel but not in the palea or the lemma. Function of MADS-box genes belonging to the *AGL2* subfamily has not been determined due to the lack of loss-of-function mutants. Therefore, it is not easy to speculate on functions of *OsMADS1*, *OsMADS24*, and *OsMADS45* simply from their expression patterns, although the distribution of their RNAs suggest their putative involvement in controlling floral development of rice.

6 Perspectives

Molecular studies of rice flowering have just begun and not much is known. However, it is not difficult to imagine the fruitful future of this field. Undoubtedly, rapid progress of rice genome research will accelerate the development of molecular study of rice flowering. In fact, a growing number of rice genes and EST clones have been isolated and found to show sequence similarity to known genes (Harushima et al. 1998). The reverse genetic method will be extremely powerful for genetic and molecular approaches to reveal the flower development of rice. Identifying mutant alleles of genes with known sequences by using endogenous retrotransposons or introduced transposons is now a reality in rice (see Izawa and Shimamoto, Chapter 4; Hirochika, Chapter 3). Furthermore, the efficiency of the rice transformation helps to study the function of isolated genes. Knowledge obtained from studies on rice floral development will be used for improvement of grain production of rice in future.

Acknowledgments. I thank Yasuro Nagato for providing pictures of *fon2* mutant used in Fig. 4.

References

Alvarez J, Smyth DR (1994) Flower development in *clavata3,* a mutation that produces enlarged floral meristems. In: Bowman J (ed) *Arabidopsis*: an atlas of morphology and development. Springer-Verlag, New York, pp 254-257

Anthony RG, James PE, Jordan BR (1993) Cloning and sequence analysis of a *flo/lfy* homologue isolated from cauliflower (*Brassica oleracea* L. var. *botrytis*). Plant Mol Biol 22:1163-1166

Bossinger G, Rohde W, Lundquist U, Salamini F (1992) Barley: genetics, biochemistry, molecular biology and biotechnology. CAB International, Wallingford, pp 231-263

Chung YY, Kim SR, Finkel D, Yanofsky M, An G (1994) Early flowering and reduced apical dominance result from ectopic expression of a rice MADS box gene. Plant Mol Biol 26:657-665

Chung YY, Kim SR, Kang HG, Noh YS, Park MC, Finkel D, An G (1995) Characterization of two rice MADS box genes homologous to *GLOBOSA*. Plant Science 109:45-56

Clark SE, Running MP, Meyerowitz EM (1993) *CLAVATA1*, a regulator of meristem and flower development in *Arabidopsis*. Development 119:397-418

Clark SE, Running MP, Meyerowitz EM (1995) *CLAVATA3* is a specific regulator of shoot and floral meristem development affecting the same processes as *CLAVATA1*. Development 121:2057-2067

Clark SE, Running MP, Meyerowitz EM (1997) The *CLAVATA1* gene encodes a putative receptor kinase that controls shoot and floral meristem size in *Arabidopsis*. Cell 89:575-595

Coen ES, Meyerowitz EM (1991) The war of the whorls: genetic interactions controlling flower development. Nature 353:31-37

Coen ES, Romero JM, Doyle S, Elliott R, Murphy G, Carpenter R (1990) *floricaula*: a homeotic gene required for flower development in *Antirrhinum majus*. Cell 63:1311-1322

Flanagan CA, Ma H (1994) Spatially and temporally regulated expression of the MADS-box gene *AGL2* in wild-type and mutant *Arabidopsis* flowers. Plant Mol Biol 26:581-595

Gaut BS, Doebley JF (1997) DNA sequence evidence for the segmental allotetraploid origin of maize. Proc Natl Acad Sci USA 94:6809-6814

Greco R, Stagi L, Colombo L, Angenent GC, Sari-Gorla M, Pe ME (1997) MADS box genes expressed in developing inflorescences of rice and sorghum. Mol Gen Genet 253:615-623

Harushima Y, Yano M, Shomura A, Sato M, Shimano T, Kuboki Y, Yamamoto T, Lin SY, Antonio BA, Parco A, Kajiya H, Huang N, Yamamoto K, Nagamura Y, Kurata N, Khush G, Sasaki T (1998) A high-density rice genetic linkage map with 2275 markers using a single F_2 population. Genetics 148:479-494

Hoshikawa K (1989) The growing rice plant. Nosan Gyoson Bunka, Kyokai (Nobunkyo), Tokyo.

Hofer J, Turner L, Hellens R, Ambrose M, Matthews P, Michael A, Ellis N (1997) *UNIFOLIATA* regulates leaf and flower morphogenesis in pea. Curr Biol 7:581-587

Huijser P, Klein J, Loonig W-E, Meijer H, Saedler H, Sommer H (1992) Bracteomania, an inflorescence anomaly, is caused by the loss of function of the MADS-box gene *squamosa* in *Antirrhinum*. EMBO J 11:1239-1249

Izawa T, Shimamoto K (1996) Becoming a model plant: the importance of rice to plant science. Trends Plant Sci 1:95-99

Kang HG, Noh YS, Chung YY, Costa M, An K, An G (1995) Phenotypicalterations of petals and sepals by ectopic expression of a rice MADS box gene in tobacco. Plant Mol Biol 29:1-10

Kelley AJ, Bonnalander MB, Meeks-Wagner R (1995) *NFL*, the tobacco homolog of *Floricaula* and *LEAFY*, is transcriptionally expressed in both vegetative and floral meristems. Plant Cell 7:225-234

Kempin SA, Savidge B, Yanofsky MF (1995) Molecular basis of the *cauliflower* phenotype in *Arabidospsis*. Science 267:522-525

Kinoshita T, Takahashi M (1991) The one hundredth report of genetical studies on rice plant. J Fac Agric Hokkaido Univ 65:1-61

Krizek BA, Meyerowitz EM (1996) Mapping the protein regions responsible for the functional specificities of the *Arabisopsis* MADS domain organ identity proteins. Proc Natl Acad Sci USA 93:4063-4070

Kyozuka J, Konishi S, Nemoto K, Izawa T, Shimamoto K (1998) Downregulation of *RFL*, the *FLO/LFY* homolog of rice, accompanied with panicle branch initiation. Proc Natl Acad Sci USA 95:1979-1982

Ma H, Yanofsky MF, Meyerowitz EM (1991) *AGL1-AGL6*, an *Arabidopsis* gene family with similarity to floral homeotic and transcription factor genes. Genes Dev 5:484-495

Ma H (1998) To be, or not to be, a flower - control of floral meristem identity. Trends Genet 14:26-32

Mackill DJ, Pinson SRM, Rutger JN (1991) Frizzy panicle, an EMS-induced mutant in the Japonica cultivar M-201. Rice Genet Newslett 9:100-102

Mandel MA, Gustafson-Brown C, Savidge B, Yanofsky MF (1992) Molecular characterzation of the *Arabidopsis* floral homeotic gene *APETALA1*. Nature 360:273-277

Mandel MA, Yanofsky M (1995) A gene triggering flower formation in *Arabidopsis*. Nature 377:522-524

Mena M, Mandel MA, Lerner D, Yanofsky M, Schmidt RJ (1995) A characterization of the MADS-box gene family in maize. Plant 8:845-854

Mena M, Ambrose BA, Meeley RB, Briggs SP, Yanofsky M, Schmidt RJ (1996) Diversification of C-function activity in maize flower development. Science 274:1537-1540

Mizukami Y, Ma H (1992) Ectopic expression of the floral homeotic gene *AGAMOUS* in transgenic *Arabidopsis* plants alters floral organ identity. Cell 71:119-131

Mizukami Y, Huang H, Tudor M, Hu Y, Ma H (1996) Functional domains of the floral regulator *AGAMOUS*: characterization of the DNA binding domain and analysis of dominant negative mutations. Plant Cell 8:831-845

Murai M, Izawa M (1994) Effects of major genes controlling morphology of panicle in rice. Breed Sci 44:247-255

Nagasawa N, Miyoshi M, Nagato Y (1996) *DL* regulates both leaf and pistil development in rice. Rice Genet Newslett 13:102-105

Nagasawa N, Miyoshi M, Kitano H, Satoh H, Nagato Y (1996) Mutations associated with floral organ number in rice. Planta 198:627-633

Purugganan MD, Rounsley SD, Schmidt RJ, Yanofsky MF (1995) Molecular evolution of flower development: diversification of the plant MADS-box regulatory gene family. Genetics 140:345-356

Riechmann JL, Meyerowitz EM (1997) MADS domain proteins in plant development. Biol Chem 378:1079-1101.

Schmidt RJ, Veit B, Mandel MA, Mena M, Hake S, Yanofsky M (1993) Identification and molecular characterization of *ZAG1*, the maize homolog of the Arabidopsi floral homeotic gene *AGAMOUS*. Plant Cell 5:729-737

Schmidt RJ, Ambrose BA (1998) The blooming of glass flower development. Curr Opin Plant Biol 1:60-67

Takahashi M, Nagasawa N, Nagato Y (1996) Mutations affecting panicle architecture in rice. Rice Genet Newslett 13:97-99

Theissen G, Strater T, Fischer A, Saedler H (1995) Structual characterization, chromosomal localization and phylogenetic evaluation of two pairs of *AGAMOUS*-like MADS-box genes from maize. Gene 156:155-166

Veit B, Schmidt RJ, Hake S, Yanofsky MF (1993) Maize flloral development: new genes and old mutants. Plant Cell 5:1005-1015

Weigel D, Alvarez J, Smyth DR, Yanofsky MF, Meyerowitz EM (1992) *LEAFY* controls floral meristem identity in *Arabidopsis. Cell* 69:843-859
Weigel D, Meyerowitz EM (1994) The ABCs of floral homeotic genes. Cell 78:203-209
Weigel D, Nilsson O (1995) A developmental switch sufficient for flower initiation in diverse plants. Nature 377:495-500
Weigel D (1995) The genetics of flower development: from floral induction to ovule morphogenesis. Annu Rev Genet 29:19-39
Yanofsky MF (1995) Floral meristems to floral organs: genes controlling early events in *Arabidopsis* flower development. Annu Rev Plant Physiol Plant Mol Biol 46:167-188

7
Analysis of the functions of rice homeobox genes

Yutaka Sato, Naoki Sentoku, and Makoto Matsuoka

BioScience Center, Nagoya University, Chikusa, Nagoya 464-8601, Japan

1 Introduction

Although most plant development occurs after embryogenesis, the basic body plan in the vascular plant is generated during embryogenesis (Meinke, 1995; Jürgens et al., 1994). Like all sexually reproducing organisms, the vascular plant begins its existence as a single cell, the fertilized egg or zygote. This cell proliferates to be an embryo with some organs and tissues. Early in the embryogenesis, axes patterns which contribute the basic organization of a plant body are formed, namely the apical-basal pattern and the radial pattern. Based on the regional information of these axes, the shoot apical meristem (SAM) and the root apical meristem are developed during embryogenesis (Jürgens, 1995).

After seed germination, the shoot and root apical meristems retain some properties of embryonic cells and continue to produce the post embryonic shoot and root systems of the adult plant. The entire ground portion of a plant body is constructed by piling up a shoot unit called phytomer which consists of an axillary bud, a leaf, and a stem. SAM continuously produces these units keeping itself as collection of indeterminate stem cell (Steeves and Sussex, 1989).

After the phase change of SAM has occurred, most flowering plants transform the vegetative meristem into the inflorescence meristem from which determinate reproductive or floral meristems are arisen. Floral meristems produce floral organs, which become sepals, petals, stamens, and carpels. Thus, plants continuously perform organogenesis throughout the life cycle, even after embryogenesis. Organogenesis after embryogenesis is a distinct feature of plant development, comparing to the development of higher animals (Walbot, 1985). In the animal embryo, all the organs and tissues are formed at least in rudimentary forms during embryogenesis.

Key words. rice, homeodomain, homeobox gene, shoot apical meristem, embryo, embryogenesis, organogenesis, in situ hybridization, retrotransposon, *Tos17*, tagging, mutant, dwarf, *d6*, internode

Only enlargement of the body and/or its maintenance in a functionally efficient state occurs in postembryonic development of animal bodies.

Extensive molecular and genetic analyses on animal embryogenesis have established the genetic programs which control patterning (Gehring et al., 1994; Lawrence and Morata, 1994), and it has been revealed that homeobox genes often act the key role in these programs as developmental regulators. Recently, homeobox genes have been cloned from various plant species, and it has been suggested that plant homeobox genes are also involved in the morphogenetic events during plant development. It is very interesting that homeobox genes act as developmental regulators both in animals and plants, which show such distinct features of body plans between them. In this chapter, recent advances in the analysis of functions of plant homeobox genes, especially focusing on the rice homeobox genes, are introduced.

2 Homeobox genes in rice

The homeodomain was originally described as a highly conserved protein motif of 60 to 63 amino acids, encoded by a short DNA fragment, the homeobox, that was found in *Drosophila* developmental control genes (McGinnis et al., 1984; Scott and Weiner, 1984). The first hint at the possible function of the homeobox genes came from computer searches through protein databases which revealed a sequence similarity to the yeast mating-type genes that have been identified as transcriptional regulators. Sequence similarities between homeodomain and the gene regulatory proteins of prokaryotes also suggested that the homeodomain might contain a helix-turn-helix DNA-binding motif. In following years, many data have confirmed that homeobox genes encode gene regulatory proteins acting as transcriptional factors which bind to specific *cis*-regulatory regions in their target genes.

Now, homeobox genes have been identified in numerous copies in a broad range of eukaryotic organisms including yeast, fungi, and plants (Scott et al., 1989). The first homeobox genes in a plant species were isolated by transposon tagging of the morphological mutant of maize, *Knotted1* (*Kn1*) mutant as *KNOTTED1* (*KN1*) gene (Vollbrecht et al., 1991). In the *Kn1* mutant, abnormal arrangements of lateral veins, sporadic outgrowths called knots, and ligule displacement are observed in leaf blades. It has been revealed that *Kn1* is a dominant mutation caused by ectopic expression of *KN1* in leaves, and that its ectopic expression results in the disorganization of the developmental program of the maize leaf blade (Smith and Hake, 1994). Analysis of the expression pattern of *KN1* by *in situ* hybridization and immunohistochemical analysis has revealed that *KN1* shows the cell type specific expression within SAM. That is, *KN1* is expressed in vegetative, inflorescence, and floral meristems and is downregulated at the site of formation of determinate lateral organs such as leaves or floral organs. Based on the expression pattern and the gain-of-function phenotype, *KN1* is thought to be involved in the maintenance of the indeterminacy within SAM and, in other words, suppression of the formation of determinate lateral organs (Smith et al., 1992).

Many homeobox-containing genes have been isolated in several higher plants including rice. Based on either sequence identity within homeodomain or conserved protein motif outside the homeodomain, plant homeobox genes have now been classified into five groups (Lu et al., 1996). These include the KNOTTED-type (KN1) homeodomain proteins, homeodomain zipper proteins (HD-ZIP), homeodomain finger proteins (PHD-finger), the GLABRA2 homeodomain proteins (HD-GL2), and the BELL1 homeodomain proteins. Among them, genes of KN1-type homeodomain proteins are most intensively analyzed. KN1-type homeobox genes are further classified to two groups, named class 1 and class 2, according to their amino acid identity (Kerstetter et al., 1994). Based on the observations, the class 1-type homeodomain proteins are expressed around SAM and overexpression of these genes affects the developmental program of the leaf blades. It has been suggested that genes of this class are involved in maintenance of SAM and/or the development of lateral organs. Also, it is assumed that the HD-GL2 genes become specialized to the establishment of the dermal layer of the shoot system, since genes of this class share a common L1 layer-specific or dermal-specific pattern of expression (Lu et al. 1996). It has not been known whether the rest of groups share their specific functions during plant development. In rice, homeobox genes from three groups, KN1-type, HD-ZIP, and HD-GL2, have been identified so far (Figure 1A). The alignment of the homeodomain regions is shown in Figure 1B.

Meijer et al. (1997) have isolated a HD-ZIP-type homeobox gene from rice. The HD-ZIP homeodomain proteins differ from other homeodomain proteins in that they contain a leucine zipper motif closely linked to the carboxyl-terminal end of the homeodomain region (Ruberti et al., 1991). The HD-ZIP motif has not been described in organisms from the animal kingdom, and it remains to be seen whether this configuration is indeed plant-specific (Schena and Davis, 1992). The rice HD-ZIP gene, *Oshox1*, is expressed in various tissues at different developmental stages, with highest levels in embryos, shoots of seedlings, and leaves of mature plants. Overexpression of the Oshox1 product in transgenic *Arabidopsis* causes to retard growth and affects leaf size and shape, indicating that its product functions as a developmental regulator. Meijer et al (1997) have also reported that the Oshox1 product interacts with the pseudopalindromic sequence, CAAT(C/G)ATTG, and its interaction with the binding sequence can repress the reporter gene activity in rice suspension cells. Through these findings, they discussed the Oshox1 function as an active repressor of target gene(s). There are at least two other HD-ZIP homeobox genes in the rice genome, as two clones encoding the consensus amino acid motif of the HD-ZIP homeodomain have been released as EST clones from the Rice Genome Project with accession number C73285 and D24543.

The second group of the rice homeobox genes is a HD-GL2 family. Very recently, Ito et al. isolated novel homeobox genes encoding the amino acid motif similar to the homeodomains of *Arabidopsis* GL2 (Rerie et al., 1994) and ATML1 (Lu et al., 1996) from rice (unpublished results). *GL2* and *ATML1*, which are the two *Arabidopsis* members of the HD-GL2 type homeobox genes, are expressed specifically in the L1 layer of SAM, the protoderm, or the developing epidermis.

A

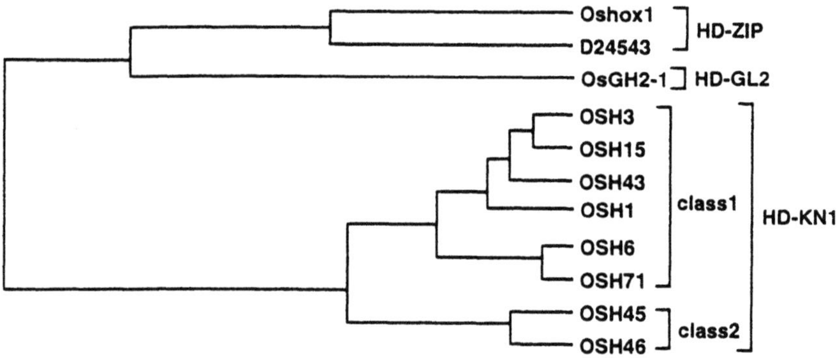

B

```
           basic      helix1            turn  helix2     turn  helix3
OSH3       KKRKKGKLPKEARQKLLTWWELHYRWPYPSEMEKIALAESTGLEQKQINNWFINQRKRHW
OSH15      KKKKKGKLPKEARQKLLHWWELHYKWPYPSETEKIALAESTGLDQKQINNWFINQRKRHW
OSH43      KRTKKGKLPKEARQKLLHWWELHYKWPYPSEMEKMTLAQTTGLDQKQINNWFINQRKRHW
OSH1       KKKKKGKLPKDARHQLLNWWELHYKWPYPSESQKVALAESTGLDLKQINNWFINQRKRHW
OSH6       KKRKKGKLPKDARSALLEWWNTHYRWPYPTEEDKLRLAARTGLDPKQINNWFINQRKRHW
OSH71      KKRKKGKLPKDARSALMDWWNTHYRWPYPTEEDKVRLAAMTGLDPKQINNWFINQRKRHW
OSH45      RKRRAGKLPGDTTSTLKAWWQSHAKWPYPTEEDKARLVQETGLQLKQINNWFINQRKRNW
OSH46      RKRRAGKLPGDTTTILKQWWQQHSKWPYPTEDDKAKLVEETGLQLKQINNWFINQRKRNW
Oshox1     GSRKKLRLSKDQAAVLEDTFKEHNT---LNPKQKAALARQLNLKPRQVEVWFQNRRARTK
D24543     GGEKKRRLSVEQVRALERXFEVENK---LXPERKXRLARXLGLQPRQVAVWFQNRRARWK
OsGH2-1    KKKRYHRHTAHQIQQMEALFKECPH---PDDKQRLKLSQELGLKPRQVKFWFQNRRTQMK
```

Fig. 1. A Phylogenetic tree of homeodomain proteins from rice. The tree reflects the degree of similarity between deduced amino acid sequences of the homeodomains from rice homeobox genes using the UPGMA method. **B** Alignment of the homeodomains of rice homeobox genes.

Another HD-GL2 homeobox gene in *Phalaenopsis*, *O39*, is also expressed specifically in placental epidermis, the protoderm of ovule primordia, and the outer cell layer surrounding the archesporial cells (Nadeau et al., 1996). While the rice HD-GL2 homeobox genes have never been characterized, it is assumed that these genes function to establish the dermal layer of the shoot system.

The third group of the rice homeobox genes is the KN1 type. This type of homeobox genes, one of the best characterized homeobox genes in plants, encodes atypical homeodomain with three extra amino acids between helix 1 and helix 2 in comparison to typical homeodomains. Initially, this type of atypical homeodomains had been considered to be unique to the plant KN1-type homeobox genes. Recently, however, Bürglin (1997) and Chen et al. (1997a) have described that such atypical homeobox genes are widespread not only in the plant kingdom but also in human, fly, *Caenorhabditis elegans*, and yeast. Thus this type of homeobox genes was already present in the common ancestor of plants, fungi, and animals and represents a branch distinct from the typical homeobox genes.

From the cDNA and genomic libraries, we have isolated eight clones encoding the KN1-type homeobox genes from rice. All these clones encoded a conserved homeodomain sequences of the KN1-type homeobox genes that consisted with 64 amino acid residues (Figure 1B). In the homeodomain, all the proteins of this type are well conserved, especially in the third helix, which is predicted as a recognition site of DNA sequences. Outside of the homeodomain, the flanking ELK (glutamate, leucine and lysine) domain, which have been often seen in the KN1 type of homeodomain proteins, is also conserved in all proteins.

Kerstetter et al. (1994) have described that the maize KN1-type homeobox genes can be divided into two classes by differences outside the conserved third helix of homeodomains. The homeodomains of the class 1 genes share 73%-89% identical residues with KN1. The class 2 genes share 55%-58% identical residues with KN1, although the conservation within each class is greater than 87%. The same situation is also observed in the rice KN1-type homeobox genes (Figure 1B). The six genes, *OSH1, 3, 6, 15, 43,* and*71*, share 75%-87% identical residues within this class and also show high similarity to the maize class 1 genes. The remaining two genes, *OSH45* and *46*, share 57%-67% identity with the rest of genes, although the identity between the genes is 87% and also show high similarity to the maize class 2 genes. The phylogenetic analysis with the rice and maize *KN1* homeobox genes demonstrates that each rice gene has one or two corresponding maize genes with higher similarity than other genes. For example, *OSH1* is paired with *KN1*, and *OSH15* is paired with the maize *RS1* and *KNOX4*. Such pairing is also seen between *OSH3* and *KNOX3*, *OSH6* and *LG3*, *OSH43* and *KNOX8*, and *OSH71* and *KNOX5* or *KNOX11*.

The relationship of these pairs between the rice and maize genes is also supported by the map positions of these genes on the rice and maize genome. Sentoku et al. have mapped the rice KN1-type homeobox genes on the rice genome by an analysis of RFLP and compared the map positions of the genes to the maize KN1-type homeobox genes by referring to the comparative linkages of RFLP markers between the rice and maize (Ahn and Tanksley, 1993). This comparative analysis has demonstrated that pairs of rice and maize homeobox genes are mapped at comparative positions on the rice or maize chromosomes (Sentoku et al., unpublished results). In some cases such as *OSH15* and *OSH71*, there are two loci on the maize genome corresponding to the single locus in rice, which is considered to be caused by duplicate chromosome segments of the maize genome (Helentjaris et al., 1988). Considering the high similarity in the primary structures and the comparative map positions, it is possible that pairs of rice and maize homeobox genes are orthologous relationships, and therefore the genes may have the same or similar biological function(s) in rice and maize.

3 Expression of rice homeobox genes in embryo and SAM

In animals, homeobox genes play important roles in cellular or regional differentiation during embryogenesis. Plant homeobox genes, by analogy of the functional roles of animal homeobox genes, have been expected to encode transcriptional regulators that mediate important developmental processes. Recently, some observations indicate the involvement of plant homeobox genes to embryogenesis and to the morphogenetic events around SAM (Smith et al., 1995; Klinge and Werr, 1995; Sato et al., 1996a; Long et al., 1996; Lu et al., 1996). In this section, current advances on the functional analyses of KN1-type homeobox genes, mainly focusing on our current research on rice KN1-type homeobox genes, which have been strongly suggested the involvement to such processes, will be introduced.

The involvement of plant homeobox genes in embryogenesis is first demonstrated by the analysis of an *Arabidopsis* embryogenesis-defective mutant, *stm* (Barton and Poethig, 1993; Long et al., 1996). *stm* (*shoot meristemless*) mutant shows the shoot meristemless phenotype of embryos, whereas other embryonic organs such as cotyledons, hypocotyl, and radicle are normally developed. It has been revealed that the *STM* gene is encoded by the KN1-type homeobox gene and *STM* is expressed as if it marks SAM during embryogenesis. This finding has demonstrated directly the involvement of the plant homeobox gene in shoot formation or maintenance during embryogenesis. To elucidate further the function of plant homeobox genes during embryo development, it is expected that much information about the function of plant homeobox genes during embryogenesis will be acquired by examining the temporal and spatial expressions during embryogenesis. In the following paragraphs, we outline the current works on the expression pattern of some KN1-type homeobox genes during embryogenesis.

As mentioned previously, the maize homeobox gene, *KN1*, is expressed at SAM in the postembryonic stages of development (Smith et al., 1992; Jackson et al., 1994). Based on this observation, Smith et al. proposed that *KN1* may act as a marker for the shoot meristem development during embryogenesis. To elucidate this idea, they tested the expression pattern of *KN1* during maize embryogenesis by *in* situ hybridization (Smith et al., 1995). They reported that expression of KN1 was not observed in the globular embryo in which no sign of the shoot differentiation was observed, while expression was observed at the same time and same place of shoot meristem differentiation. Based on these observations, they have discussed that *KN1* is a useful marker for the shoot development during embryogenesis.

We have also investigated the expression pattern of two KN1-type homeobox genes, *OSH1* and *OSH15*, during rice embryogenesis and have proposed the possible involvement of these genes in the establishment of cellular identities during embryogenesis (Sato et al., 1996a; Sato et al., 1998). Rice embryos develop much quickly than those of other cereal plants such as maize and barley. They complete most of the morphogenetic events within 1 week under normal conditions (Figure 2). In the developmental course of rice embryo, Sato et al. performed in situ hybridization analyses on embryos at 2 DAP through 7 DAP with the digoxygenin- or fluorescein-labeled *OSH1* and *OSH15* antisense RNAs as probes.

After fertilization, rice embryos grow in a globular shape with relatively irregular cell divisions. In the early globular stage embryo (2 DAP), embryos are composed of about 100 cells and are about 50 μm in length. Although no sign of organ differentiation can be observed at this stage, the expression of *OSH1* has already been observed in the ventral and basal part of the embryo where SAM would de-

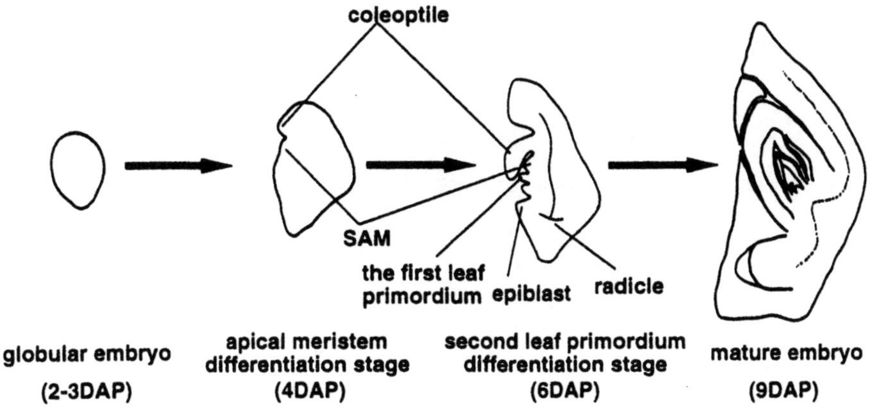

Fig. 2. Schematic representation of a developmental course of wild-type rice embryogenesis.

velop later in embryogenesis (Figure 3A). In the same section, however, *OSH15* expression was not detected (Figure 3B). In the late globular stage embryo (3 DAP), a dorsoventral polarity within the embryo becomes evident from the gradient of cell size although no organogenetic events are yet observed, and *OSH1* mRNA accumulation, as indicated by a pink signal from its fluorescein-labeled probe, continued in the ventral and basal part of the embryo (Figure 3C). In the same area, the first appearance of *OSH15*, as indicated by a digoxygenin-labeled probe, was observed (Figure 3D). Thus, the timing of the initiation of gene expression during the embryogenesis may be different for the two genes, but both genes are expressed in the same region, that is, the future shoot region, before any visible organ differentiation.

Such localized expression of *OSH1* and *OSH15* before morphological differentiation indicates that both genes function at an earlier stage than the morphological differentiation of SAM. To clarify the relationship between expression of *OSH1* and *OSH15* and shoot differentiation, Sato et al. conducted in situ hybridization using the embryos of the rice embryogenesis-defective mutant, *organless1* (*orl1*), which is caused by a single recessive mutation (Hong et al., 1995). *orl1* lacks embryonic organs including shoot and radicle, although the embryo grows relatively large (larger than 1000 μm). Before the late globular stage, it is impossible to distinguish the wild type from *orl1* embryos under the microscope because both embryos show a globular shape without any organs. At 4 DAP and later stages, *orl1* embryos are morphologically distinguishable from the wild-type embryos by the absence of the shoot and the radicle. Sato et al. (1996a) examined the expression of *OSH1* and *OSH15* in embryos of the plants that were heterozygous for *orl1*. At the globular stage, the section of mutant embryo could not be identified. (The expected frequency of homozygous *orl1* embryo is 25% in the panicle of heterozygous plants.)

More than 50 early globular embryos examined showed the same hybridization

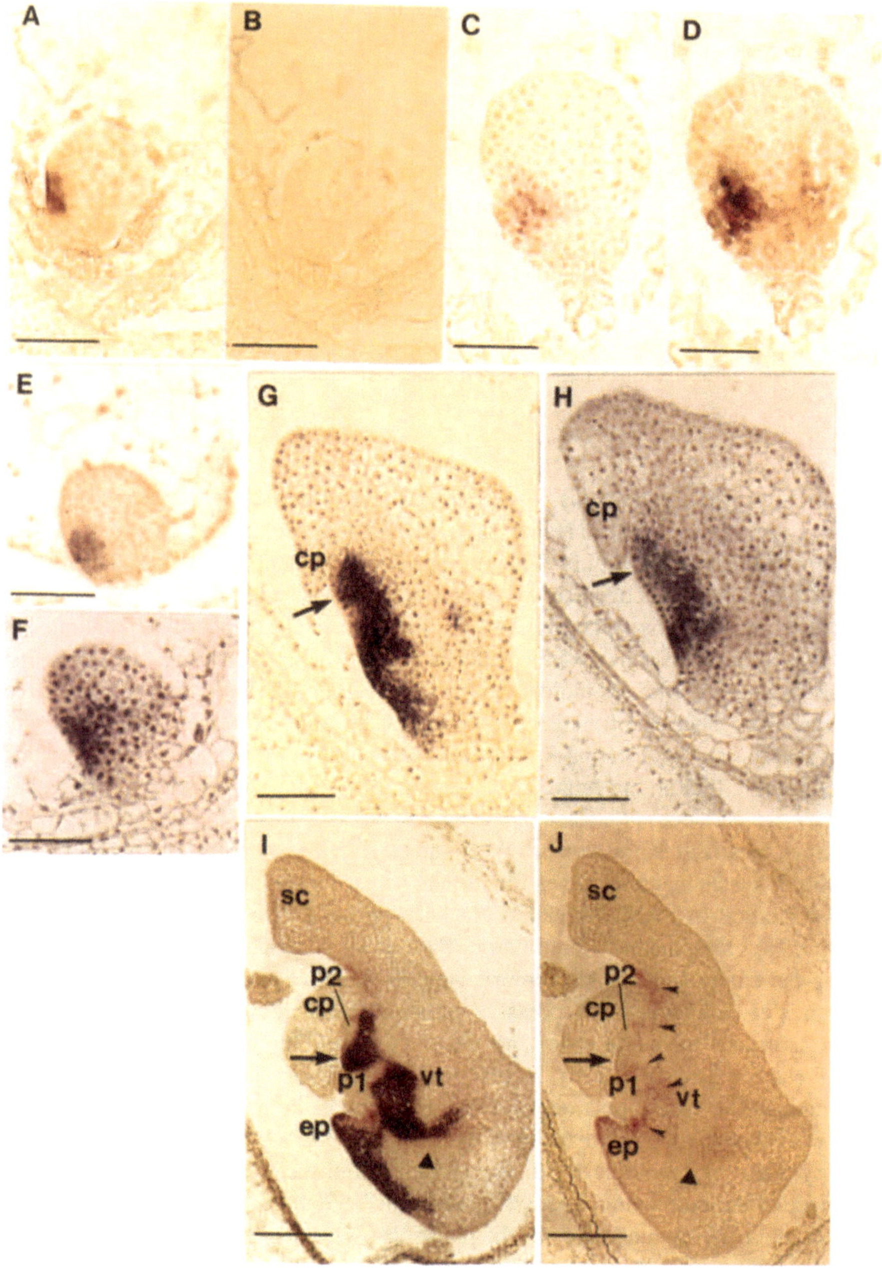
A
B
C
D
E
G
H
cp
cp
F
I
J
sc
p2
cp
p1
vt
ep
sc
p2
cp
p1
vt
ep

pattern as the wild type that marked the future shoot region, indicating that both *OSH1* and *OSH15* expression is normally regulated in the *orl1* embryo (Figure 3E,F). These observations indicate that the expression of *OSH1* and *OSH15* is exactly maintained even in the *orl1* mutant embryos lacking SAM. This confirms that, in the wild-type embryo at the globular stage, *OSH1* and *OSH15* are first expressed earlier than the onset of morphological differentiation of the shoot. The localized expression at the shoot and/or to the zone where it will develop early in embryogenesis is also observed in maize *KN1*, *Arabidopsis STM*, and tobacco *NTH15*. These findings indicate the involvement of this class of homeobox genes in the formation of shoot and/or its maintenance during embryogenesis and the later stages of plant development including vegetative and reproductive stages. However, the timing of the appearance of *OSH15* expression was later than that of *OSH1*. Thus, there may be an epistatic relationship between these two genes. It is noteworthy that some homeobox genes which are thought to be involved in shoot formation during embryogenesis such as *KN1*, *OSH1*, *OSH15*, *NTH15*, *KNAT1*, and some other class 1 genes produce ectopic shoots on the leaves of transgenic tobacco or *Arabidopsis* plants overproducing these genes (Sinha et al., 1993; Kano-Murakami et al., 1993; Sato et al., 1996b; Chuck et al., 1996; Müller et al., 1995; Tamaoki et al., 1997). It is possible that this formation of ectopic shoots may mimic the process of shoot formation in the natural context of embryogenesis.

As rice embryogenesis proceeds, differences in the expression pattern between *OSH1* and *OSH15* begin to be observed. At 5 DAP, which had a developing co-

◄

Fig. 3. In situ mRNA localization of *OSH1* and *OSH15* in developing embryos of wild-type and in *orl1* mutant embryo. Longitudinal sections through the embryo were hybridized with an *OSH15* and/or an *OSH1* antisense probe. **A–D**, **I** and **J** are double labeling of the same sections for *OSH1* and *OSH15*. Probes were labeled with digoxigenin-UTP or fluorescein-UTP, and the transcript-specific hybridization signal is visualized as either *purple* or *pink* color, respectively. **E–H** are single labeling for *OSH1* (**E** and **G**) or *OSH15* (**F** and **H**). Probes were labeled with digoxigenin-UTP, and the transcript-specific hybridization signal is visualized as *purple* color. **A** Double staining of mixtures of fluorescein labeled *OSH15* (*pink*) and digoxigenin-labeled *OSH1* (*purple*) antisense probe in the globular stage embryo (2DAP). **B** Single staining of fluorescein-labeled *OSH15* antisense probe (*pink*) in the same section as in **A**. In this stage, no hybridization signal of *OSH15* was observed. **C** Single staining of fluorescein-labeled *OSH1* antisense probe (*pink*) in the late globular stage embryo (3DAP). **D** Double staining of mixtures of fluorescein-labeled *OSH1* (*pink*) and digoxigenin-labeled *OSH15* (*purple*) antisense probe in the same section as in **C**. **E** Single staining of digoxigenin-labeled *OSH1* antisense probe (*purple*) in the *orl1* mutant embryo (4DAP). **F** Single staining of digoxigenin-labeled *OSH15* antisense probe (*purple*) in the *orl1* mutant embryo (4DAP). **G** Single staining of digoxigenin-labeled *OSH1* antisense probe (*purple*) in the coleoptilar stage embryo (4DAP). **H** Single staining of digoxigenin-labeled *OSH15* antisense probe (*purple*) in the coleoptilar stage embryo (4DAP). **I** Double staining of mixtures of fluorescein-labeled *OSH15* (*pink*) and digoxigenin-labeled *OSH1* (*purple*) antisense probe in the second leaf primordium differentiation stage. **J** Single staining of fluorescein-labeled *OSH15* antisense probe (*pink*) in the same section as in **I**. *Bars* in **A**, **B**, **C**, **D**, **E**, and **F**, 100 μm; *bars* in **G** and **H**, 160 μm; *bars* in **I** and **J**, 200 μm. *cp*, coleoptile; *ep*, epiblast; *sc*, scutellum; *vt*, vascular tissue; *p1*, the first leaf primordium; *p2*, the second leaf primordium; *arrow*, shoot apex; *filled triangle*, radicle apex.

leoptile, SAM, and provascular tissue, *OSH15* was expressed at the ventral side of the embryo, but the expression within SAM was downregulated (Figure 3H), whereas the expression of *OSH1* within SAM continued high (Figure 3G). Rice embryos develop successively two primordia of embryonic leaves at the periphery of SAM in a distichous phyllotaxis, and the root cap differentiates at the tip of the radicle primordium by 7 DAP. In the embryo at this stage, differences of the expression patterns between *OSH1* and *OSH15* became clearer (Figure 3I,J). *OSH1* expression as stained with the digoxygenin-labeled probe was observed in SAM, ventral side around the provascular tissue, epiblast (bud scale), and below it (Figure 3I). High-level expression of *OSH1* in SAM during embryogenesis leads to the possibility that *OSH1* is involved in the maintenance of the indeterminate state of the shoot apical meristem as is the case for the *KN1* in maize (Smith et al., 1992). On the other hand, the expression of *OSH15* on the same section stained with the fluorescein-labeled probe was restricted to the boundaries between the embryonic organs but not within SAM. Expression was detected between the scutellum and the coleoptile, the coleoptile and the second leaf primordium, the shoot apical meristem and the first leaf primordium, the first leaf primordium and the coleoptile, and the coleoptile and the epiblast (indicated by arrowheads in Figure 3J). Thus, the expression pattern of *OSH15* changes dramatically before and after formation of SAM during embryogenesis. Such dramatic change in the *OSH15* expression suggested the hypothesis that *OSH15* may have different functions before and after formation of SAM. *OSH15* may first function in the formation of SAM during embryogenesis. After shoot formation, *OSH1* is involved in the maintenance of SAM whereas *OSH15* is not. The function of *OSH15* after shoot formation in embryos is probably similar to those after seed germination, because the pattern and the site of the expression of *OSH15* are quite similar to those around the vegetative shoot.

After seed germination, expression of the both homeobox genes continues around SAM of the vegetative and reproductive shoots (Figure 4A,B). Figure 4A shows the *OSH15* expression at a near-median longitudinal section of a rice vegetative shoot. *OSH15* expression appears as pairs of signals on opposite flanks of SAM. The cross section through the plane X in Figure 4A revealed that those pairs of signal came from a ring shape (Figure 4B). In the same section, the expression of *OSH1* was observed as disc shaped (Figure 4C) because *OSH1* was expressed continuously in the meristem dome of shoot apex (Sentoku et al., unpublished results). The expression of both genes was also observed in the axillary buds at the base of the second leaf primordium (indicated by arrowheads in Figure 4B,C). The earliest appearance of *OSH15* expression in SAM was observed in two to four cells before any visible differentiation of P0 leaf (indicated by an arrowhead in Figure 4A). From these observations, *OSH15* expression in vegetative shoots appears to represent the boundary of each phytomer as proposed by Schneeberger et al. for *RS1* gene from maize, which may be an orthologous gene of *OSH15* in maize (Schneeberger et al., 1995). Alternatively, *OSH15* expression may mark the future internodes in the postembryonic stages of development and mesocotyls during embryogenesis and may be involved in their differentiation.

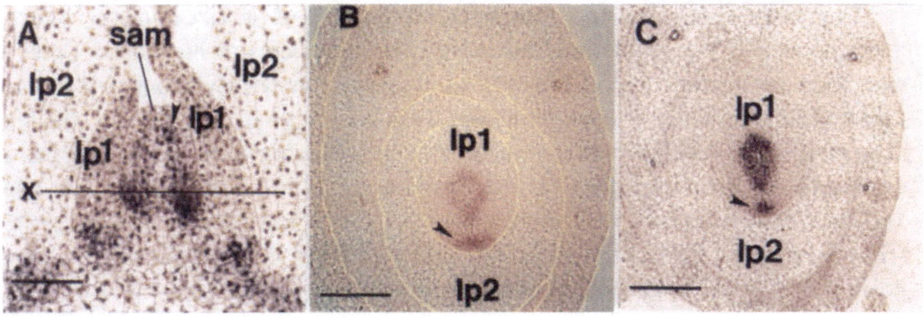

Fig. 4. In situ mRNA localization of *OSH15* around the vegetative and inflorescence shoot apical meristem and the floral meristem. **A** Single staining of digoxigenin labeled *OSH15* antisense probe *(purple)* in the longitudinal section of the vegetative shoot apical meristem. *Line x* indicates the approximate plane of bisection shown in **B** and **C**. **B** Single staining of fluorescein labeled *OSH15* antisense probe *(pink)* in the cross section around the vegetative shoot apical meristem. The *yellow outlines* show the border of leaves. *Arrowhead* indicates the expression of *OSH15* in the axillary bud. **C** Double staining of mixtures of fluorescein-labeled *OSH15* *(pink)* and digoxigenin-labeled *OSH1* *(purple)* antisense probe in the same section as in **B**. *Arrowhead* indicates the expression of *OSH15* in the axillary bud. *sam*, shoot apical meristem; *lp1*, leaf primordium 1; *lp2*, leaf primordium 2. *Bar* in **A**, 50 μm; *bars* in **B** and **C**, 100 μm

4 Analysis of the function of homeobox genes in rice

The KN1-type homeobox genes often cause severe morphological abnormalities when overexpressed in tobacco, *Arabidopsis*, and tomato, and in dominant mutations in maize, barley, or tomato (Smith and Hake, 1994; Sinha et al., 1993; Matsuoka et al., 1993; Tamaoki et al., 1997; Chuck et al., 1996; Hareven et al., 1996; Schneeberger et al., 1995; Müller et al., 1995; Chen et al., 1997b, Lincoln et al., 1994). However, it is difficult to determine the function(s) of these genes in the wild-type context from the phenotype of gain-of-function transgenic plants or dominant neomorphic mutations.

Recently, loss-of-function alleles of the maize *KN1* gene were reported (Kerstetter et al., 1997). Although some of the phenotypes exhibited poor penetrance, perhaps due to genetic redundancy in maize, these mutations have revealed a role for *KN1* in meristem maintenance as previously expected. In animals, families of related homeobox genes are often involved in related developmental processes (Manak and Scott, 1994). Based on the observation that most of the KN1-type homeobox genes are expressed around SAM (Smith et al., 1992; Jackson et al., 1994; Lincoln et al., 1994; Tamaoki et al., 1997), and from the phenotypes of the loss-of-function mutations in *KN1* and *STM* (Kerstetter et al., 1997; Long et al., 1996), it is believed that genes of this class are involved in the maintenance of SAM and/or the development of lateral organs from SAM, although the biological event(s) in which each KN1-type homeobox gene is engaged is still unclear except for *KN1* and *STM*.

To gain an understanding on the bona fide function of the rice homeobox genes, a crucial step is the isolation of recessive mutation alleles of homeobox genes in rice. Rice has become a model plant for the study of monocotyledonous plants because of the feasibility of transformation, its relatively small genome size, the high saturation of molecular markers on the rice genome, and large-scale analysis of expressed sequence tags (Izawa and Shimamoto, 1996). Furthermore, the gene knockout system using a rice retrotransposon, *Tos17*, has been recently exploited (Hirochika et al., 1996; Hirochika, 1997; Hirochika et al., in preparation). Transpositions of *Tos17* are activated during tissue culture and several copies of *Tos17* can be inserted into the rice genome. In this system, it is possible to screen for mutants with insertion of *Tos17* into genes of interest from a large pool of plants regenerated after a few months of tissue culture.

Using this system, Sato et al. have succeeded in isolating and characterizing the loss-of-function mutants of the rice homeobox gene *OSH15* (Sato et al. unpublished results). Rare insertions of a transposable element into a gene can be detected by PCR-based screening (Ballinger and Benzer 1989; Kaiser and Goodwin 1990). A possible approach for screening loss-of-function mutations in the rice homeobox gene is depicted in Figure 5A. Sato et al. designed transposon-specific primers from the long terminal repeat (LTR) of *Tos17* and a gene-specific primer from the conserved region among the rice KN1-type homeodomain. If an insertion of *Tos17* occurs in the rice homeobox gene, a suitable template will be generated which can be amplified exponentially by PCR using gene-specific and transposon-specific primers. For efficient screening, Sato et al. adopted a pool sampling system in which plants were arranged in a two-dimensional matrix. Genomic DNA was isolated from leaves sampled in pools consisting of either a column or a row. In this way, the genomic DNA from each plant is represented in a unique combination of a row and a column. By screening DNA pools that contained genomic DNA from about 550 plants in which transposition of the rice retrotransposon *Tos17* had been induced and random insertion had occurred, a plant with insertion of *Tos17* into *OSH15* was identified. Since *Tos17* was inserted into the fourth exon of *OSH15*, just upstream of the homeodomain region (Figure 5B), this transposon-tagged gene was predicted to be a loss-of-function allele.

Northern blot analysis using a *OSH15* gene-specific probe and total RNAs from plants with homozygous for *Tos17* insertion into *OSH15* revealed that no normal *OSH15* transcript was generated from the mutant allele. This also suggests that the *Tos17* insertion represents a loss-of-function allele. All the plants homozygous for the insertion displayed a dwarf phenotype (Figure 5C). This indicates that the insertion of *Tos17* into *OSH15* is genetically linked to the dwarf phenotype. To confirm that the dwarf phenotype is caused by the loss-of-function of *OSH15*, Sato et al. sought additional mutant alleles of *OSH15* by identifying mutants with similar phenotype which were mapped near *OSH15*.

Because of their agronomic importance, a large number of dwarf mutants of rice have been isolated and characterized. These mutants are categorized into six groups based on the elongation pattern of the upper four to five internodes (Takeda, 1977). To fit the *OSH15* loss-of-function mutant into this classification, Sato et al.

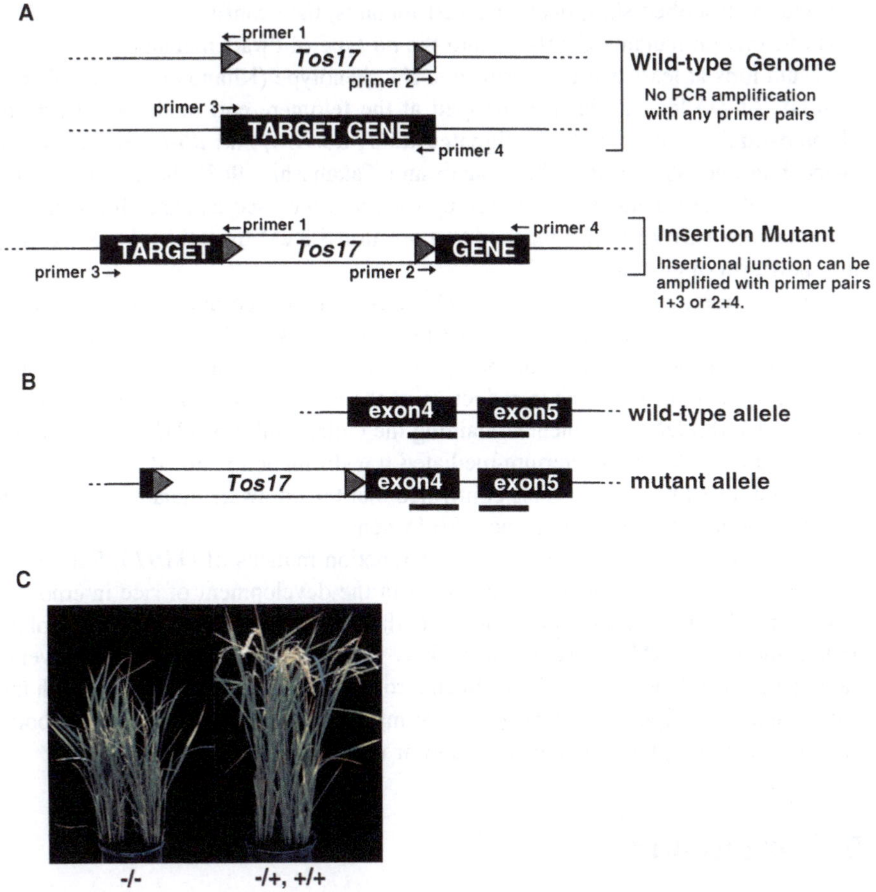

Fig. 5. A General approach for detecting transposon insertions. An insertion of the rice retrotransposon *Tos17* into a target gene can be detected by PCR amplification using primer pairs 1+3 or 2+4. *Triangles* indicate the long terminal repeat (LTR). **B** Insertion site of the rice retrotransposon *Tos17* into the fourth exon of the rice homeobox gene, *OSH15*. *Filled boxes* indicate exons; *bold underline* indicates the approximate position of the homeodomain. The length of *Tos17* in not drawn to scale: *Tos17* is 4.3 kbp whereas the homeodomain is about 190 bp. **C** The appearance of homozygous (-/-), heterozygous (-/+), and wild-type (+/+) plants. Homozygous plants exhibited a dwarf phenotype (compare the height of the positions of the panicles).

measured the length of the upper four internodes and panicle of the loss-of-function mutants and compared them with those of wild-type plants. Although almost no difference was observed in the length of the panicle and the first internode between the mutant and the wild-type plant, the lengths of the second, third, and fourth internodes in the mutant were shortened to approximately 25%, 27%, and 15% of the wild-type, respectively. By comparing this internode elongation pattern to the previously described six groups of dwarf mutants, they confirmed that the *OSH15* loss-of-function mutants clearly fit into the *d6* type of dwarf mutants.

Mutations at least two loci result in a *d6* phenotype (Kitano et al. unpublished results). At *D6* locus, which is mapped at the telomere end of the short arm of chromosome 7, three alleles designated *d6-1*, *d6-ID6*, and *d6-tankanshirasasa* (Kinoshita and Shinbashi, 1982; Nagao and Takahashi, 1963), have been identified. All of these mutants are recessive, and are often used as tester lines for this chromosome. *OSH15* is also mapped to a position 8.3 cM from the telomere end of the short arm of chromosome 7.

Furthermore, all the alleles of the *d6* locus contained complete or partial deletions in the *OSH15* gene. This strongly suggests that loss-of-function of the *OSH15* gene causes the *d6*-type dwarf phenotype in rice. Sato et al. also tried complementation analysis of *d6-1* by the introduction of the wild-type *OSH15* gene. Introduction of a 14-kbp DNA fragment containing the entire wild-type *OSH15* gene to the *d6-1* mutant by the agrobacterium-mediated transformation resulted in the restoration of the dwarf phenotype. This confirmed that the *d6* dwarf mutant is caused by the loss-of-function mutation in the *OSH15* gene.

Based on the phenotype of the loss-of-function mutants of *OSH15*, Sato et al. have demonstrated that *OSH15* is involved in the development of rice internodes. The internodes and leaves, which constitute the most of the ground part of the plant axis, come from SAM. Therefore, it is likely that *OSH15* is involved in the events having a close relationship to SAM. Such a role of *OSH15* is in agreement with the notion that KN1-type homeobox genes are master regulators that specify the body plan as controlling the organogenetic and/or morphogenetic events in SAM.

5 Conclusion

Many plant homeobox genes have been cloned from various plant species since the first plant homeobox gene, *KN1*, was cloned. Among these, the roles in plant development have been elucidated for only a few genes so far. Identification of the recessive mutation alleles of genes of interest is one of the powerful ways for understanding its function. In the plant homeobox genes, phenotypes of the loss-of-function mutants were reported on *KN1, STM, BELL1, GL2,* and *OSH15* (Kerstetter et al., 1997; Long et al., 1996; Reiser et al., 1995; Rerie et al., 1994). Among them, *STM, BELL1*, and *GL2* genes were cloned from recessive mutation alleles and their functions were discussed based on the phenotypes of the mutants. This kind of approach is called "forward genetics." On the other hand, the recessive mutation alleles of the *KN1* were produced genetically. Intragenic suppressors were identified from the heterozygous dominant *Kn1* mutants, and among the progeny of these suppressors, the recessive *KN1* alleles were selected. Identification of recessive mutation alleles of *OSH15* is the first example in which the knockout line was screened based on the DNA sequence of a gene. This kind of approach is called "reverse genetics." Recently, targeted gene inactivation in plants has become practical.

Insertionally mutagenized lines with T-DNA or (retro)transposons have been generated in *Arabidopsis*, maize, *Petunia*, and rice. Using these lines, it is possible to screen knockout lines of genes of interest systematically by the PCR. Another possibility to obtain knockout lines is the homologous recombination method. Indeed, some examples of gene disruption based on the homologous recombination have also been reported in *Arabidopsis* (Miao and Lam, 1995). Establishment of the methods or the materials for targeted gene inactivation enables further the analysis of functions of plant homeobox genes by reverse genetics.

Acknowledgments. We are grateful to Drs. Yasuo Nagato and Hidemi Kitano for the use of rice mutants and Dr. Hirohiko Hirochika for kindly providing the materials for the screening of loss-of-function mutants. We also thank Akemi Tagiri for her advice and help on *in situ* hybridization work.

References

Ahn S, Tanksley SD (1993) Comparative linkage map of the rice and maize genomes. Proc Natl Acad Sci USA 90: 7980-7984

Ballinger DG, Benzer S (1989) Targeted gene mutations in *Drosophila*. Proc Natl Acad Sci USA 86: 9402-9406

Barton MK, Poethig RS (1993) Formation of the shoot apical meristem in *Arabidopsis thaliana*-an analysis of development in the wild type and in the *shootmeristemless* mutant. Development 119: 823-831

Bürglin TR (1997) Analysis of TALE superclass homeobox genes (MEIS, PBC, KNOX, Iroquois, TGIF) reveals a novel domain conserved between plants and animals. Nucleic Acids Res 25: 4173-4180

Chen H, Rossier C, Nakamura Y, et al (1997a) Cloning of a novel homeobox-containing gene, *PKNOX1*, and mapping to human chromosome 21q22.3. Genomics 41: 193-200

Chen JJ, Janssen BJ, Williams A, et al (1997b) A gene fusion at a homeobox locus: alterations in leaf shape and implications for morphological evolution. Plant Cell 9: 1289-1304

Chuck G, Lincoln C, Hake S (1996) *KNAT1* induces lobed leaves with ectopic meristems when overexpressed in *Arabidopsis*. Plant Cell 8: 1277-1289

Gehring WJ, Affolter M, Bürglin T (1994) Homeodomain proteins. Annu Rev Biochem 63: 487-526

Hareven D, Gutfinger T, Parnis A, et al (1996) The making of compound leaf: genetic manipulation of leaf architecure in tomato. Cell 84: 735-744

Helentjaris T, Weber D, Wright S (1988) Identification of the genomic locations of duplicate nucleotide sequences in maize by analysis of restriction fragmet length polymorphisms. Genetics 118: 353-363

Hirochika H (1997) Retrotransposons of rice: their regulation and use for genome analysis. Plant Mol Biol 35: 231-240

Hirochika H, Sugimoto K, Otsuki Y et al (1996) Retrotransposons of rice involved in mutations induced by tissue culture. Proc Natl Acad Sci USA 93: 7783-7788

Hong SK, Aoki T, Kitano H, et al (1995) Phenotypic diversity of 188 rice embryo mutants.

Dev Genet 16: 298-310
Izawa T, Shimamoto K (1996) Becoming a model plant: the importance of rice to plant science. Trends Plant Sci 1: 95-99
Jackson D, Veit B, Hake S (1994) Expression of maize *KNOTTED1* related homeobox genes in the shoot apical meristem predicts patterns of morphogenesis in the vegetative shoot. Development 120: 405-413
Jürgens G (1995) Axis formation in plant embryogenesis: cues and clues. Cell 81: 467-470
Jürgens G, Ruiz T, Berleth T (1994) Embryonic pattern formation in flowering plants. Annu Rev Genet 28: 351-371
Kaiser K, Goodwin SF (1990) "Site-selected" transposon mutagenesis of *Drosophila*. Proc Natl Acad Sci USA 87: 1686-1690
Kano-Murakami Y, Yanai T, Tagiri A, et al (1993) A rice homeobox gene, *OSH1*, causes unusual phenotype in transgenic tobacco. FEBS Lett 334: 365-368
Kerstetter RA, Laudencia-Chingcuanco D, Smith LG, et al (1997) Loss-of-function mutations in the maize homeobox gene, *knotted1*, are defective in shoot meristem maintenace. Development 124: 3045-3054
Kerstetter R, Vollbrecht E, Lowe B, et al (1994) Sequence analysis and expression patterns divide the maize *knotted1*-like homeobox genes into two classes. Plant Cell 6: 1877-1887
Kinoshita T, Shinbashi N (1982) Identification of dwarf genes and their character expression in the isogenic background. Japan J Breed 32: 219-231
Klinge B, Werr W (1995) Transcription of the *Zea mays* homeobox (*ZmHox*) genes is activated early in embryogenesis and restricted to meristems of the maize plant. Dev Genet 16: 349-357
Lawrence PA, Morata G (1994) Homeobox genes: their function in *Drosophila* segmentation and pattern formation. Cell 78: 181-189
Lincoln C, Long J, Yamaguchi J, et al (1994) A *knotted1*-like homeobox gene in *Arabidopsis* is expressed in the vegetative meristem and dramatically alters leaf morphology when overexpressed in trangenic plants. *Plant Cell* 6: 1859-1876
Long JA, Moan EI, Medford JI, et al (1996) A member of the KNOTTED class of homeodomain proteins encoded by the *STM* gene of *Arabidopsis*. Nature 379: 66-69
Lu P, Porat R, Nadeau JA, et al (1996) Identification of a meristem L1 layer-specific gene in *Arabidopsis* that is expressed during embryonic pattern formation and defines a new class of homeobox genes. Plant Cell 8: 2155-2168
Manak JR, Scott MP (1994) A class act: conservation of homeodomain protein functions. Development (Suppl) 1994: 61-77
Matsuoka M, Ichikawa H, Saito A, et al (1993) Expression of a rice homeobox gene causes altered morphology of transgenic plants. Plant Cell 5: 1039-1048
McGinnis W, Levine MS, Hafen E, et al (1984) A conserved DNA sequence in homeotic genes of the *Drosophila* antennapedia and bithorax complexes. Nature 308: 428-433
Meijer AH, Scarpella E, Dijk EL,et al (1997) Transcriptional repression by Oshox1, a novel homeodomain leucine zipper protein from rice. Plant J 11: 263-276
Meinke DW (1995) Molecular genetics of plant embryogenesis. Annu Rev Plant Physiol Plant Mol Biol 46: 369-394
Miao ZH, Lam E (1995) Targeted disruption of the TGA3 locus in *Arabidopsis thaliana*. Plant J 7: 359-365
Müller KJ, Romano N, Gerstner O, et al (1995) The barley *Hooded* mutation caused by a duplication in a homeobox gene intron. Nature 374: 727-730

Nadeau JA, Zhang XS, Li J, et al (1996) Ovule development: identification of stage-specific and tissue-specific cDNAs. Plant Cell 8: 213-239

Nagao S, Takahashi M (1963) Trial construction of twelve linkage groups in Japanese rice. Genetical studies on rice plant, XXVII. J Fac Agric Hokkaido Univ Sapporo 53: 72-130

Reiser L, Modrusan Z, Margossian L, et al (1995) The *BELL1* gene encodes a homeodomain protein involved in pattern formation in the *Arabidopsis* ovule primordium. Cell 83: 735-742

Rerie WG, Feldmann KA, Marks DM (1994) The *GLABRA2* gene encodes a homeo domain protein required for normal trichome development in *Arabidopsis*. Genes Dev 8: 1388-1399

Ruberti I, Sessa G, Lucchetti S, et al (1991) A novel class of plant proteins containing a homeodomain with a closely linked leucine zipper motif. EMBO J 10: 1787-1791

Sato Y, Hong SK, Tagiri A, et al (1996a) A rice homeobox gene, *OSH1*, is expressed before organ differentiation in a specific region during early embryogenesis. Proc Natl Acad Sci USA 93: 8117-8122

Sato Y, Sentoku N, Nagato Y, et al (1998) Isolation and characterization of a rice homeobox gene, *OSH15*. Plant Mol Biol: in press

Sato Y, Tamaoki M, Murakami T, et al (1996b) Abnormal cell divisions in leaf primordia caused by the expression of the rice homeobox gene, *OSH1*, lead to altered morphology of leaves in transgenic tobacco. Mol Gen Genet 251: 13-22

Schena M, Davis RW (1992) HD-zip proteins: members of an *Arabidopsis* homeodomain protein superfamily. Proc Natl Acad Sci USA 89: 3894-3898

Schneeberger RG, Becraft PW, Hake S, et al (1995) Ectopic expression of the *knox* homeobox gene *rough sheath1* alters cell fate in the maize leaf. Genes Dev 9: 2292-2304

Scott MP, Weiner AJ (1984) Structural relationships among genes that control development: sequence homology between the *Antennapedia*, *Ultrabithorax*, and *fushi tarazu* loci of Drosophila. Proc Natl Acad Sci USA 81: 4115-4119

Scott MP, Tamkun JW, Hartzell GWIII (1989) The structure and function of the homeodomain. Biochem Biophys Acta 989: 25-48

Sinha NR, Williams RE, Hake S (1993) Overexpression of the maize homeo box gene, *KNOTTED-1*, causes a switch from determinate to indeterminate cell fates. Genes Dev 7: 787-795

Smith LG, Hake S (1994) Molecular genetic approaches to leaf development: *Knotted* and beyond. Can J Bot 72: 617-625

Smith LG, Greene B, Veit B, Hake S (1992) A dominant mutation in the maize homeobox gene, *Knotted-1*, causes its ectopic expression in leaf cells with altered fates. Development 116: 21-30

Smith LG, Jackson D, Hake S (1995) Expression of *knotted1* marks shoot meristem formation during maize embryogenesis. Dev Genet 16: 344-348

Steeves TA, Sussex IM (1989) Patterns in plant development. Cambridge University Press.

Takeda K (1977) Internode elongation and dwarfism in some gramineous plants. Gamma Field Symp 16: 1-18

Tamaoki M, Kusaba S, Kano-Murakami Y, et al (1997) Ectopic expression of a tobacco homeobox gene, *NTH15*, dramatically alters leaf morphology and hormone levels in transgenic tobacco. Plant Cell Physiol 38: 917-927

Vollbrecht E, Veit B, Sinha N, Hake S (1991) The developmental gene Knotted-1 is a member of a maize homeobox gene family. Nature 350: 241-243

Walbot V (1985) On the life strategies of plant and animals. Trends Genet 1: 165-169

Part 3

Regulation of gene expression

8
Regulation of gene expression by abscisic acid in rice

Tsukaho Hattori and Tokunori Hobo

Center for Molecular Biology and Genetics, Mie University, 1515 Kamihama-cho, Tsu 514-8507, Japan

1 Introduction

Two major roles of the phytohormone abscisic acid (ABA) are known. First, ABA mediates responses of plants to environmental stress, such as high salinity, drought, and low temperature. Second, ABA is required for various events during seed maturation, which include the accumulation of storage proteins and other reserve materials, the acquisition of desiccation tolerance, the arrest of embryonic development, and achievement of dormancy (Finkelstein et al. 1985; McCarty 1995). In most cases, these processes are associated with gene expression that is induced by ABA. Stomatal closure in response to water stress is the only known ABA-regulated process in which gene expression is not required. During the late phase of seed development, the embryo accumulates characteristic sets of mRNAs and proteins, which are called LEA (late embryogenesis abundant) mRNAs and proteins (Baker et al. 1988; Skriver and Mundy 1990; Dure 1993a,b). In many cases, ABA has been shown to be involved in the induction of LEA genes (Berge et al. 1989; Espelund et al. 1992; Marcotte et al. 1989; Mundy and Chua 1988; Pla et al. 1989, 1991; Skriver and Mundy 1990; Ulrich et al. 1990; Yamaguchi-Shinozaki et al. 1989).

Some, but not all, genes for seed storage proteins that accumulate in the embryo and aleurone tissues are also shown to be regulated by ABA (Bray and Beachy 1985; Sun et al. 1996; Kriz et al. 1990; Heck et al. 1993). Various genes are induced by low temperature, drought and high salinity stresses. Although there are ABA-dependent and ABA-independent signaling pathways that lead to gene expression in response to low temperature and drought (Shinozaki and Yamaguchi-

---

Key words. abscisic acid response element (ABRE), coupling element, stress, VP1, seed maturation

Shinozaki 1996), many of known ABA-regulated genes are those originally identified as low temperature and/or drought-inducible genes. Many of LEA and their related genes are also shown to be expressed in vegetative tissues in response to the environmental stresses mentioned above (Dure 1993a). Since the seed maturation and the responses of plant to drought, low temperature and high salinity both involve processes to tolerate with dehydration, LEA proteins are proposed to have functions in protecting cellular components and processes under dehydrated conditions. Predicted structures of LEA proteins also support the proposal (Dure et al., 1989; Dure 1993b).

Since ABA-inducible genes were isolated mainly from cereal plants in earlier times, more information about the regulation of gene expression by ABA is available from the studies using these species. Among them, the studies with rice have greatly contributed to elucidation of the mechanism of ABA-regulated gene expression. For example, *Rab16A* (formerly *Rab21*) from rice is one of the ABA-inducible genes cloned at the earliest time and studied in detail for their regulation by ABA (Mundy and Chua 1988). In addition, rice cells have been used for a heterologous expression system to characterize the promoters of ABA-inducible genes from other monocot species (Marcotte et al. 1988; Marcotte et al. 1989; Pla et al. 1993). Furthermore, the importance of the rice system in the study of regulation of gene expression by ABA has become more evident as transgenic rice can now be easily produced by the *Agrobacterium*-mediated method (Hiei et al. 1994).

Recently, the current knowledge of ABA-regulated gene expression was excellently reviewed by Busk and Pages (1998). The reviews by Giraudat et al. (1994), McCarty (1995), and Leung and Giraudat (1998) on ABA signaling and gene expression from a genetic point of view are also valuable. Therefore, in this review, we focus on the studies using rice as the source of genes or an experimental system.

2 ABA-regulated genes from rice

2.1 Stress-induced and LEA genes

ABA-regulated genes cloned from rice are summarized in Table 1. As mentioned, many ABA-inducible genes have been cloned as those induced by stresses such as low temperature and water stress by differential screening and then confirmed to be induced by exogenous ABA. For example, *osr40cl* (Moons et al. 1997b) and *SalT* (Claes et al. 1990) were isolated as high salinity-induced genes, *lip5* and *19* were isolated as low temperature-induced cDNAs (Aguan et al. 1991), and *WSI5* and *WSI724* were isolated as water stress-induced cDNAs (Takahashi et al. 1994). On screening for such stress-induced cDNAs, those which are not subject to ABA regulation have been also identified (Aguan et al. 1991; Takahashi et al. 1994), indicating that there are both ABA-dependent and -independent signaling pathways for stress-induced gene expression, as has been well demonstrated for other plant systems (Shinozaki and Yamaguchi-Shinozaki 1996).

The products of *Rab16A, B, C,* and *D,* (Mundy et al. 1990; Yamaguchi-Shinozaki et al. 1989) and *Rab25* (Kusano et al. 1992) belong to the group 2 LEA protein family or alternatively are called dehydrins (Close et al. 1993). The four *Rab16* genes, *A* through *D,* are tandemnly located in a short region of the same chromosome (Yamaguchi-Shinozaki et al. 1989). Interestingly, another rice dehydrin, *lip9,* is induced by low temperature but not by ABA, at least in the callus (Aguan et al. 1991). The generally assumed function of LEA proteins such as dehydrins is protection of cellular components from osmotic stress (Dure et al. 1989; Dure 1993b; Close et al. 1993). However, they may have a more specific function since a maize dehydrin, Rab17, has been shown to be localized both in the nucleus and cytoplasm and has an NLS- (nuclear localization signal-) binding activity (Goday et al. 1994). The products of *Osem* (Hattori et al. 1995), and *Oslea3* (Moons et al. 1997a) and *lip5* (Aguan et al. 1991), belong to the group 1 and group 3 LEA protein families, respectively.

2.2 Tissue/organ-dependent regulation

Tissue/organ-dependent regulation is known for several rice ABA-responsive genes. The expression of *Osem* is seed specific and is not induced in shoots and roots of young plants (Hattori et al. 1995). It can be induced by exogenous ABA in imbibed dry-mature embryos (Hattori et al. 1995). In situ hybridization has revealed that the expression of *Osem* is localized in the shoot, radicle, vascular bundles, and epidermal cells of the embryo and in the aleurone layers (Miyoshi et al., unpublished results). An embryo globulin gene, *REG2,* is also seed (the embryo and aleurone layers) specific and induced by ABA only weakly in imbibed embryos but not at all in suspension-cultured cells, in which *Osem* is strongly induced (Sun et al. 1996). The spatial pattern of *REG2* expression in the embryo is almost the opposite to that of *Osem* (Miyoshi et al., unpublished results). *Rab16A* displays a spatial pattern of expression in embryo distinct from that of either *Osem* or *REG2* (Miyoshi et al., unpublished results). In contrast to *Osem, REG2, Rab16A, B,* and *C* are expressed in vegetative tissues in response to water stresses and exogenous ABA, whereas *Rab16D* is induced exclusively in vegetative tissues (Yamaguchi-Shinozaki et al. 1989). An interesting organ-dependent regulation has been reported for *osr40cl* (Moons et al. 1997b). Salt stress or exogenous ABA induces a marked increase in *osr40cl* mRNA in the root, while its expression in the shoot is constitutive (Moons et al. 1997b).

2.3 Antagonistic regulation against gibberellin-induced gene expression

In addition to positive regulation of gene expression during seed development and stress responses, ABA is known to suppress gene expression induced by gibberellin (GA). In particular, the suppression of GA-induced expression of genes for α-

Table 1. ABA-regulated genes from rice

Gene/clone	Protein product/homology	Function	Comment	Reference
Genes upregulated by ABA				
Osem	group 1 LEA protein	unknown	(H)[a] seed specific	Hattori et al. (1995)
Rab16A	group 2 LEA protein (dehydrin)	unknown	(A)[a]	Mundy and Chua (1988)
Rab16B, C, D	group 2 LEA protein (dehydrin)	unknown	(H) Rab16 A, B, C and C are located in a 30-kb region of a chromosome	Yamaguchi-Shinozaki et al. (1989)
Rab25	group 2 LEA protein (dehydrin)	unknown	(A)	Kusano et al. (1992)
Oslea3	group 3 LEA protein	unknown	(O)[a]	Moons et al. (1997a)
REG2	embryo globlulin-2	storage protein	(O) embryo specific	Sun et al. (1996)
lip5	=WSI724, related to Arabidopsis cold-regulated gene Cor11	unknown	(L) [a] weak ABA-induction	Aguan et al. (1991)
lip9	group 2 LEA protein (dehydrin)	unknown	(L) weak ABA-induction	Aguan et al. (1991)
lip19	bZIP protein	transcription factor	(L) basic region is similar to that of maize OCSBF-1	Aguan et al. (1993)
WSI18	related to group 3 LEA protein	unknown	(W)[a] weak ABA-induction	Takahashi et al. (1994)
WSI724	=lip5		(W)	Takahashi et al. (1994)
EFA27	EF-hand calcium binding protein	signalin molecule?	(A)	Frandsen et al. (1996)
Sodcc2	cytosolic copper/zinc-superoxide dismutase	enzyme	(O)	Sakamoto et al. (1995)

OsMT-2	metallothionein-like protein	metal binding	(H) weak ABA-induction	Hsieh et al. (1996)
LTP2, 3, 4	Lipid transfer protein	lipid exchange	(H) tissue/organ dependent regulation	Vignol et al. (1997)
osr40cl	osr40	unknown	(S) [a] tissue/organ dependent regulation	Moons et al. (1997b)
rTip1	intrinsic membrane protein	water channel		Liu et al. (1994)
Osdrr	PR10 (pathogenesis-related protein 10)	unknown	(O) salt an JA-induckble	Moons et al. (1997c)
HRGP	hydroxyproline-rich glycoportein	cell wall component	(H)	Guo et al. (1994)
SalT	SalT	unknown	(S) Tissue/organ dependent regulation; complex interaction with JA	Claes et al. (1990) Moons et al. (1997c)
Genes repressed by ABA				
Ramy1A	α-amylase	enzyme	induced by GA	Itoh et al. (1995)
α-amy8=Ramy3E	α-amylase	enzyme	induced by GA	Wang et al. (1996)
β-amylase	β-amylase	enzyme	not induced by GA	Wang et al. (1996)
cbp3	type 3 carboxypeptidase	enzyme	induced by GA	Washio and Ishikawa (1992)
pRP60	REP-1, cysteine-endopeptidase	enzyme	induced by GA	Shintani et al. (1997)

[a]Isolated based on homology (H), or as a gene induced by low temperature (L), water stress (W), salt stress (S), or other (O).

amylase and other hydrolytic enzymes during cereal seed germination has been well documented (see Chapter 9). In rice, ABA suppression of GA-induced gene expression has been reported for α-amylase genes *α-amy8* (*Ramy3E*) (Wang et al. 1996) and *Ramy1A* (Itoh et al. 1995), *cpb3* (type III carboxypeptidase) (Washio and Ishikawa 1992), and REP-1 (cysteine endopeptidase) (Shintani et al. 1997). The regulation of the rice β-amylase gene is different from that of these hydrolytic enzyme genes (Wang et al. 1996). The germination-induced expression of the β-amylase gene does not appear to depend on GA. However, the induction is suppressed by ABA. Interestingly, the suppressive effect of ABA can be antagonized by exogenous GA (Wang et al. 1996).

The GA induction of α-amylase mRNAs appears to require newly synthesized transcription factor(s) (see Chapter 9). In addition, the ABA suppression of the GA-induced transcription has been shown to be achieved through GA-responsive *cis*-elements (GARE) (Skriver et al. 1991). From these results, it has been postulated that the antagonistic effect of ABA on the GA-responsive transcription is primarily due to a decrease in the amount or activity of the factor interacting with GARE (see Chapter 9).

2.4 Interaction with jasmonic acid regulation

Complex interactions between jasmonic acid (JA) and ABA have been reported for the expression of salt stress-induced genes (Moons et al. 1997c). Salt stress and ABA induce the accumulation of *oslea3* mRNA. The ABA-induced accumulation of *oslea3* mRNA is suppressed by JA. In contrast, both ABA and JA can induce mRNA of *salT*, which was originally identified as a high-salinity-inducible gene. However, simultaneous application of JA and ABA at certain concentration ranges inhibits the accumulation of *salT* mRNA. In addition, ABA has a negative effect on salt stress-induced accumulation of *salT* mRNA, while JA has a moderate synergistic effect on it. Furthermore, the accumulation of protein products of these and other salt stress-regulated genes does not always correlate with that of their mRNAs, suggesting further complex interacting regulations by these hormones and stress at a translational or posttranslational level (Moons et al. 1997c).

3 Analyses of *cis*-acting elements involved in ABA-regulated transcription

3.1 Identification of ABA-responsive elements

ABA-regulated promoter activity was first reported for the wheat *Em* gene by Marcotte et al. (1988). They demonstrated it by transiently transforming rice cultured-cell protoplasts with promoter-GUS fusion constructs by the PEG-mediated method. With this transient assay system, a very clear induction by ABA was demonstrated. The level of induction was 15- to 30 fold for the *Em* promoter. Such a clear induction system allowed further analyses of *cis*-acting elements required for ABA-regulated transcription. Soon after this report, Mundy et al. (1990) reported a similar result for *Rab16A* promoter using a rice homologous system (a rice promoter expressed in rice cells). In the same report, they first suggested putative *cis*-acting elements from the results with a promoter deletion analysis, and electrophoretic mobility shift assays and DNase I footprint experiments using nuclear extracts in addition to the identification of sequence motifs conserved among ABA-regulated promoters. These putative *cis*-acting elements, denoted motif I and motif II, contained consensuses of PyACGTGGC and CG G/C CGCGCT, respectively. Putative *cis*-acting elements [Em1a (GACACGTGGC) and Em1b (CACACGTGCC)] similar to motif I, were also pointed out in the study of the wheat *Em* promoter. Function of *Em*1a as an ABA-responsive element (ABRE) was demonstrated by a loss-of-function type experiment; a base-change mutation of this sequence resulted in a great loss of ABA inducibility (Guiltinan et al. 1990). Conclusive evidence that the putative ABRE, ACGTGGC, mediates ABA-inducible transcription was provided by a gain-of-function type experiment, which demonstrated that artificially hexamerized copies of 11-bp oligonucleotides containing motif I from *Rab16A* can confer ABA responsiveness to a heterologous minimal promoter (Skriver et al. 1991).

As summarized in Table 2, functional analyses of other ABA-inducible promoters also indicate that the sequences containing ACGTGGC or ACGTGTC function as an ABRE (Pla et al. 1993; Michel et al. 1993; Hattori et al. 1995; Shen and Ho 1995; Shen et al. 1996; Ono et al. 1996; Busk et al. 1997). Analyses of the promoter of the rice *Osem* have revealed that it contains at least three functional ABREs, namely, motif A, motif A', and motif B, although motif A plays the most critical role in ABA-regulated transcription (Hattori et al. 1995). One commonly recognized feature of ABRE sequences is that they contain ACGT, which is a core sequence of binding sites for various kinds of plant bZIP-type DNA-binding proteins (see Meshi and Iwabuchi 1995 for review). The cloning of EmBP1 that binds to Em1a and Em1b is one of the earliest reports for plant bZIP proteins (see below; Guiltinan et al. 1990). The palindromic sequence C<u>ACGTG</u> is defined as a G-box (Giuliano et al. 1988; Izawa et al. 1993). As the 3' 5-bp sequence of G-box is contained in the ABREs, they are referred to as "G-box-like ABREs." Since the ACGT core sequence is conserved in most of the functionally defined ABREs, its presence has been often taken as one of the criteria for putative identification of an ABRE. However, the complete conservation of the ACGT sequence does not appear essential for the functionality of ABRE (see following).

3.2 Coupling elements

Although the functional importance of G-box-like sequences in ABA-regulated transcription has been clearly demonstrated, there have been two puzzles. First,

Table 2. Functionally identified ABREs, experimentally determined sequence requirement of ABRE, and the preferred binding sequences of GBFs

	-4	-3	-2	-1	0	0	+1	+2	+3	+4
Functionally identified ABREs										
ACGT-type										
Osem motif A (Hattori et al. 1995)	C	G	T	A	C	G	T	G	T	C
Osem motif B (Hattori et al. 1995)	T	C	G	A	C	G	T	G	T	C
Rab16A motif I (Skriver et al. 1991)	C	G	T	A	C	G	T	G	G	C
Rab16B motif I (Ono et al. 1996)	A	G	T	A	C	G	T	G	G	C
HVA22 ABRE2(R) (Shen and Ho 1995)	C	G	C	A	C	G	T	G	T	C
HVA22 ABRE3 (Shen and Ho 1995)	T	G	T	A	C	G	T	G	G	C
HVA1 A2 (Shen et al. 1996)	C	C	T	A	C	G	T	G	G	C
Em Em1a (Guiltinan et al. 1990)	G	A	C	A	C	G	T	G	G	C
Em Em1b (R) (Vasil et al. 1995)	G	G	C	A	C	G	T	G	T	G
Rab17 ABRE 1 (Busk et al. 1997)	G	A	G	A	C	G	T	G	G	C
Rab17 ABRE4 (Busk et al. 1997)	A	C	T	A	C	G	T	G	G	G
consensus	***N***	***N***	***T/G/C***	***A***	***C***	***G***	***T***	***G***	***G/T***	***C/(G)***
CE3-type										
Osem CE3 (Hobo et al. 1998)	G	A	C	G	C	G	T	G	T	C
HVA1 CE3 (Shen et al. 1996)	A	A	C	G	C	G	T	G	T	C
Rab16B motif III (Ono et al. 1996)	G	C	C	G	C	G	T	G	G	C
hex3 (tobbaco) (Lam and Chua 1991)	G	A	C	G	C	G	G	G	G	C
consensus	***G/(A)***	***C/A***	***C***	***G***	***C***	***G***	***T***	***G***	***G/T***	***C***
ABRE requirement (ACGT-type)[a]	n.t.	n.t.	T>G=C>A	n.t. (A)	n.t(C)	n.t. (G)	n.t. (T)	G>>>C, T, G	G=T>>C, T	C>A,C,G
Strong GBF binding Sequences[b]	G	C/A	C	A	C	G	T	G	G/T	C

[a]Experimentally examined by changing each base at each position of motif A in the 55-bp context of the *Osem* promoter by the protoplast transient assay (Totsuka et al, unpublished results; see text). n.t.: not tested.
[b]Based on the results by Izawa et al (1993); the bases of the "strong half site" sequences on [+] side (3' side) are underlined.

only multiple copies but not a single copy of G-box-like ABRE can confer ABA responsiveness to a heterologous minimal promoter (Skriver et al. 1991; Vasil et al. 1995). Second, the G-box or G-box-like sequences similar or identical to the G-box-like ABRE have been identified as *cis*-acting elements in the promoters of various genes which are regulated by other signals such as cell cycle (Nakayama et al. 1989), white light (Giuliano et al. 1988; Kao et al. 1996), UV light (Schulze-Lefert et al. 1989), auxin (Liu et al. 1994; Ulmasov et al. 1995), anaerobic conditions (McKendree and Ferl 1992), and jasmonic acid (Mason et al. 1993). The presence of a second sequence element in the vicinity of the G-box element has been demonstrated to play an important role in specifying the light-responsiveness of the parsley *CHS* promoter (Block et al. 1990). Shen and Ho (1995) carried out a linker-scan analysis of a 44-bp fragment from the barley *HVA22* promoter that can confer ABA responsiveness to a heterologous minimal promoter and have identified a second *cis*-acting element (TGCCACCGG) named CE1 (for coupling element 1), which functions with a single copy of G-box-like ABRE to constitute a minimal unit (ABA-response complex; ABRC) for ABA responsiveness. Similar analyses with the barley *HVA1* promoter have identified another coupling element, CE3 (ACGCGTGTC; see Table 2) distinct from CE1 (Shen et al. 1996). These coupling elements have been proposed to specify ABA responsiveness in combination with G-box-like ABREs (Shen and Ho; 1995; Shen et al. 1996).

A CE3-like sequence is also present in the *Osem* promoter approximately 40 bp downstream of motif A (a G-box-like ABRE). A 55-bp fragment containing both motif A and the CE3-like element can confer ABA responsiveness to the CaMV 35S minimal promoter (Fig. 1). Linker-scan analysis of this 55-bp sequence revealed that the CE3-like element is required for ABA responsiveness together with motif A as in the barley *HVA1* promoter (Fig. 1; Hobo et al., unpublished results). Since the sequences of motif A and CE3 were found to be similar (see Table 2), motif-exchange experiments were carried out. CE3, as well as CE1, has been originally identified as a second *cis*-acting element distinct from G-box-like ABRE. However, the CE3 and motif A sequences are very similar because the 6 bp of the 3'-portion (CGTGTC) of the CE3 sequence was identical to that of motif A. In fact, they were shown to be functionally equivalent by motif-exchange experiments us-

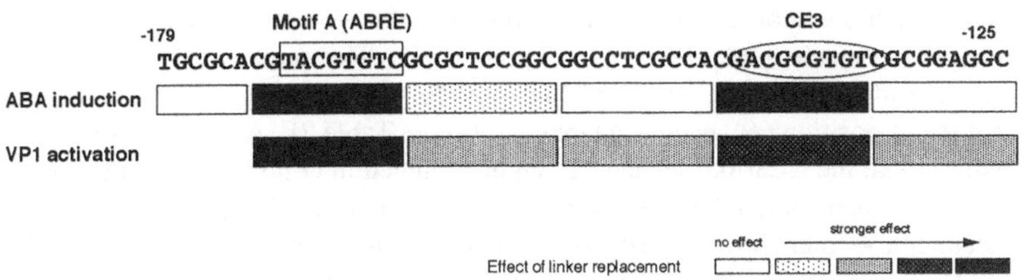

Fig. 1. The nucleotide sequence of the 55-bp fragment of Osem promoter that can confer ABA responsiveness and the ability to be activated by VP1 to a heterologous minimal promoter. Underneath the sequence, the effect of linker-scan mutations on ABA induction and VP1 activation (Hobo et al., unpublished results) is schematically represented.

ing the 55-bp fragment of *Osem*; either of mutant 55-bp fragments in which motif A was replaced with CE3 (CE3-CE3), CE3 with motif A (motif A-motif A), or both replaced by each other (5'-CE3-motif A-3'), exhibited a high level of ABA inducibility comparable to that of the wild-type 55-bp fragment (5'-motif A-CE3-3') (Hobo et al., unpublished results). Functional equivalency of CE3 and motif A (G-box-like ABRE) was also demonstrated by the comparable binding of a bZIP factor, OSBZ8 (see below; Hobo et al., unpublished results).

As mentioned earlier, functionally identified ABRE sequences have the consensus of $A_{-1}C_{0}G_{0}T_{+1}G_{+2}(G/T)_{+3}C_{+4}$ (see Table 2), and most of the proposed potential ABREs in ABA-inducible promoters also fall into this consensus (Michel et al. 1993; Hattori et al. 1995; Shen et al. 1996; Vasil et al. 1995; Busk et al. 1997; Pla et al. 1993; Ono et al. 1996). It should be noted that the consensus contains not only the ACGT core but also the pentanucleotide GTGGC or GTGTC, which is the strong half-site of the palindromic G-box sequence for binding of GBF-type bZIP proteins (Niu and Guiltinan 1994; Izawa et al. 1993; see Table 2). Several cloned GBF-type bZIP proteins have been shown to bind these ABREs (Guiltinan et al. 1990; Lu et al. 1996; Oeda et al. 1991; Nakagawa et al. 1996b). Although the CE3 sequence $(A/G)_{-4}A_{-3}C_{-2}\mathit{G}_{-1}C_{0}G_{0}T_{+1}G_{+2}T_{+3}C_{+4}$ does not maintain the ACGT core, it can be regarded as a near-complete palindrome of the strong half-site, in which the -1 position only differs from the complete palindrome.

Ono et al. (1996) have reported a GCCGCGTGGC sequence in the rice *Rab16B* promoter, which they called motif III, as an ABA response element distinct from G-box-like ABRE. They have shown that a tetramer of motif III can confer ABA responsiveness to the CaMV 35S minimal promoter. However, this sequence also can be regarded as an incomplete palindrome of the strong half-site, which also only differs from the complete palindrome at the -1 position, with the same nucleotide substitution as that in the CE3 sequence (see Table 2). Furthermore, they have also demonstrated that a 40-bp fragment of the *Rab16B* promoter, which contains single copies of motif I (a G-box-like ABRE) and motif III, can confer a good ABA responsiveness to the minimal promoter and that both elements are essential to this function. Thus, motif III is considered to be equivalent to the CE3 sequence. A tetramer of the *hex*3 sequence GAC*G*CGTGGC, which was originally created as a mutant sequence of the hexamer sequence found in the wheat histone gene promoters, can also confer ABA responsiveness to the -90 promoter of CaMV 35S gene in transgenic tobacco (Lam and Chua, 1991). This sequence can be regarded as a hybrid of two different strong half-sites for GBF binding with the exception that the -1 position (A) again is changed to G (see Table 2). All these observations indicate that the CE3-like sequences can be included in G-box-like ABREs. Thus, a coupling element to ABRE is not necessarily to be one distinct from ABRE; it can be either a second copy of ABRE or a completely different element such as CE1.

A coupling element was first identified as such that functions with a gibberellin-responsive element (GARE) to form a gibberellin-responsive *cis*-element complex (GARC) (Rogers and Rogers 1992). It has been shown that GARC can be converted into ABRC, which still depends on the same coupling element, by replacing a GARE with an ABRE (Rogers and Rogers 1992). This suggests that a coupling element can be a general or constitutive element rather than an element that is specific to each signal. A minimal *cis*-element complex for auxin response (AuRC) of the GH3 promoter also consists of at least two elements; one is specific to the auxin signal and the other is constitutively positive (Ulmasov et al. 1995). An artificially manipulated AuRC, in which the constitutive element is replaced with a GAL4 binding site, can still respond to auxins when a GAL4-rel activator is provided by cotransfection (Ulmasov et al. 1995). These observations suggest that the involvement of at least two elements, either similar or dissimilar, as a minimal unit may be a widely existing mechanism for transcriptional induction in response to various hormonal and other signals. In other words, a single signal-regulated transcription factor molecule on a promoter is not sufficient; a second factor, either signal regulated as in the case for the ABRE/CE3 complex, or constitutive as in the case for the ABRE/CE1 complex, GARC and AuRC, is required to achieve a significant transcriptional activation in response to a signal.

4 Tissue/organ-dependent expression of ABA-responsive gene expression

As already mentioned, some of LEA and other ABA-regulated genes are expressed specifically in embryonic tissues, while others are induced also in vegetative tissues by treatment with ABA, dehydration, or cold stress. For example, *Osem* is specifically expressed in embryo and aleurone tissues (Hattori et al. 1995); *Rab16A-C* are induced by ABA in vegetative tissues as well as in seed tissues; *Rab16D* is induced only in vegetative tissues (Yamaguchi-Shinozaki et al. 1989). However, all these genes with different tissue/organ specificities appear to be regulated by ABA through the same type of G-box-like ABRE. There must be some mechanisms that define the tissue/organ specificities of ABA-regulated transcription.

4.1 Expression of ABA-responsive genes in seed tissues

The embryos of *viviparous-1* (*vp1*) mutants in maize germinate precociously on the ear and are insensitive to ABA (Robertson 1955; Robichaud et al. 1980). The expression of at least some sets of ABA-regulated genes in embryonic tissues have been shown to require functional products of *Vp1* (VP1) (McCarty et al. 1989; McCarty et al. 1991). VP1 is expressed specifically in seed tissues and shown to be a transcriptional activator by domain-swapping experiments with other well-characterized eukaryotic transcription factors (McCarty et al. 1989; McCarty et al. 1991). The product of the *Arabidopsis* orthologous locus, *ABI3*, is shown to function similarly (Giraudat et al. 1992; Finkelstein 1993; Parcy et al. 1994). This VP1/ABI3 factor is proposed to function to specify the ABA-regulated transcription in seed tissue (Hattori et al. 1992; Parcy et al. 1994). The strongest evidence for this hypothesis is that transgenic *Arabidopsis* ectopically expressing ABI3 can express seed-specific genes in nonseed tissues in response to ABA (Parcy et al. 1994). We have also produced transgenic rice plants that express OSVP1 (the rice homologue of VP1) in vegetative tissues and shown that they express normally seed-specific *Osem* in root tissues in response to ABA (Hoshino et al., unpublished results).

Cotransfection of VP1 expression vector into protoplasts can transactivate the promoters of ABA-regulated genes such as *Em* and *Osem* (McCarty et al. 1991; Hattori et al. 1992; Hattori et al. 1994; Hattori et al. 1995). *Cis*-acting elements required for VP1 regulation in the *Osem* and *Em* promoters have been determined by such transactivation experiments (Hattori et al. 1995; Vasil et al. 1995). The 55-bp fragment of the *Osem* promoter that can confer ABA responsiveness to a heterologous minimal promoter, as mentioned above, was also shown to be activated by VP1 (Hobo et al., unpublished results). Experiments with various base-change mutation in the natural or the 55-bp context revealed that the same sequence elements as those for ABA, namely motif A and CE3, are required for the VP1 activation (Hattori et al. 1995; Hobo et al., unpublished results). Motif A and CE3 were again shown to be interchangeable with respect to VP1 activation, as well as for ABA response (see Fig. 1; Hobo et al., unpublished results). Similar results were obtained by the analyses of the *Em* promoter assayed in maize protoplasts (Vasil et al. 1995). The experiments using the maize protoplasts also demonstrated that a tetramer of the G-box sequence derived from the light-responsive CHS promoter can confer both ABA responsiveness and VP1 activation to a heterologous minimal promoter (Vasil et al. 1995). Thus, these results indicate that two or more copies of G-box-related ABRE/CE3 elements can constitute a VP1-regulated ABA-responsive *cis*-element complex.

Because of the overlap of the *cis*-acting elements for ABA and VP1 regulation, it has been postulated that VP1 complexes with a factor that directly binds to ABRE/CE3 (Hattori et al. 1995; Vasil et al. 1995; Hill et al. 1996). This hypothesis implies that the hormonal signal and the tissue/organ or developmental signal are integrated by physical interaction between the terminal molecules of each signaling pathway. Although the possibility has not been excluded that VP1 itself directly binds to ABRE/CE3 elements and mediates the ABA signal, the former hypothesis is more likely, considering the failure so far in demonstrating a specific binding activity to ABRE/CE3 as well as the fact that the G-box ABREs have been shown to be bound by various bZIP-type transcription factors (Guiltinan et al. 1990; Oeda et al. 1991; Nakagawa et al. 1996b). In fact, we have recently cloned a rice bZIP protein that interacts with VP1 (see below). While VP1 functions through the G-box-like ABREs on one hand, it also interacts with another *cis*-element called Sph box (Hattori et al. 1992; Kao et al. 1996). Sph box is identified as a *cis*-acting element that is responsible for the VP1 regulation in the maize *C1* promoter (Hattori et al. 1992). The Sph box is considered to be the same element as that designated legumin box or RY-repeat, which is found in many seed-specific genes (Helmut et al. 1992). An Sph-box-like sequence is also found in the *Osem* promoter and appears to contribute quantitatively to the VP1 regulation (Hattori et al. 1995). In contrast to the interaction with G-box-like ABREs, VP1 directly binds to the Sph box of the *C1* promoter (Suzuki et al. 1997). This Sph-box-binding activity is mapped at the C-terminal region called B3 domain, which is highly conserved among the VP1 homologues of various plant species. Although the B3 domain is essential for regulation through the Sph box, it is not required for the activation through the G-box-like ABREs (Carson et al. 1997; Hobo et al., unpublished results).

4.2 Expression of ABA-responsive genes in vegetative tissues

Very little is known about the mechanism that specifies ABA-responsive transcription in vegetative tissues. An element called GRA, which is present in the vicinity of G-box ABREs in the maize *Rab17* promoter, was shown to enhance the ABA-induced transcription in the leaves but not in embryo by particle bombardment transient expression assays (Busk et al. 1997). We have recently started analyses of *cis*-acting elements involved in tissue/organ-specific regulation of ABA-responsive transcription using transgenic rice. Preliminary results of such analyses revealed the presence of *cis*-elements that specify or enhance the ABA-responsive transcription in vegetative tissues. Transgenic rice plants carrying *Osem* promoter (up to the -176 position relative to the transcription start site)::GUS fusion gene were found to exhibit preferential GUS expression in embryos and aleurone layers; however, when a fragment of the *Rab16A* promoter (-188 to -693) was placed in front of the *Osem* -176 promoter, a significant level of GUS mRNA is induced in root tissues in response to ABA (Kagaya et al., unpublished results).

5 *Trans*-acting factors involved in ABA regulation

5.1 bZIP-type factors that bind to ABREs

As already mentioned, several bZIP-type transcription factors have been shown to bind to G-box-like ABRE. Tobacco TAF-1 and wheat EmBP1 have been cloned by the Southwestern screening method using the ABREs of *Rab16A* (motif I) and *Em* (Em1a and Em1b), respectively, as the probes (Guiltinan et al. 1990; Oeda et al. 1991). Although these proteins were cloned with the intention to isolate the transcription factors responsible for ABA-regulated transcription, only insufficient evidence for their involvement in ABA regulation has been obtained beside their binding to the G-box-like ABREs. Both EmBP1 and TAF-1 are classified into the GBF-type (class I) bZIP proteins according to their binding specificities and structural characteristics (Izawa et al. 1993). There have been a number of GBF-type bZIP proteins cloned to date, most of which have been obtained by the Southwestern screening using the G box or its related sequences derived from the promoters under various regulation such as visible and UV light, auxin, cell cycle, and anaerobic conditions (Schindler et al. 1992a; Weisshaar et al. 1991; Tabata et al. 1989, Hong et al. 1995; de Vetten and Ferl 1995). It has been difficult to conclusively determine the actual functions of these GBFs because of the similarity in the binding specificities as well as in the *cis*-element sequences.

Several GBF-type bZIP proteins have been cloned from rice, which include OSBZ6, OSBZ8, osZIP-1a, osZIP-2a, and osZIP-2b (Nakagawa et al. 1996b; Nantel and Quatrano 1996; Nakagawa and Hattori, unpublished results). Among them, OSBZ8 is of particular interest with respect to ABA-regulated transcription (Nakagawa et al. 1996b). OSBZ8 was isolated from a rice embryo cDNA library by use of degenerate oligonucleotide probes that corresponded to the amino acid sequences conserved among the basic regions of plant GBF-type bZIP proteins. OSBZ8 has structural features typical of the GBF-type bZIP proteins and binds to the G-box and G-box-like sequences that include ABREs such as motif A, motif I, and Em1a. The observation that relates this factor to ABA-regulated transcription is that the accumulation of *OSBZ8* mRNA was induced by ABA, which preceded the induction of *Osem* and *Rab16A* mRNAs and was not inhibited by an inhibitor of protein synthesis, cycloheximide (Nakagawa et al. 1996b). By contrast, the induction of *Osem* and *Rab16A* was at least partially inhibited by cycloheximide. Thus, OSBZ8 appears to be involved in the ABA-regulated transcription in some way. However, no further conclusive evidence has been obtained that OSBZ8 primarily participates in ABA-regulated transcription, although it was shown to transactivate the 55-bp ABRC of the *Osem* promoter to some extent (about 2.5-fold) in a cotransfection experiment (Hobo et al., unpublished result). osZIP-1a might be also involved in the ABA regulation since it was demonstrated to transactivate the *Em* promoter slightly but to a statistically significant level in an ABA-dependent manner (Nantel and Quatrano 1996). In contrast to those two GBF-type bZIP proteins, another recently cloned bZIP protein TRAB1 has been more conclusively shown to be involved in ABA-regulated transcription (see section 5.3).

5.2 The binding characteristics of GBFs and the sequence requirements of ABRE

As already mentioned, GBF-type bZIP proteins (homodimers) strongly bind to palindromic G-box sequences with the strong half-site, either GTGGC or GTGTC, or the sequence with combination of the two (Schindler et al. 1992b; Izawa et al. 1993). All the functionally identified ABREs including CE3 contain either of the strong half-sites although some variations occur at position +4 (see Table 2). Some ABREs have the G-box sequences with two strong half-sites, either palindromic (e.g., CHS G box) or in the combination of two (e.g., Em1a), shown to function as ABREs. However, other ABREs comprise a strong half-site and a nonstrong half-site or non-G-box half-site although the ACGT core is preserved (see Table 2). Although GBFs bind to such nonpalindromic G-box-like ABREs, the binding affinities to them are much lower compared to those to the sequences with the strong half-sites on both sides (Nakagawa et al. 1996b; Schindler et al. 1992b; Izawa et al. 1993; Nakagawa and Hattori, unpublished observation).

We made an attempt to derive some useful information about the transcription factors that actually bind to ABRE and mediate the ABA signal in vivo by examining the sequence requirement of G-box-like ABRE (Totsuka and Hattori, unpublished results). We systematically examined it by substituting each base at position, -2, +2, +3 or +4 of motif A ($T_{-2}A_{-1}C_0G_0T_{+1}G_{+2}T_{+3}C_{+4}$) in the 55-bp context using the protoplast transient assay (Table 2). Such analyses revealed that the sequence requirement of ABRE almost completely coincides with that of the GBF binding so long as the 3' half-site ($G_0T_{+1}G_{+2}T_{+3}C_{+4}$) is considered. However, no strong base preference at +2 position was found although the original "T" appeared to be optimal for ABA responsiveness. For example, the substitution of T_{-2} with C, which renders motif A to the palindromic G-box (core), did not increase but rather decreased the level of ABA responsiveness. Furthermore, when motif A was changed to a high-affinity site for GBFs comprising the two strong half-sites, either palindromic or in combination, the level of induction as judged by fold-induction decreased due to the elevated basal level of expression. Thus the high-affinity GBF sites do not serve as good ABREs as the original motif A does. It is puzzling why only the sequence requirement of the 3' half-site matches with the half-site specificity of GBF binding. One likely possibility is that a heterodimer composed of a GBF type and a different type of bZIP proteins binds to ABREs and mediates the ABA signal. Alternatively, the binding affinities measured in vitro might be different from those in vivo.

5.3 A transcription factor that interacts with VP1 can mediate the ABA signal

As mentioned, VP1 is postulated to interact or complex with a transcription factor that directly binds to G-box-like ABRE. If the hypothesis is true, the factor that specifically interacts with VP1 should be able to mediate the ABA signal. Recently, we have succeeded in cloning a rice bZIP factor that specifically interacts with OSVP1 by yeast two-hybrid screening. This bZIP protein, designated TRAB1, contains a basic region similar to those of the *Arabidopsis* GBF4 (Menkens and Cashmore 1994) and the *Helianthus* DPBF-1 and DPBF-2 (Kim et al. 1997), which appear to belong to another class of plant bZIP factors distinct from the above mentioned GBF class. A specific interaction between TRAB1 and OSVP1 was confirmed not only in yeast but also in rice protoplast cells. Furthermore, when it was fused to the GAL4 DNA-binding domain and cotransfected with a GUS reporter construct carrying GAL4-binding sites, the expression of GUS was found to be inducible by ABA. This indicates that TRAB1 is responsible for ABA-regulated transcription. Although TRAB1 can clearly mediate ABA signal, heterodimer formation with GBFs such as OSBZ8 and osZIP-1a might be still important, be-

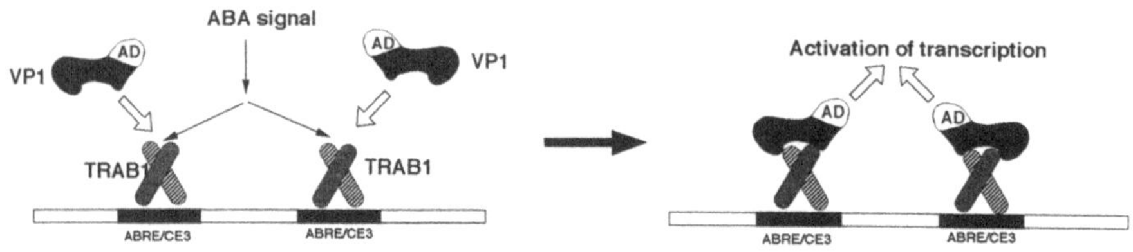

Fig. 2. A model for VP1-dependent ABA-induced transcriptional activation. As a result of ABA signal transduction, the ABA-responsive bZIP factor (TRAB1) is somehow modified so that it can interact with VP1, which provides a transcriptional activation function. The bZIP factors may be heterodimers. Synergistic action of two VP1/ABF complexes may be necessary to achieve a significant transcriptional activation. *AD*, activation domain of VP1.

cause of the strong coincidence between the sequence requirement of ABRE and the GBF strong half-site sequence. Considering our results and the fact that VP1 has a transcriptional activation domain, the interaction between TRAB1 and VP1 might be regulated by the ABA signal, as in a model illustrated in Fig. 2. The model is also based on the observation that the amount and subcellular localization of VP1 are not affected by ABA (Nakagawa et al. 1996a). The report that the in vivo footprints on ABREs are not affected by ABA is consistent with this model (Busk et al. 1997).

6 Signal transduction leading to ABA-regulated gene expression

The mechanism of signal transduction has been studied by various methods in various plant systems. They include genetic studies using *Arabidopsis*, those utilizing the inhibitors of various signaling molecules and cloning of potential signal transduction components. However, most of these studies have been carried out using plant species other than rice.

Two ABA-insensitive loci in *Arabidopsis*, *abi1* and *abi2* have been cloned, both of which are found to encode protein phosphatases of type 2C (Leung et al. 1994; Meyer et al. 1994; Leung et al. 1997). Other studies also indicate the involvement of protein phosphorylation-dephosphorylation in ABA signaling that leads to gene expression (Knetsch et al. 1996; Shen et al. 1996; Heimovaara-Dijkstra et al. 1996). Calcium ion (Sheen 1996) and cyclic ADP-ribose (Wu et al. 1997) are implicated as second-messenger molecules. Recently, the cloning of another ABA-insensitive locus, *abi4*, has been reported; it encodes a transcription factor containing an AP2 domain (Finkelstein et al. 1998). However, the direct target genes and *cis*-acting element(s) are not known. Ishitani et al. (1997) have reported a comprehensive isolation of *Arabidopsis* mutants that have defects in ABA or stress signaling by screening a mutagenized population of transgenic plants carrying a luciferase fusion gene with a stress/ABA-responsive promoter from the *Arabidopsis RD29B* gene. The screening that directly scores ABA/stress-regulated gene expression may be advantageous over the conventional screening of ABA-insensitive mutants, which depends on seed germination assays, for the isolation of downstream mutants of the ABA-signaling pathway. A genetic study using these mutants has suggested the presence of an intricate network of signaling pathways for ABA and distinct stresses. Cloning of these genes will greatly facilitate understanding the mechanism of ABA signaling that leads to the induction of gene expression.

7 Conclusions and perspectives

As summarized in the beginning of this review, various ABA-regulated genes have been isolated from rice. Many of them have been isolated as those induced by various stresses. The rice systems have greatly contributed to the current knowledge about the transcriptional regulation by ABA. The analyses of *cis*-acting elements for the ABA regulation have revealed a synergistic action of at least two elements, a G-box-like ABRE (including the CE-like sequence) and either a second copy of the ABRE/CE3 or a coupling element distinct from ABRE. Tissue/organ-dependent regulation of ABA-responsive genes has been demonstrated. The VP1/ABI3 factors, which were originally cloned from maize and *Arabidopsis* based on genetic information, are responsible, although not solely, for specifying the expression of at least some ABA-regulated genes in the seed tissues. VP1/ABI3 can act through the G-box-like ABREs, although it does not appear to bind directly to it. Several bZIP proteins have been cloned as candidates for the functional *trans*-acting factors for ABRE. The rice TRAB1, a bZIP protein which has been cloned as a factor that interacts with VP1 by yeast two-hybrid screening, has been shown to be able to mediate ABA-induced transcriptional activation. In contrast to the studies on *cis*-acting elements and *trans*-acting factors, the mechanisms of ABA-signal transduction have been mainly investigated with plant species other than rice. However, the cloning of TRAB1 has now provided the firm endpoint from which we can now trace the pathway.

References

Aguan K, Sugawara K, Suzuki N, et al (1993) Low-temperature dependent expression of a rice gene encoding a protein with a leucine-zipper motif. Mol Gen Genet 240:1-8

Aguan K, Sugawara K, Suzuki N, et al (1991) Isolation of genes for low temperature-induced proteins in rice by a simple subtractive method. Plant Cell Physiol 32:1285-1289

Baker J, Steele C, Dure L III (1988) Sequence and characterization of 5 *Lea* proteins and their genes from cotton. Plant Mol Biol 11:277-291

Berge S K, Bartholomow DM, Quatrano RS (1989) Control of the expression of wheat embryo genes by abscisic acid. In: Goldberg R (ed) The molecular basis of plant development. Liss, New York, pp 193-201

Block A, Dangl JL, Hahlbrock K, et al (1990) Functional borders, genetic fine structure, and distance requirements of cis elements mediating light responsiveness of the parsley chalcone synthase promoter. Proc Natl Acad Sci USA 87:5387-5391

Bray EA, Beachy RN (1985) Regulation by ABA of β-conglycinin expression in cultured developing soybean cotyledons. Plant Physiol 79:746-750

Busk PK, Jensen AB, Pages M (1997) Regulatory elements in vivo in the promoter of the abscisic acid responsive gene rab17 from maize. Plant J 11:1285-1295

Busk PK, Pages M (1998) Regulation of abscisic acid-induced transcription. Plant Mol Biol 37:425-435

Carson CB, Hattori T, Rosenkrans L, et al (1997) The quiescent/colorless allele of viviparous1 show that the conserved B3 domain of VP1 is not essential for ABA-regulated gene expression in the seed. Plant J 12:1231-1240

Claes B, Dekeyser R, Villarroel R, et al (1990) Characterization of a rice gene showing organ-specific expression in response to salt stress and drought. Plant Cell 2:19-27

Close TJ, Fenton RD, Yang A, et al (1993) Dehydrin: the protein. In: Close TJ, Bray EA (eds) Plant responses to cellular dehydration during environmental stress. American Society of Plant Physiologists, Rockville, MD, pp 104-118

de Vetten NC, Ferl RJ (1995) Characterization of a maize G-box binding factor that is induced by hypoxia. Plant J 7(4):589-601

Dure L III (1993a) The Lea proteins of higher plants. In: Verma DPS (ed) Control of plant gene expression. CRC Press, Boca Raton, FL, pp 325-335

Dure L III, Crouch M, Harada J, et al (1989) Common amino acid sequence domains among the *Lea* proteins of higher plants. Plant Mol Biol 12:475-486

Dure L III (1993b) A repeating 11-mer amino acid motif and plant desiccation. Plant J 3:363-369

Espelund M, Saeboe-Larssen S, Hughes DW, et al (1992) Late embryogenesis-abundant genes encoding proteins with different numbers of hydrophilic repeats are regulated differentially by abscisic acid and osmotic stress. Plant J 2:241-252

Finkelstein RR, Wang ML, Lynch T J, et al (1998) The *Arabidopsis* abscisic acid response locus ABI4 encodes an APETALA2 domain protein. Plant Cell 10:1043-1054

Finkelstein RR, Tenberg KM, Shumway JE, et al (1985) Role of ABA in maturation of rapeseed embryos. Plant Physiol 78:630-636

Finkelstein RR (1993) Abscisic acid-insensitive mutations provide evidence for stage specific signal pathways regulating expression of an *Arabidopsis* late embryogenesis-abundant (*lea*) gene. Mol Gen Genet 238:401-408

Frandsen G, Muller-Uri F, Nielsen M, et al (1996) Novel plant Ca^{2+}-binding protein expressed in response to abscisic acid and osmotic stress. J Biol Chem 271:343-348

Giraudat J, Hauge BM, Valon C, et al (1992) Isolation of the *Arabidopsis ABI3* gene by positional cloning. Plant Cell 4:1251-1261

Giraudat J, Parcy F, Bertauche N, et al (1994) Current advances in abscisic acid action and signaling. Plant Mol Biol 26:1557-1577

Giuliano G, Pichersky E, Malik VS, et al (1988) An evolutionarily conserved protein binding sequence upstream of a plant light-regulated gene. Proc Natl Acad Sci USA 85:7089-7093

Goday A, Jensen AB, Culianez-Maca FA, et al (1994) The maize abscisic acid-responsive protein Rab17 is located in the nucleus and interacts with nuclear localization signals. Plant Cell 6:351-360

Guiltinan MJ, Marcotte WR, Quatrano RS (1990) A plant leucine zipper protein that recognizes an abscisic acid response element. Science 250:267-270

Guo Y, Delseny M, Puigdomenech P (1994) mRNA accumulation and promoter activity of the gene coding for a hydroxyproline-rich glycoprotein in *Oryza sativa*. Plant Mol Biol 25:159-165

Hattori T, Terada T, Hamasuna S (1994) Sequence and functional analyses of the rice gene homologous to the maize Vp1. Plant Mol Biol 24:805-810

Hattori T, Terada T, Hamasuna S (1995) Regulation of the *Osem* gene by abscisic acid and the transcriptional activator VP1: analysis of *cis*-acting promoter elements required for regulation by abscisic acid and VP1. Plant J 7:913-925

Hattori T, Vasil V, Rosenkrans L, et al (1992) The *Viviparous-1* gene and abscisic acid activate the *C1* regulatory gene for anthocyanin biosynthesis during seed maturation in maize. Genes Dev 6:609-618

Heck GR, Chamberlain AK, David Ho T-H (1993) Barley embryo globulin 1 gene, *Beg1*: Characterization of cDNA, chromosome mapping and regulation of expression. Mol Gen Genet 239:209-218

Heimovaara-Dikstra S, Nieland GJF, van der Meulen RM, et al (1996) Abscisic acid-induced gene expression requires the activity of protein(s) sensitive to the protein-tyrosine phosphatase inhibitor phenylarsine oxide. Plant Growth Regul 18:115-123

Helmut B, Nagy I, Villarroel R, et al (1992) *Cis*-analysis of a seed protein gene promoter: the conservative RY repeat CATGCATG within the legumin box is essential for tissue-specific expression of legumin gene. Plant J 2:233-239

Hiei Y, Ohta S, Komari T, et al (1994) Efficient transformation of rice (*Oryza sativa* L.) mediated by *Agrobacterium* and sequence analysis of the boundaries of the T-DNA. Plant J 6:271-282

Hill A, Nantel A, Rock CD, et al (1996) A conserved domain of the viviparaous-1 gene product enhances the DNA binding activity of the bZIP protein EmBP-1 and other transcription factors. J Biol Chem 271:3366-3374

Hong JC, Cheong YH, Nagao RT, et al (1995) Isolation of two soybean G-box binding factors which interact with a G-box sequence of an auxin-responsive gene. Plant J 8:99-211

Hsieh HM, Liu WK, Chang A, et al (1996) RNA expression patterns of a type 2 metallothionein-like gene from rice. Plant Mol Biol 32:525-529

Ishitani M, Xiong L, Stevenson B, et al (1997) Genetic analysis of osmotic and cold stress signal transduction in *Arabidopsis*: interactions and convergence of abscisic acid-dependent and abscisic acid-independent pathways. Plant Cell 9:1935-1949

Itoh K, Yamaguchi J, Huang N, et al (1995) Developmental and hormonal regulation of rice α-amylase (*Ramy1A*)-gusA fusion genes in transgenic rice seeds. Plant Physiol 107:25-31

Izawa T, Foster R, Chua N-H (1993) Plant b-ZIP protein DNA-binding specificity. J Mol Biol 230:1131-1144

Kao CY, Cocciolone SM, Vasil IK, et al (1996) Localization and interaction of the *cis*-acting elements for abscisic acid, VIVIPAROUS1, and light activation of the *C1* gene of maize. Plant Cell 8:1171-1179

Kim SY, Chung H-J, Thomas TL (1997) Isolation of a novel class of bZIP transcription factors that interact with ABA-responsive and embryo-specification elements in the Dc3 promoter using yeast one-hybrid system. Plant J 11:1237-1251

Knetsch MLW, Wang M, Snaar-Jagalska, BE, et al (1996) Abscisic acid induces mitogen-activated protein kinase activation in barley aleurone protoplasts. Plant Cell 8:1061-1067

Kriz AR, Wallace MS, Pavia R (1990) Globulin gene expression in embryos of maize viviparous mutants. Plant Physiol 92:538-542

Kusano T, Aguan K, Abe M, et al (1992) Nucleotide sequence of a rice rab16 homologue gene. Plant Mol Biol 18:127-129

Lam E, Chua N-H (1991) Tetramer of a 21-base pair synthetic element confers seed expression and transcriptional enhancement in response to water stress and abscisic acid. J Biol Chem 266:17131-17135

Leung J, Bouvier-Durand M, Morris PC, et al (1994) *Arabidopsis* ABA response gene ABI1: features of a calcium-modulated protein phosphatase Science 264:1448-1452

Leung J, Merlot S, Giraudat J (1997) The *Arabidopsis* ABSCISIC ACID-INSENSITIVE2 (ABI2) and ABI1 genes encode homologous protein phosphatases 2C involved in abscisic acid signal transduction. Plant Cell 9:759-771

Leung JM, Giraudat J (1998) Abscisic acid signal transduction. Annu Rev Plant Physiol Plant Mol Biol 49:199-222

Liu Q, Umeda M, Uchimiya H (1994) Isolation and expression analysis of two rice genes encoding the major intrinsic protein. Plant Mol Biol 26:2003-2007

Lu G, Paul AL, McCarty DR, et al (1996) Transcription factor veracity: is GBF3 responsible for ABA-regulated expression of Arabidopsis Adh? Plant Cell 8:847-857

Marcotte WR, Baylely CC, Quatrano RS (1988) Regulation of a wheat promoter by abscisic acid in rice protoplasts. Nature 335:454-457

Marcotte WR Jr, Russel SH, Quatrano RS (1989) Abscisic acid-responsive sequences from the Em gene of wheat. Plant Cell 1:969-976

Mason HS, DeWald DB, Mullet JE (1993) Identification of a methyl jasmonate-responsive domain in the soybean vspB promoter. Plant Cell 5:241-251

McCarty DR (1995) Genetic control and integration of maturation and germination pathways in seed development. Annu Rev Plant Physiol Plant Mol Biol 46:71-93

McCarty DR, Hattori T, Carson CB, et al (1991) The Viviparous-1 developmental gene of maize encodes a novel transcriptional activator. Cell 66:895-905

McCarty DR, Carson CB, Stinard PS, et al (1989) Molecular analysis of *viviparous-1* : an abscisic acid-insensitive mutant of maize. Plant Cell 1:523-532

McCarty DR, Hattori T, Carson CB, et al (1991) The Viviparous-*1* developmental gene of maize encodes a novel transcriptional activator. Cell 66:895-905

McKendree WL, Ferl RJ (1992) Functional elements of the *Arabidopsis Adh* promoter include the G-box. Plant Mol Biol 19:859-862

Menkens AE, Cashmore AR (1994) Isolation and characterization of a fourth *Arabidopsis* thaliana G-box-binding factor, which has similarities to Fos oncoprotein. Proc Natl Acad Sci USA 91:2522-2526

Meshi T, Iwabuchi M (1995) Plant transcription factors. Plant Cell Physiol. 36: 405-1420

Meyer K, Leube MP, Grill E (1994) A protein phosphatase 2C involved in ABA signal transduction in *Arabidopsis thaliana*. Science 264:1452-1455

Michel D, Salamini F, Bartels D, et al (1993) Analysis of a desiccation and ABA-responsive promoter isolated from the resurrection plant *Craterostigma plantagineum*. Plant J 4:29-40

Moons A, De Keyser A, Van Montagu M (1997a) A group 3 LEA cDNA of rice, responsive to abscisic acid, but not to jasmonic acid, shows variety-specific differences in salt stress response. Gene 191:197-204

Moons A, Gielen J, Vandekerckhove J, et al (1997b) An abscisic-acid- and salt-stress-responsive rice cDNA from a novel plant gene family. Planta 202:443-454

Moons A, Prinsen E, Bauw G, et al (1997c) Antagonistic effect of abscisic acid and jasmonates on salt stress-inducible transcripts in rice roots. Plant Cell 9:2243-3359

Mundy J, Chua N-H (1988) Abscisic acid and water-stress induce the expression of a novel rice gene. EMBO J 7:2279-2286

Mundy J, Yamaguchi-Shinozaki K, Chua N-H (1990) Nuclear proteins bind conserved elements in the abscisic acid-responsive promoter of a rice rab gene. Proc Natl Acad Sci USA 87:1406-1410

Nakagawa H, Ohkura E, Ohmiya Y, et al (1996a) The seed-specific transcription factor VP1 (OSVP1) is expressed in rice suspension-cultured cells. Plant Cell Physiol 37:355-362

Nakagawa H, Ohmiya K, Hattori T (1996b) A rice bZIP protein, designated OSBZ8, is rapidly induced by abscisic acid. Plant J 9:217-227

Nakayama T, Ohtsubo N, Mikami K, et al (1989) *Cis*-acting sequences that modulate transcription of wheat histone H3 gene and 3' processing of H3 premature mRNA. Plant Cell Physiol 30:825-832

Nantel A, Quatrano RS (1996) Characterization of three rice basic/leucine zipper factors, including two inhibitors of EmBP-1 DNA binding activity. J Biol Chem 271:31296-31305

Niu X, Guiltinan MJ (1994) DNA binding specificity of the wheat bZIP protein EmBP-1. Nucleic Acids Res 22:4969-4978

Oeda K, Salinas J, Chua N-H (1991) A tobacco bZip transcription activator (TAF-1) binds to G-Box-like motif conserved in plant genes. EMBO J 10:1793-1802

Ono A, Izawa T, Chua N-H, et al (1996) The *rab16B* promoter of rice contains two distinct abscisic acid-responsive elements. Plant Physiol 112:483-491

Parcy F, Valon C, Raynal M, et al (1994) Regulation of gene expression programs during *Arabidopsis* seed development: roles of the *ABI3* locus and of endogenous abscisic acid. Plant Cell 6:1567-1582

Pla M, Goday A, Vilardell J, et al (1989) Differential regulation of ABA-induced 23-25 kDa proteins in embryo and vegetative tissues of the viviparous mutants of maize. Plant Mol Biol 13:385-394

Pla M, Gomez J, Pages M (1991) Regulation of the abscisic acid-responsive gene *rab28* in maize *viviparous* mutants. Mol Gen Genet 230:394-400

Pla M, Vilardell J, Guiltinan MJ, et al (1993) The cis-regulatory element CCACGTGG is involved in ABA and water-stress responses of the maize gene rab28. Plant Mol Biol 21:259-266

Robertson DS (1955) The genetics of vivipary in maize. Genetics 40: 745-760

Robichaud CS, Wong J, Sussex IM (1980) Control *in vitro* growth of viviparous embryo mutants of maize by abscisic acid. Dev Genet 1:325-330

Rogers JC, Rogers SW (1992) Definition and functional implications of gibberellin and abscisic acid *cis*-acting hormone-response complexes. Plant Cell 4:1443-1451

Sakamoto A, Okumura T, Kaminaka H, et al (1995) Structure and differential response to abscisic acid of two promoters for the cytosolic copper/zinc-superoxide dismutase genes, SodCc1 and SodCc2, in rice protoplasts. FEBS Lett 358:62-66

Schindler U, Menkens AE, Beckmann H, et al (1992a) Heterodimerization between light-regulated and ubiquitously expressed *Arabidopsis* GBF bZIP proteins. EMBO J 11:1261-1273

Schindler U, Terzaghi W, Beckman H, et al (1992b) DNA-binding site preferences and transcriptional activation properties of the *Arabidopsis* transcription factor GBF1. EMBO J 11:1275-1289

Schulze-Lefert P, Dangl JL, Becker-Andre M, et al (1989) Inducible *in vivo* DNA footprints define sequences necessary for UV light activation of the parsley chalcone synthase gene. EMBO J 8:651-656

Sheen J (1996) Ca^{2+}-dependent protein kinases and stress signal transduction in plants. Science 274:1900-1902

Shen Q, Ho T-H D (1995) Functional dissection of an abscisic acid (ABA)-inducible gene reveals two independent ABA-responsive complexes each containing a G-box and novel *cis*-acting element. Plant Cell 7:295-307

Shen Q, Zhang P, Ho T-H (1996) Modular nature of abscisic acid (ABA) response complexes: composite promoter units that are necessary and sufficient for ABA induction of gene expression in barley. Plant Cell 8:1107-1119

Shinozaki K, Yamaguchi-Shinozaki K (1996) Molecular responses to drought and cold stress. Curr Opin Biotechnol 7:161-167

Shintani A, Kato H, Minamikawa T (1997) Hormonal regulation of expression of two cysteine endopeptidase genes in rice seedlings. Plant Cell Physiol 38:1242-1248

Skriver K, Mundy J (1990) Gene expression in response to abscisic acid and osmotic stress. Plant Cell 2:503-512

Skriver K, Olsen FL, Roger J, et al (1991) *Cis*-acting DNA elements responsive to gibberellin and its antagonist abscisic acid. Proc Natl Acad Sci USA 88:7266-7270

Sun J-L, Nakagawa H, Karita S, et al (1996) Rice embryo globulins: N-terminal amino acid sequences, cDNA cloning and expression. Plant Cell Physiol 37:612-620

Suzuki M, Kao CY, McCarty DR (1997) The conserved B3 domain of VIVIPAROUS1 has a cooperative DNA binding activity. Plant Cell 9:799-807

TabataT, Takase H, Takayama S, et al (1989) A protein that binds to a *cis*-acting element of wheat histone gene has a leucine zipper motif. Science 245:965-967

Takahashi R, Joshee N, Kitagawa Y (1994) Induction of chilling resistance by water stress, and cDNA sequence analysis and expression of water stress-regulated genes in rice. Plant Mol Biol 26:339-352

Ulmasov T, Liu ZB, Hagen G, Guilfoyle TJ (1995) Composite structure of auxin response elements. Plant Cell 7:1611-1623

Ulrich TU, Wurtele ES, Nikolau BJ (1990) Sequence of *EmB-1*, an mRNA accumulating specifically in embryos of carrot. Nucleic Acids Res 18:2826

Vasil V, Marcotte WR, Rosenkrans L Jr, et al (1995) Overlap of Viviparous1 (VP1) and abscisic acid response elements in the *Em* promoter: G-box elements are sufficient but not necessary for VP1 transactivation. Plant Cell 7:1511-1518

Vignols F, Wigger M, Garcia-Garrido JM, et al (1997) Rice lipid transfer protein (LTP) genes belong to a complex multigene family and are differentially regulated. Gene 195:177-186

Wang SM, Lue WL, Eimert K, et al (1996) Phytohormone-regulated b-amylase gene expression in rice. Plant Mol Biol 31:975-982

Washio K, Ishikawa K (1992) Structure and expression during the germination of rice seeds of the gene for a carboxypeptidase. Plant Mol Biol 19:631-640

Weisshaar B, Armstrong GA, Block A, et al (1991) Light-inducible and constitutively expressed DNA-binding proteins recognizing a plant promoter element with functional relevance in light responsiveness. EMBO J 10:1777-1786

Wu Y, Kuzma J, Marechal E, et al (1997) Abscisic acid signaling through cyclic ADP-ribose in plants. Science 278:2126-2130

Yamaguchi-Shinozaki K, Mundy J, et al (1989) Four tightly linked *rab* genes are differentially expressed in rice. Plant Mol Biol 14:29-39

9 Regulation of α-amylase gene expression

Su-May Yu

Institute of Molecular Biology, Academia Sinica, Nankang, Taipei 11529, Taiwan, R.O.C.

1 Introduction

α-Amylases are endo-amylolytic enzymes which catalyze the hydrolysis of α-1,4-linked glucose polymers and play an important role in the degradation of starch and glycogen in higher plants, animals, and many microorganisms. During germination of cereal grains, α-amylases are the primary enzymes in the conversion of endospermal starch to simple sugars that provide energy for the growing seedling. The α-amylases have been of considerable interest to plant biologists due to their importance to both agriculture and industry. In agriculture, α-amylase activity that contributes to high seedling vigor is an essential agronomic trait in cereal crops. In the brewing industry, the germinating cereal grains which contain high amounts of α-amylase malt are necessary and desirable to ensure maximum production of fermentable sugars. In addition, the induction of α-amylase synthesis in aleurone cells by the phytohormone gibberellins has been a classical model system for studying the hormonal regulation of gene expression in plants (Jacobsen et al. 1995).

Most studies of α-amylases in plants have focused on hormonal regulation of gene expression and protein secretion in germinating cereal grains. However, α-amylases are also present in many other tissues of monocotyledonous and dicotyledonous plants. Knowledge of the physiological roles of α-amylases in plants, in particular regulated starch mobilization, has been scarce (Beck and Ziegler 1989; Steup 1988). The existence and potential importance of α-amylase gene regulation by sugars became apparent only in the last few years (Yu et al. 1991, 1996). Previous reviews have discussed the hormonal (Fincher 1989; Jacobsen et al. 1995) and metabolite (Thomas and Rodriguez 1994) regulation of α-amylase gene expression in germinating cereal grains and seedlings. Recently, the gene and enzyme struc-

---

Key words. rice, cereal, germination, α-amylase gene, gibberellin, sugar, osmotic stress, gene regulation, feedback regulation, differential regulation, signal transduction, embryo, endosperm, aleurone, cell culture

ture and the regulation of synthesis and secretion of α-amylases have also been discussed (Mitsui and Itoh 1997). Here I present some recent advances in our knowledge of α-amylase gene regulation by various factors in cereal plants, in particular the regulation of α-amylase gene expression in rice will be deliberated.

2 Regulation by gibberellin (GA)

The amylolytic breakdown of stored starch is one of the central biochemical events in the germination of cereal grains. In the early stages of seed germination, the embryo synthesizes gibberellins which diffuse to the aleurone cells where gibberellins act as signals to initiate the synthesis and secretion of α-amylases and other hydrolases (Fincher 1989). These enzymes digest the stored starch and other nutrients in the endosperm to provide sugars for the growth of the young seedling. α-Amylases are the most abundant hydrolases produced in barley aleurone tissue in response to GA_3 (one of the gibberellins) during seed germination. The appearance of α-amylases in germinating seeds is due to the de novo synthesis of these enzymes, which is accompanied by a dramatic increase in the amount of α-amylase mRNA (Jacobsen et al. 1995). Abscisic acid (ABA) is another phytohormone that is present in developing grains and antagonizes the action of GA_3. In principle, GA_3 and ABA act in similar but opposing ways at the level of transcription (Jacobsen et al. 1995).

Little is known of the molecular mechanisms that connect perception of GA signal with the activation of α-amylase gene expression. What we know about the GA regulation of α-amylase gene expression is mostly based on studies with germinating barley and wheat seeds. However, several studies suggest that similar principles and modes of hormonal regulation of α-amylase gene expression also occur in germinating rice seeds (Itoh et al. 1995; Thomas and Rodriguez 1994; Yu et al. 1992, 1996). One observed difference is that aleurones of many cereal species, including barley, exhibit considerable α-amylase gene expression in the absence of added GA (Fincher 1989). However, no α-amylase gene expression is detected in the rice aleurones unless GA is added (Yu et al. 1991). It is not clear whether this is due to a difference in endogenous low GA concentrations or GA-independent regulation. Nevertheless, studies on various cereal species have indicated that GA regulation of α-amylase expression involves transcriptional control (Jacobsen and Beach 1985; Zwar and Hooley 1986), leading to the belief that control is probably exerted through interaction between hormone response elements located in the gene promoter and protein transcription factors. This belief was confirmed when chimeric genes containing promoters of wheat and barley α-amylase genes fused to reporter genes were introduced into barley or oat aleurone protoplasts, and the expression of the reporter genes was shown to be regulated by GA_3 and ABA in a dose-dependent manner (Huttly and Baulcombe 1989; Jacobsen and Close 1991; Salmenkallio et al. 1990; Skriver et al. 1991). Similar results were obtained later by introducing the α-amylase gene promoter-reporter gene into barley or rice intact aleurone layers using particle bombardment (Lanahan et al. 1992; Tanida et al. 1994).

2.1 Promoter *cis*-acting elements involved in GA regulation

Nucleotide sequence comparisons of several rice, barley, and wheat α-amylase genes by computer analysis revealed three highly conserved sequence motifs, C/TCTTTTC/T (pyrimidine box), TAACAA/GA, and TATCCAC/T boxes (Huang et al. 1990b). Functional analysis of a barley high-pI α-amylase gene (*Amy6-4*) promoter by transient expression in barley aleurone protoplasts revealed that the TAACAAA box can act alone as a GA response element (GARE) (Skriver et al. 1991). In addition, both the TAACAAA and TATCCAC box act cooperatively, and together they may form a gibberellin-response complex (GARC) (Gubler and Jacobsen 1992). However, analysis of a barley low-pI α-amylase gene (*Amy32b*) promoter in barley aleurone tissue suggests that all three conserved boxes are required for a full GA_3 response and comprise the GARC (Lanahan et al. 1992). GARC in some α-amylase genes may include additional sequences, as an Opaque-2 like element upstream of the pyrimidine box and another sequence (-192 to -158) of the *Amy32b* promoter are also essential for high-level expression in response to GA_3 (Lanahan et al. 1992; Rogers and Rogers 1992). It seems that composition of the GARC varies among different members of the α-amylase gene family. Promoter analyses have also investigated whether ABA antagonism of the GA-regulated α-amylase gene expression is mediated via an ABA response element. Evidence suggests that ABA acts via elements in the GARC (Skriver et al. 1991; Lanahan et al. 1992). Whether ABA acts directly on transcription factors which bind to GARC or on the signal transduction pathway is not known.

Transgenic approaches have been applied to investigate the function of the α-amylase gene promoter in germinating rice seeds. Expression of two promoter fragments, -232 to +31 and -748 to +31 relative to transcription start sites, of a rice α-amylase gene *AmylA(αAmy7)* fused to GUS in germinating transgenic rice seeds were both found to be subject to similar temporal, spatial, and hormonal regulation (Itoh et al. 1995). GUS expression was first detected in the scutellar epithelium and then in the aleurone layer. The pattern of temporal and spatial expression of the *AmylA(αAmy7)* promoter was consistent with the studies using RNA gel blot and in situ hybridization analyses (Karrer et al. 1991; Ranjhan et al. 1992; Yu et al. 1996). Within the 232-bp *AmylA(αAmy7)* promoter region, the pyrimidine box, the TAACAAA box, and the TATCCAT box are all included. It would be interesting to determine how the various *cis*-elements in the α-amylase gene promoter work together or separately to achieve hormonal as well as temporal and spatial expression of α-amylases in cereals.

It is worthwhile to note that although the GARE box is essential for GA-dependent expression of several α-amylase genes in aleurone cells, it is not present in several rice α-amylase gene promoters. For instance, the GARE box is present in the promoter of *αAmy7* but not in those of *αAmy3(Amy3D)*, *αAmy6(Amy3B)* and *αAmy8 (Amy3E)* (Huang et al. 1990a; Lu et al. 1998), although *αAmy6* and *αAmy8* as well as *αAmy7* are highly expressed and *αAmy3* is weakly expressed in the aleurone of germinating rice seeds (Yu et al. 1996, unpublished results). One explanation is that GA is not the only initiating factor or only an initiating factor, but is perhaps more importantly an accelerating factor, in the expression of α-amylase genes in aleurone cells of germinating seeds. However, GA is required for initiating the expression of α-amylase genes in aleurones of embryoless rice half-seeds, as accumulation of α-amylase mRNA in this tissue is not detected unless GA_3 is added (Yu et al. 1991). Thus, an alternative explanation is that the function of the GARE box could be replaced by other yet unidentified sequences in the promoters of some α-amylase genes, such as *αAmy6* and *αAmy8*, or the regulatory mechanism of α-amylase gene expression in the intact germinating seeds is somewhat different from that in the embryoless half-seeds in terms of GA dependency.

2.2 *Trans*-acting factors involved in GA regulation

Hormonal regulation of transcription is likely to involve *trans*-acting factors which interact with *cis*-acting elements in the GARC. The characterization of *trans*-acting factors involved in α-amylase gene expression has only recently been initiated. Nuclear proteins from barley aleurone layers that bind to the GARE and TATCCAT box of the *Amy32b* promoter in a GA_3-dependent manner have been identified by gel retardation and DNase I footprint analyses (Sutliff et al. 1993). DNA-binding proteins from rice germinating seeds or GA_3-treated embryoless half-seeds have also been shown to interact with the pyrimidine box in the promoter of two rice α-amylase genes, *OSamy-c(Amy1B)* and *Amy3C* (Ou-Lee et al. 1988; Goldman et al. 1994). Whether or not these DNA-protein interactions play roles in GA-regulated α-amylase gene expression remains to be determined.

A study carried out by Gubler et al. (1995) demonstrated the function of a GARE-binding protein as a part of the GA response pathway. It was first found that the central GARC element, TAACAAA in the *Amy6-4* promoter and TAACAGA in the *Amy32b* promoter, resemble the c-Myb and v-Myb consensus binding sequence. Myb-related proteins are a group of highly conserved transcriptional activators involved in the regulation of growth, development, and metabolism of a wide range of organisms (Luscher and Eisenman 1990; Martin and Paz-Ares 1997). A cDNA (*HvGAMyb*) encoding the GA-inducible, Myb-related protein (GAMyb) was later isolated from a barley aleurone cDNA library. GAMyb can bind to the TAACAAA box in vitro as well as activate the transcription of a barley high pI α-amylase gene (*Amy pHV19)* promoter fused to a reporter gene in the absence of GA in the barley aleurone transient expression system. Recently, a cDNA clone, *OsGAMyb*, which encodes a homologue to the barley HvGAMyb has also been isolated from a rice endosperm cDNA library (Gubler et al. 1997). Expression of *OsGAMyb* in aleurone cells is stimulated by GA_3, and the gene product transactivates the *Amy32b* promoter both in the presence and absence of GA_3 in transient expression analysis of bombarded barley aleurones (Gubler et al. 1997). These findings shed a new light on the nature of the interaction between *cis*-acting elements and *trans*-acting factors in GA-regulated α-amylase gene expression.

2.3 Signal transduction of GA

Evidence from several studies suggests that the site of GA perception is at the plasma membrane (Hooley et al. 1991; Gilroy and Jones 1994) and that GA-induced elevations of cytosolic calcium (Bush 1996) and calmodulin (Schuurink et al. 1996) are early events in signal transduction. Recently, cGMP was shown to play an important role in GA signaling in barley aleurone cells, as a reduction in the cGMP level prevented GA-induced accumulation of α-amylase and GAMyb mRNAs (Penson et al. 1996).However, cGMP cannot substitute for GA in the induction of α-amylase gene expression.

A cDNA (*OsCa-atpase*) encoding GA-responsive Ca^{2+}-ATPase has been isolated from rice aleurone cDNA library and has been suggested to play an important role in the GA-dependent signal transduction pathway (Chen et al. 1997). In a transient assay in rice aleurone cells, expression of this Ca^{2+}-ATPase bypassed the GA requirement for stimulation of *OSamy-c* expression. In the presence of GA, the rice Ca^{2+}-ATPase and the Ca^{2+} channels appeared to coregulate the local concentration of cytosolic Ca^{2+}, which in turn activated the expression of downstream a-amylase genes (Chen et al. 1997).

2.4 Tissue-specific regulation by GA

One interesting aspect regarding GA regulation is the tissue-specific response of α-amylase genes to GA. The expression of an α-amylase gene *Amy3D(αAmy3)* in rice embryos does not respond to addition of GA (Karrer and Rodriguez 1992). Insensitivity of α-amylase production to exogenously applied GA_3 was also observed in the embryos of sorghum (Aisien and Palmer 1983) and barley (MacGregor and Marchylo 1986). Saturation of the GA level in the embryo is suggested as a cause for the GA insensitivity in stimulating the expression of α-amylases in barley embryos, as the level of α-amylases reduced by treatment of embryos with a GA biosynthesis inhibitor can be restored by application of exogenous GA_3 (Perata et al. 1997). However, expression of *Amy3D(αAmy3)*, *Amy3E(αAmy8)*, and total α-amylase genes in rice cell cultures are not enhanced by GA or reduced by a GA biosynthesis inhibitor (Simmons and Rodriguez 1989; Yu et al. 1991). The GA insensitivity in tissues other than aleurone cells, at least in cultured cells, could be caused by a lack of GA receptor(s), a deficiency in the GA signal transduction pathway, or an incapability of expressing GA-responding transcription factors. These findings also suggest that the expression of α-amylase genes can be regulated by other mechanism, as is discussed in the next section.

3 Regulation by sugars

A few years ago expression of the α-amylase gene family in cultured rice suspension cells was found to be repressed by sugars present in the culture medium and induced by their absence (Yu et al. 1991, 1992). Additional physiological changes induced by sucrose starvation that accompany the induction of α-amylase gene expression have also been observed, including rapid consumption of cellular sugars and starch, reduction in metabolic activity, and extensive autophagic activity (increase in vacuolar volume and degradation of cytoplasmic constituents) (Chen et al. 1994). It is unclear how all of these physiological changes are coordinately regulated by a sugar starvation signal at the molecular level. Due to the ease of culturing suspension cells and in manipulating sugar levels in medium, the sugar-dependent repression of α-amylase gene expression became an ideal model system for studying the mechanism of sugar repression in plants. In addition, the coordinated expression of α-amylases and degradation of starch in cultured cells is useful for studying the mechanism of regulated starch degradation under defined conditions.

The mechanism of sugar repression of α-amylase gene expression in cultured rice suspension cells has been explored and has been found to involve both transcriptional and posttranscription control (Sheu et al. 1994, 1996). Both the transcription rate and mRNA stability increase as cells are starved of sucrose. A protein synthesis inhibitor, cycloheximide (CHX), induces massive accumulation of α-amylase mRNA in cells provided with sucrose, suggesting that de novo synthesized proteins are involved in the sugar-dependent regulation (Sheu et al. 1994, 1996). Chimeric genes carrying promoters from *αAmy3* and *αAmy8*, fused to a reporter gene have been introduced into rice and expression of the chimeric genes in transformed suspension cells was shown to be repressed by sucrose (Chan et al. 1993, 1994; Huang et al. 1993). Expression of the *αAmy8-GUS* chimeric gene is regulated similarly in transgenic potato and tobacco, indicating that the regulatory sequences within the rice α-amylase gene promoter could be recognized by these two plant species (Chan et al. 1994). This findings suggest that the pathway for transducing signals of sugar and the regulatory mechanism of sugar repression on the α-amylase gene promoter are conserved and retain common features during evolutionary divergence of monocots and dicots.

3.1 Differential expression in germinating seeds

α-Amylase activity has been shown to correlate with rice seedling vigor (Lee et al. 1986; Sasahara et al. 1986; Williams and Peterson 1973) and that it is essential for rapid germination and early seedling growth, especially in rice-growing regions (e.g., California, Louisiana, and Texas) that practice sowing partially germinated seeds in preflooded fields. Rice α-amylase isozymes are encoded by approximately ten genes (Huang et al. 1990b), and preferential expression of *Amy1A(Amy7)* was shown to be an important factor in determining seedling vigor in rice (Karrer et al. 1993). Coordinated but differential expression of α-amylase genes in rice has also been observed in the embryo and endosperm of germinating seeds (Karrer et al. 1991, 1992; Yu et al. 1992, 1996), other tissues (Huang et al. 1990a), and cultured cells (Sheu et al. 1996; Simmons et al. 1991; Yu et al. 1992). In general, the overall expression of α-amylase genes is more abundant in endosperm of germinating seeds and in cultured cells (Thomas and Rodriguez 1994). In germinating seeds, the coordinated expression of several α-amylase isozymes may represent dosage requirements necessary to provide correct amounts of gene products during rapid germination and early seedling growth, while the differential expression of α-amylase isozymes could be due to their distinct substrate specificities, necessary to hydrolyze starch with various structures, during different stages of germination and seedling development (Thomas and Rodriguez 1994). Hence, the significance and mechanism of coordinated and differential regulation of α-amylase gene expression during germination became a subject of interest.

3.2 Differential regulation by sugars

Early studies observed that synthesis of α-amylase in germinating barley embryos could be repressed by sugars (Briggs and Clutterbuck 1973; Smith and Briggs 1980). Later a study showed that expression of *Amy3D(Amy3)* in germinating rice embryos was transient and could be repressed by endosperm extract of germinating rice seeds or by pure sugars (Karrer and Rodriguez 1992). Recently, we demonstrated that sugars regulate the accumulation of gibberellin as well as differential expression of α-amylase genes in germinating rice seeds in a concentration-dependent and tissue-specific manner (Yu et al. 1996). A remarkably different sensitivity to changes in tissue sugar levels is suggested as a mechanism for differential expression of α-amylase genes in the rice embryo and aleurone cells (Yu et al. 1996). In the aleurone cells, expression of *αAmy3*, *αAmy6*, *αAmy7*, and *αAmy8* is regulated by gibberellins produced in the embryo and by osmotically active sugars. In the embryo, expression of *αAmy3* and *αAmy8* and production of gibberellins is transient, presumably induced by depletion of sugars in the embryo upon imbibition and repressed by sugar influx from the endosperm as germination proceeds (Yu et al. 1996). In the embryo, the transient expression of *αAmy3* and *αAmy8* precedes the expression of *αAmy7* (Thomas and Rodriguez 1994; Yu, unpublished result), which suggests possible distinct roles played by the individual α-amylase isozymes in starch hydrolysis. An interactive loop among sugars, gibberellins, and α-amylase gene expression in the germinating cereal grain has been proposed (Fig. 1).

3.3 Promoter *cis*-acting elements involved in sugar regulation

The mechanism of sugar regulation of α-amylase gene expression has been further studied using the rice suspension cell system (Sheu et al. 1996). By using the α-amylase gene-specific DNA fragments as probes, mRNA levels of eight α-amylase genes were shown to increase in response to sucrose starvation. Levels of the *αAmy3* and *αAmy8* mRNAs were distinctly higher and constituted 90% of total α-amylase mRNA accumulated after sucrose starvation. Nuclear run-on transcription analysis revealed that transcription rates of the eight α-amylase genes increased by sucrose starvation and that a positive correlation existed between the transcription rates and the steady-state mRNA levels, suggesting that transcriptional regulation

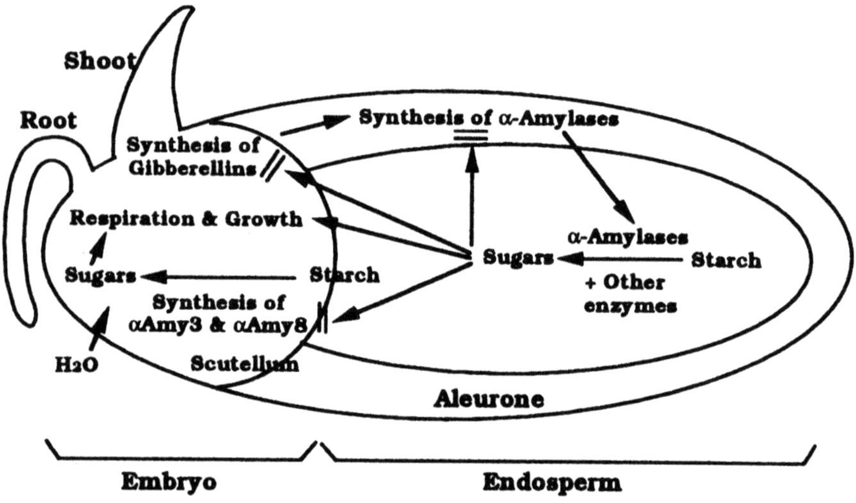

Fig.1. A scheme illustrating the interactions among sugars, gibberellins, and α-amylases in rice grains during germination and seedling growth. The *parallel bars* indicate the presence of negative regulation by sugars. Early in seed germination, active metabolism commences and respiration rate rises rapidly following hydration of the dry seed, which causes a rapid drop of sugar level in the scutellum. Depletion of sugars in the scutellum may trigger the expression of *αAmy3* and *αAmy8* and degradation of starch in this tissue, and onset of gibberellin production and release into the endosperm. It is also possible that the pre-existing gibberellins in the embryo trigger the expression of α-amylase genes in the scutellum. The breakdown of endosperm starch begins initially in the region immediately adjacent to the scutellar epithelium and later in regions under the aleurone layer. As hydrolysis products from the endosperm reach the embryo, expression of *αAmy3* and *αAmy8* and production of gibberellin are ceased. Endosperm starch is gradually hydrolyzed by α-amylases synthesized and secreted by aleurone cells, and a high concentration of hydrolysis products accumulated in the subaleurone space in turn suppress the expression of α-amylase genes in aleurone cells. (From Yu et al. 1996)

plays an important role in the coordinated and differential expression of individual α-amylase genes. *αAmy3* has been chosen as a model gene for study because it constitutes approximately 60% of total α-amylase mRNAs in rice cells starved of sucrose (Sheu et al. 1996) and because its expression in germinating rice embryo is regulated by sugars (Karrer and Rodriguez 1992; Yu et al. 1996).

The *αAmy3* promoter has been shown to mediate sugar-dependent repression of reporter gene expression in transgenic rice suspension cells (Huang et al. 1993). In a recent study using a rice protoplast transient assay system, we analyzed the *cis*-acting elements in the *αAmy3* promoter by both loss- and gain-of-function analyses (Lu et al. 1998). The major sugar response sequence (SRS) is located between 186 and 82 bp upstream of the transcription start site. The SRS confers sugar responsiveness to a minimal promoter in an orientation-independent manner. It also converts a sugar-insensitive rice actin gene promoter into a sugar-sensitive promoter in a dose-dependent manner. A linker-scan mutation study identified three essential motifs: the GC box, the G box, and the TATCCA element, within the SRS. Sequences containing either the GC box plus G box or the TATCCA element each can mediate sugar response; however, they act synergistically to give a high-level glucose starvation-induced expression. Nuclear proteins from rice suspension cells binding to the TATCCA element in a sequence-specific and sugar-dependent manner have also been identified.

Examination of promoter sequences of eight rice, six barley, and five wheat α-amylase genes reveals that all except four of these genes have TATCCA variants, and that they all contain a TATCCA element at proximity upstream of transcription start sites (Table 1). This observation suggests that the TATCCA element may play a role in the regulation of α-amylase gene expression. Mutations of the TATCCA element in the promoters of both *Amy-pHV19* (Gubler and Jacobsen 1992) and *Amy32b* (Lanahan et al. 1992) were found to lower expression to about 20% of maxima but maintained GA responsiveness in the barley aleurone. The TATCCA element is duplicated in the *αAmy3* promoter (Table 1), and mutation of each of the duplicated TATCCA element also reduced the *αAmy3* promoter activity to 12% and 8%, respectively, of the wild-type sequence but maintained sugar responsiveness in the rice protoplasts (Lu et al. 1998). These results suggest that the TATCCA element enhances transcription in concert with the hormone- or sugar-regulated interactions on other sequences to promote the transcription of α-amylase promoters in a signal- and tissue-dependent manner. Since the TATCCA element is an important component of the GARC of α-amylase genes, these findings suggest that the regulation of α-amylase gene expression by sugar and hormone signals may share common regulatory machinery.

3.4 mRNA stability regulated by sugars

As previously mentioned, a decrease in transcript stability is one of the important mechanisms that control the sugar repression of α-amylase gene expression in rice

Table 1. Sequence alignment of the α-amylase gene promoter regions containing the TATCCA box at proximity upstream of the transcription start sites

Gene	Plant	Position[a]	Sequence[b]
Amy1A(αAmy7)	Rice	-156	GGCTGTTATCCATCCAATC
Amy1B(Osamy-c)	Rice	-154	CAGTGTTATCCATCCAATC
Amy2A	Rice	-320	AATTAGTATCCAAACATTC
Amy3A	Rice	-107	GGCCGTTATCCCCGCACTA
Amy3B(αAmy6)	Rice	-89	CGAACTTATCCAACAACGA
Amy3C(Osamy-b)	Rice	-103	TGAACTTATCCAACGGAAT
Amy3D(αAmy3)	Rice	-126	TTGCCTTATCCATATCCAC
Amy3E(αAmy8)	Rice	-139	CACCTTTATCCATGTTGCT
Amy1A	Barley	-129	CCGACATATCCATCGATGC
Amy46	Barley	-129	CCGACATATCCATCGATGT
gRAm56	Barley	-129	CCGACATATCCATCGATGT
Amy6-4	Barley	-132	CCGACATATCCACTGGCCC
Amy2A	Barley	-184	CCTGTCTATCACACCACAG
Amy32b	Barley	-114	GTCTGGTATCCATGCAGTG
Amy1/18	Wheat	-93	CCGACATATCCACTGGCCC
Amy2/53	Wheat	-115	GTCTGGTATCCATGCAGTG
Amy2/54	Wheat	-116	GTCTGGTATCCATGCAGTG
Amy2/8	Wheat	-114	GTCTGGTATCCATGCAGTG
Amy3	Wheat		GGTCCTTATCCATGCGGAT

[a] The negative numbers refer to the distance between the transcription start site and the first nucleotide T in the TATCCA box. The exact transcription start sites of wheat *Amy3* have not yet been determined.
[b] Source of the α-amylase gene sequences is GenBank.

suspension cells (Sheu et al. 1994, 1996). To elucidate the mechanism of sugar-dependent mRNA turnover, we recently examined the effect of α*Amy3* 3'-UTR on mRNA abundance and stability by functional analyses in rice protoplasts and in transformed rice suspension cells. By examining the transient expression of chimeric genes in rice protoplasts, the α*Amy3* 3'-UTR was shown to mediate the sugar-dependent repression of fused heterologous gene expression (Chan and Yu 1998a). The same kinetics of accumulation of endogenous α*Amy3* mRNA and reporter mRNA carrying the α*Amy3* 3'-UTR in response to glucose deprivation was observed, suggesting that the α*Amy3* 3'-UTR is sufficient, and probably the major determinant, for controlling the abundance of these transcripts (Chan and Yu 1998a). Further study demonstrated that the entire α*Amy3* 3'-UTR and two of its subdomains can independently mediate sugar-dependent repression of reporter mRNA accumulation (Chan and Yu 1998b). Analysis of reporter mRNA half-lives showed that the entire α*Amy3* 3'-UTR and the two subdomains each functioned as a sugar-dependent destabilizing determinant in the turnover of mRNA (Chan and Yu 1998b). The identification of sequence elements in the α-amylase mRNA that dictate differential stability has very important implications for the study of sugar-dependent mRNA decay mechanisms.

4 Regulation by osmotic stress

Although it is generally accepted that gibberellins that diffuse from the embryo activate the expression of α-amylase genes in aleurone cells at early stages of germination, little is known of the regulation of α-amylase gene expression in aleurone cells at later stages. An early study observed that 6 days beyond germination, concentration of solutes arose from hydrolysis of endosperm and accumulated in the endosperm was high (570 mM) and might have caused an osmotic stress, which in turn repressed the biosynthesis of α-amylases in the aleurone layer at later stages of germination (Jones and Armstrong 1971). Later, a study demonstrated that high concentrations of polyethylene glycol or mannitol inhibited the synthesis of α-amylase in isolated barley aleurone layers, and suggested that osmotic stress results in a general inhibition of protein synthesis, including α-amylase production (Chrispeels 1973). An inhibition of ribosome binding to the endoplasmic reticulum has been found to associate with the inhibition of α-amylase synthesis under osmotic stress (Armstrong and Jones 1973).

4.1 Osmotic regulation in germinating seeds

In the aleurone cells of germinating rice seeds, levels of *αAmy3*, *αAmy6*, and *αAmy8* mRNA increase with time and reach a first peak at day 5, decline significantly at day 6, then rise again and reach a second peak at day 8 (Yu et al. 1996). The fluctuation in mRNA levels suggests that some mechanism exists that operates to induce or repress α-amylase gene expression in aleurone cells and which is controlled at the level of either transcription and/or mRNA stability. Five days after rice seed germination, the concentration of sugars in the endosperm can reach approximately 550 mM, which is high enough to cause an osmotic stress and repress the expression of *αAmy8* in the aleurone cells (Yu et al. 1996). In the embryo of germinating rice seeds, level of gibberellins increase with time, reach a peak at day 3, and decline thereafter (Yu et al. 1996). Because the peak of gibberellin level precedes that of α-amylase mRNA levels, leading to a hypothesis that the levels of α-amylase mRNA in the endosperm fluctuate in response to fluctuation in concentrations of gibberellins transported from the embryo and hydrolysis products accumulated in the subaleurone space. In the embryo, low concentrations (25-50 mM) of glucose effectively repress both α-amylase gene expression and gibberellin production but, in the endosperm, only high concentrations (≥300 mM) of glucose repress α-amylase gene expression (Yu et al. 1996). Therefore, it is proposed that sugars regulate the expression of α-amylase genes and synthesis of gibberellins in the embryo via a metabolite feedback control mechanism, but regulate the expression of α-amylase genes in the aleurone via an osmotic control mechanism.

4.2 Osmotic regulation in cultured suspension cells

In a recent study, expression of α-amylase genes in cultured rice suspension cells was also shown to be subject to osmotic regulation (Lu and Yu, unpublished result). At a concentration of 30 mM, glucose or sucrose, but not mannitol, reduces the accumulation of α-amylase mRNA. At concentrations ≥300 mM, glucose or sucrose further reduces the accumulation of α-amylase mRNA. At equivalent concentrations (≥300 mM), mannitol also reduces the accumulation of α-amylase mRNA. These observations suggest that low concentrations of metabolizable sugars repress the expression of α-amylase genes via a metabolite feedback control mechanism, and high concentrations of sugars or solutes repress the expression of α-amylase genes via an osmotic control mechanism. It is not known whether identical or separate *cis*-acting elements in the α-amylase gene promoters act to mediate the sugar repression and osmotic control. In addition to the TATCCA element, a putative G-box sequence was also identified as an essential component of the sugar response complex in the *αAmy3* promoter (Lu et al. 1998). The G-box is present in the 5' upstream regions of many plant genes exhibiting regulation by a variety of environmental signals and physiological cues (DeVetten and Ferl 1994). Function of the putative G-box in the osmotic regulation of α-amylase gene expression remained to be determined. However, very likely the regulation of α-amylase gene expression by environmental factors, i.e., osmotic stress, and by physiological cues, i.e., gibberellin or sugar signal, may share some common regulatory machinery. These findings provide new insight into the mechanisms involved in the regulation of α-amylase gene expression in cereals.

5 Concluding remarks

It is now clear that expression of α-amylase genes in rice is under different modes of tissue-specific regulation: in the embryo of germinating seeds and in cultured suspension cells, expression is activated by sugar deprivation and repressed by sugar provision (Karrer and Rodriguez 1992; Yu et al. 1991, 1992, 1996); in the endosperm of germinating seeds, expression is activated by gibberellic acid and repressed by abscisic acid and osmotic stress (Itoh et al. 1995; Yu et al. 1996). Evidence suggests that similar modes of regulation on α-amylase genes also operate in other cereal plant species (Briggs 1992; Perata et al. 1997). Study on the mechanisms of coordinated and/or differential regulation of α-amylase gene expression in germinating cereal grains is complicated by the differential sensitivity of individual genes to the various physiological and environmental regulatory signals. The full story of regulation of α-amylase gene expression by these signals is unlikely to unfold unless individual genes are studied. Differences in the promoter or mRNA structures may provide a rational basis for understanding the differential regulation of α-amylase genes. To date, the function of relatively few *cis*-acting elements in the α-amylase gene promoters have been analyzed; however, complex-

Table 2. Comparison of functional conserved sequences in the promoters of α-amylase genes

Consensus sequence	Barley *Amy6-4*	Barley *Amy pHV19*	Barley *Amy32b*	Rice *αAmy3*
O2S[a]	None	-185/-177 (+)	-149/-136 (+)	None
CTTTT[b]	-167/-163 (ND)	-168/-164 (-)	-128/-124 (+)	-311/-307 (ND)
TAACAA/GA[c]	-142/-136 (+)	-143/-137 (+)	-119/-113 (+)	None
TATCCA	-123/-118 (ND)	-124/-119 (+)	-106/-101 (+)	-116/-105 (+)

Function of *cis*-acting elements in the promoters of barley *Amy6-4* (Skriver et al. 1991), *Amy pHV19* (Gubler and Jacobsen 1992), and *Amy32b* (Lanahan et al. 1992) were determined for GA-dependent regulation and that of rice *αAmy3* (Lu et al. 1998) was determined for sugar-dependent regulation. The negative numbers refer to the distance between the transcription start site and the first and last nucleotides of the consensus sequence. ND, not determined by function analysis; +, important for high level expression; -, no effect on expression.
[a] Opaque-2 binding sequence.
[b] Pyrimidine box.
[c] GARE box.

ity in the composition of these regulatory elements has already been observed (Table 2). Apparently, considerably more functional analyses of promoter and mRNA sequences are required before a clear picture of the mechanisms controlling the response to various signals can be expected to emerge. Furthermore, many important and challenging questions remained unanswered:

- How are gibberellins, sugar, or osmotic stress signals transduced from receptor molecules to the nucleus to trigger α-amylase gene transcription, or to the cytoplasm to control the α-amylase mRNA stability? What are the components involved in these signal transduction pathways?
- Do the multiple signaling pathways 'crosstalk' with each other in regulating the expression of α-amylase genes?
- Are the *cis*-acting elements involved in the response to various signals identical, overlapping, or separate from each other within a single α-amylase gene promoter? Will different combinations of these *cis*-acting elements determine the signal-specific expression of an α-amylase gene promoter?
- How many and what are the *trans*-acting factors that regulate the expression of α-amylase genes?
- Do the *trans*-acting factors that regulate α-amylase gene expression interact and act cooperatively, or do they act independently, in response to the various physiological and environmental signals?

Rice suspension cells are ideal for the study of intracellular signal transduction, transcriptional, and posttranscriptional regulation of α-amylase gene expression.

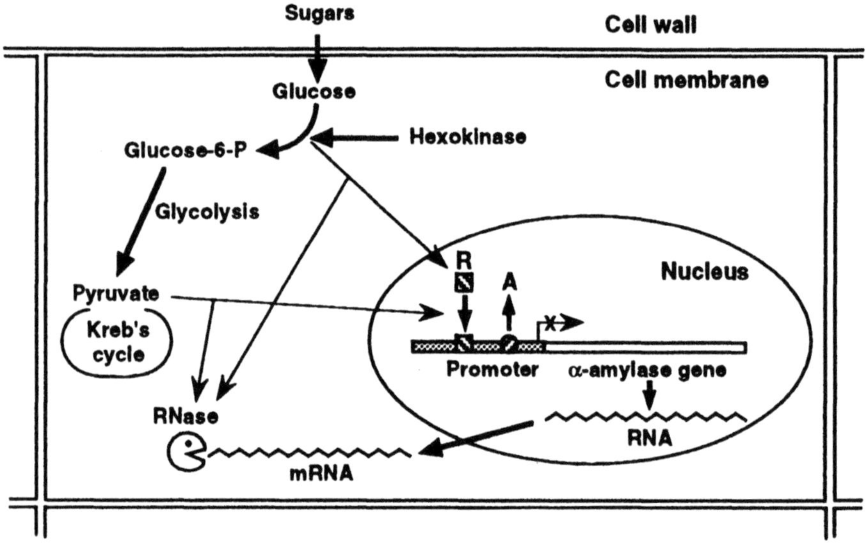

Fig. 2. A hypothetical model for the sugar signal transduction pathway and repression of α-amylase gene expression, via inhibition of transcription and degradation of mRNA, in rice cells. **A**, transcriptional activator; **R**, transcriptional repressor. The signal transduction pathway in the scheme is proposed based on the review by Jang and Sheen (1997) and a study by Aubert et al. (1996). The signal transduction pathway is branched off to separately regulate the transcription of α-amylase gene promoter in the nucleus and the stability of α-amylase mRNA in the cytoplasm.

In the past few years, significant progress on the understanding of the regulatory mechanism of sugar-dependent α-amylase gene expression (Fig. 2) has been made based on studies using the rice cell culture system (Chan et al. 1993, 1994; Chen et al. 1994; Huang et al. 1993; Sheu et al. 1994; Thomas et al. 1994; Yu et al. 1991, 1992). However, differences in the mechanisms controlling α-amylase gene expression must exist between the germinating seeds and the suspension cell culture systems, especially when the developmental and spatial regulatory factors are also involved. In addition, to date, the GA regulation of α-amylase gene expression can only be studied in seeds. Therefore, the new insight and concepts generated in the cell culture system, and the interaction among regulatory factors that respond to various signals, have to be further analyzed in transgenic plants.

The study of signal transduction pathways and mechanisms of GA, sugar, and osmotic control of α-amylase gene expression are critical to our understanding of sink (the embryo axis) and source (the scutellum, aleurone, and endosperm) interactions that control the germination process and development of cereal seedlings (Fig. 1). Additional knowledge in these areas will also enhance our ability to manipulate seed germination for biotechnological applications.

Acknowledgment. I thank Chung-An Lu, Lin-Tze Yu, and Douglas Platt for help in preparing the manuscript. The research in the author's laboratory is supported by grants from Academia Sinica and the National Science Council of the Republic of China.

References

Aisien AO, Palmer GH (1983) The sorghum embryo in relation to the hydrolysis of the endosperm during germination and seedling growth. J Sci Food Agric 34:113-121

Armstrong JE, Jones RL (1973) Osmotic regulation of α-amylase synthesis and polyribosome formation in aleurone cells of barley. J Cell Biol 59:444-455

Aubert S, Gout E, Bligny R, et al. (1996) Ultrastructural and biochemical characterization of autophagy in higher plant cells subjected to carbon deprivation: control by the supply of mitochondria with respiratory substrates. J Cell Biol 133:1251-1263.

Beck E, Ziegler P (1989) Biosynthesis and degradation of starch in higher plants. Annu Rev Plant Physiol Plant Mol Biol 40:95-117

Briggs DE, Clutterbuck VJ (1973) Generation of α-amylase in germinating *Hordeum distichon*. Phytochemistry 12:1047-1050

Briggs DE (1992) Barley germination: Biochemical changes and hormonal control. In: Shewry PR (ed) Barley: Genetics, Biochemistry, Molecular Biology and Biotechnology. CAB International, Oxon, UK, pp 369-401

Bush DS (1996) Effects of gibberellic acid and environmental factors on cytosolic calcium in wheat aleurone cells. Planta 199:88-89

Chan M-T, Chang H-H, Ho S-L, et al. (1993) *Agrobacterium*-mediated production of transgenic rice plants expressing a chimeric α-amylase promoter/β-glucuronidase gene. Plant Mol Biol 22:491-506

Chan M-T, Chao Y-C, Yu S-M (1994) Novel gene expression system for plant cells based on induction of α-amylase promoter by carbohydrate starvation. J Biol Chem 269:17635-17641

Chan M-T, Yu S-M (1998a) The 3' untranslated region of a rice α-amylase gene mediates sugar-dependent abundance of mRNA. Plant J 15:685-695

Chan M-T, Yu S-M (1998b) The 3' untranslated region of a rice α-amylase gene functions as a sugar-dependent mRNA stability determinant. Proc Natl Acad Sci USA 95:6543-6547

Chen M-H, Liu L-F, Chen Y-R, et al. (1994) Expression of α-amylases, carbohydrate metabolism, and autophagy in cultured rice cells are coordinately regulated by sugar nutrient. Plant J 6:625-636

Chen X, Chang M, Wang B, Wu R (1997) Cloning of a Ca^{2+}-ATPase gene and the role of cytosolic Ca^{2+} in the gibberellin-dependent signaling pathway in aleurone cells. Plant J 11:363-371

Chrispeels MJ (1973) Mechanism of osmotic regulation of hydrolase synthesis in aleurone cells of barley: inhibition of protein synthesis. Biochem Biophys Res Commun 53:99-104

DeVetten NC, Ferl RJ (1994) Transcriptional regulation of environmentally inducible genes in plants by an evolutionary conserved family of G-box binding factors. Int J Biochem 26:1055-1068

Fincher GB (1989) Molecular and cellular biology associated with endosperm mobilization in germinating cereal grains. Annu Rev Plant Physiol Plant Mol Biol 40:305-346

Gilroy S, Jones RL (1994) Perception of gibberellin and abscisic acid at the external face of barley (*Hordeum vulgare* L.) aleurone protoplasts. Plant Physiol 104:1185-1192

Goldman S, Mawal YR, Tanida I, et al. (1994) Studies of a gibberellin-dependent DNA-binding protein related to the expression of a rice α-amylase gene. Plant Sci 99:75-88

Gubler F, Jacobsen JV (1992) Gibberellin-responsive elements in the promoter of a barley high-pI α-amylase gene. Plant Cell 4:1435-1441

Gubler F, Kalla R, Roberts JK, et al. (1995) Gibberellin-regulated expression of a *myb* gene in barley aleurone cells: evidence for Myb transactivation of a high-pI α-amylase gene promoter. Plant Cell 7:1879-1891

Gubler F, Watts RJ, Kalla R, et al. (1997) Cloning of a rice cDNA encoding a transcription factor homologous to barley GAMyb. Plant Cell Physiol 38:362-365

Hooley R, Beale MH, Smith SJ (1991) Gibberellin perception at the plasma membrane of *Avena fatua* aleurone protoplasts. Planta 183:274-280

Huang N, Koizumi N, Reinl S, et al. (1990a) Structural organization and differential expression of rice α-amylase genes. Nucleic Acids Res 18:7007-7014

Huang N, Sutliff TD, Litts JC, et al. (1990b) Classification and characterization of the rice α-amylase multigene family. Plant Mol Biol 14:655-668

Huang N, Chandler J, Thomas BR, et al. (1993) Metabolic regulation of α-amylase gene expression in transgenic cell culture of rice (*Oryza sativa* L.). Plant Mol Biol 23:737-747

Huttly AK, Baulcombe DC (1989) A wheat α-*amy2* promoter is regulated by gibberellin in transformed oat aleurone protoplasts. EMBO J 8:1907-1913

Itoh K, Yamaguchi J, Huang N, et al. (1995) Developmental and hormonal regulation of rice α-amylase (*RAmy1A*)-*gusA* fusion genes in transgenic rice seeds. Plant Physiol 107:25-31

Jacobsen JV, Beach LR (1985) Evidence for control of transcription of α-amylase and ribosomal RNA genes in barley aleurone protoplasts by gibberellic and abscisic acid. Nature 316:275-277

Jacobsen JV, Close TJ (1991) Control of transient expression of chimaeric genes by gibberellic acid and abscisic acid in protoplasts prepared from mature barley aleurone layers. Plant Mol Biol 16:713-724

Jacobsen JV, Gubler F, Chandler PM (1995) Gibberellin action in germinated cereal grains. In: Davies PJ (ed) Plant Hormones: Physiology, Biochemistry, and Molecular Biology. Kluwer Academic Publishers, Dordrecht, pp 246-271

Jang J-C, Sheen J (1997) Sugar sensing in higher plants. Trends Plant Sci 2:208-214

Jones RL, Armstrong JE (1971) Evidence for osmotic regulation of hydrolytic enzyme production in germinating barley seeds. Plant Physiol 48:137-142

Karrer EE, Litts JC, Rodriguez RL (1991) Differential expression of α-amylase genes in germinating rice and barley seeds. Plant Mol Biol 16:797-805

Karrer EE, Rodriguez RL (1992) Metabolic regulation of rice α-amylase and sucrose synthase genes in *planta*. Plant J 2:517-523

Karrer EE, Chandler JM, Foolad MR, et al. (1993) Correlation between α-amylase gene expression and seedling vigor in rice. Euphytica 66:163-169

Lanahan MB, Ho T-HD, Rogers SW, et al. (1992) A gibberellic response complex in cereal α-amylase gene promoters. Plant Cell 4:203-211

Lee CC, Li CC, Sung FJM (1986) Physiological and genetic studies on seedling vigor in rice (*Oryza sativa* L.). II. Inheritances of α-amylase activity and seedling vigor in rice. J Agric Assoc China 135:17-24

Lu C-A, Lim, E-K, Yu S-M (1998) Sugar response sequence in the promoter of a rice α-amylase gene serves as a transcriptional enhancer. J Biol Chem 273:10120-10131

Luscher B, Elsenman RN (1990) New light on Myb and Myc. Part II. Myb. Genes Dev 4:2235-2241

MacGregor AW, Marchylo BA (1986) α-Amylase components in excised, incubated barely embryos. J Inst Brew 92:159-161

Martin C, Paz-Ares J (1997) Myb transcription factors in plants. Trends Genet 13:67-73

Mitsui T, Itoh K (1997) The α-amylase multigene family. Trends Plant Sci 2:255-261

Ou-Lee T-M, Turgeon R, Wu R (1988) Interaction of gibberellin-induced factor with the upstream region of an α-amylase gene in rice aleurone tissue. Proc Natl Acad Sci USA 85:6366-6369

Perata P, Matsukura C, Vernieri P, et al. (1997) Sugar repression of a gibberellin-dependent signaling pathway in barley embryos. Plant Cell 9:2197-2208

Penson SP, Schuurink RC, Fath A, et al. (1996) cGMP is required for gibberellic acid-induced gene expression in barley aleurone. Plant Cell 8:2325-2333

Ranjhan S, Karrer EE, Rodriguez RL (1992) Localizing α-amylase gene expression in germinated rice grains. Plant Cell Physiol 33:73-79

Rogers JC, Rogers SW (1992) Definition and functional implications of gibberellin and abscisic acid *cis*-acting hormone response complexes. Plant Cell 4:1443-1451

Salmenkallio M, Hannus R, Teeri TH, et al. (1990) Regulation of α-amylase promoter by gibberellic acid and abscisic acid in barley protoplasts transformed by electroporation. Plant Cell Rep 9:352-355

Sasahara T, Ikarashi H, Kambayashi M (1986) Genetic variations in embryo and endosperm weights, seedling growth parameters and α-amylase activity of the germinated grains in rice (*Oryza sativa* L.). Japan J Breed 36:248-261

Schuurink RC, Chan PV, Jones RL (1996) Modulation of calmodulin mRNA and protein levels in barley aleurone. Plant Physiol 111:371-380

Sheu J-J, Jan S-P, Lee H-T, et al. (1994) Control of transcription and mRNA turnover as mechanisms of metabolic repression of α-amylase gene expression. Plant J 5:655-664

Sheu J-J, Yu T-S, Tong W-F, et al. (1996) Carbohydrate starvation stimulates differential expression of rice α-amylase genes that is modulated through complicated transcriptional and posttranscriptional processes. J Biol Chem 271:26998-27004

Simmons CR, Rodriguez RL (1989) High-level synthesis and secretion of α-amylase from rice callus. In: Whitaker JR, Sonnet PE (eds) Biocatalysis in Agricultural Biotechnology, American Chemical Society, Washington, DC, pp 202-214.

Simmons CR, Huang N, Cao Y, et al. (1991) Synthesis and secretion of α-amylase by rice callus: evidence for differential gene expression. Biotechnol Bioeng 38:545-551

Skriver K, Olsen FL, Rogers JC, et al. (1991) *cis*-Acting DNA elements responsive to gibberellin and its antagonist abscisic acid. Proc Natl Acad Sci USA 88:7266-7270

Smith MT, Briggs DE (1980) Externally applied gibberellic acid and α-amylase formation in grains of barley (*Hordeum distichon*). Phytochemistry 19:1025-1033

Steup M (1988) Starch degradation. In: Preiss J (ed) Biochemistry of Plants, Vol 14. Academic Press, New York, pp 255-296

Sutliff TD, Lanahan MB, Ho T-HD (1993) Gibberellin treatment stimulates nuclear factor binding to the gibberellin response complex in a barley α-amylase promoter. Plant Cell 5:1681-1692

Tanida I, Kim J-K, Wu R (1994) Functional dissection of a rice high-pI α-amylase gene promoter. Mol Gen Genet 244:127-134

Thomas BR, Rodriguez RL (1994) Metabolite signals regulate gene expression and source/sink relations in cereal seedlings. Plant Physiol 106:1235-1239

Williams JF, Peterson ML (1973) Relations between α-amylase activity and growth of rice seedlings. Crop Sci 13:612-615

Yu S-M, Kuo Y-H, Sheu G, et al. (1991) Metabolic derepression of α-amylase gene expression in suspension-cultured cells of rice. J Biol Chem 266:21131-21137

Yu S-M, Tzou W-S, Lo W-S, et al. (1992) Regulation of α-amylase-encoding gene expression in germinating seeds and cultured cells of rice. Gene 122:247-253

Yu S-M, Lee Y-C, Fang S-C, et al. (1996) Sugars act as signal molecules and osmotica to regulate the expression of α-amylase genes and metabolic activities in germinating cereal grains. Plant Mol Biol 30:1277-1289

Zwar JA, Hooeley R (1986) Hormonal regulation of α-amylase gene transcription in wild oat (*Avena fatua* L.) aleurone protoplasts. Plant Physiol 80:459-463

10
Structure and expression of rice seed storage protein genes

Fumio Takaiwa

Department of Biotechnology, National Institute of Agrobiological Resources, Kannondai 2-1-2, Tsukuba, Ibaraki 305-8602, Japan

1 Introduction

Cereal seeds accumulate storage proteins, accounting for about 5%-15% of mature grains, compared to up to 40% in legume seeds. Rice grain has the lowest protein content (about 5%-8%) of the major cereals. According to the soluble properties of Osborne, seed storage proteins are classified into water-soluble alubumin, salt-soluble globulin, and alcohol-soluble prolamins and acidic or basic solution soluble glutelins. Rice and oat accumulate glutelins and globulins as a major seed storage protein, which is in remakable contrast to that prolamins in other cereal seeds represent major seed storage proteins accounting for about half of the total grain proteins. These storage proteins are deposited into separate protein bodies that are morphologically distinct. It has been shown that the spherical-type protein bodies store prolamines, whereas the irregularly shaped, electron-dense protein bodies contain glutelin.

Seed storage proteins are synthesized in abundance in the embryo or the endosperm through a complex plant developmental program. These proteins are utilized as an essential nutrient source to support plant seedling growth upon germination. Seed storage proteins also have become the major source of nutrition of monogastric animals, including humans. Rice provides an important source of protein, especially for many people in Asia, Africa, and South America. However, the amino acid composition of rice storage proteins is not well balanced due to the lightly low content of lysine and threonine. Moreover, a considerable percentage of the rice storage proteins are not digested by monogastric animals (Tanaka et al.

Key words. endosperm, *cis*-elements, transcription factors, glutelin, prolamin, alubumin, globulin, GCN4 motif, AACA motif, transgenic plants, Myb protein, bZIP protein

1975). Some globulins and alubumins are known to cause allergy in hypersensitive patients (Limas et al. 1990; Matsuda et al. 1991; Shibasaki et al. 1977).

2 Structure of rice seed storage proteins

2.1 Glutelins

Glutelin is the most abundant storage protein in rice, constituting 60%-80% of total endosperm protein in mature seeds. Regardless of its insoluble properties, the rice glutelin shares many properties with the 11S globulins of dicotyledonous plants. Glutelin is composed of two 35- to 37-kDa acidic and 20- to 22-kDa basic subunits. Both subunit types show a heterogeneity. When glutelins were subjected to two-dimensional polyacrylamide gel electrophoresis, acidic and basic subunits were resolved into at least 16 and 9 spots, respectively (Wen and Luthe 1985). These groups of polypeptides are synthesized as a larger precursor (57 kDa) and then processed into two major mature subunits by posttranslational cleavage. Such a posttranslational processing was confirmed by pulse-chase labeling studies of *in vivo* labeling of seed proteins and *in vitro* translation studies of endosperm membrane-bound polysomes (Yamagata et al. 1982; Luthe 1983). The posttranscriptional processing site (Asn/Gly) is completely conserved in 11S globulins of higher plants. The amino acid sequence deduced from glutelin cDNA is about 30%-35% homologous to those of leguminous 11-12S globulin (Higuchi and Fukazawa 1987; Takaiwa et al. 1986, 1987a). A direct structural relationship between the rice glutelin and 11S globulins was identified by the partial N-terminal amino acid sequences of their basic subunits (Zhao et al. 1983). The positions of cysteines involved in linking of two subunits of soybean glycinin are also conserved in the glutelins. Glutelin is accumulated in protein body type II (PB-II) through the Golgi apparatus, which is large and uniformly stained.

To date, eight types of DNA sequences for glutelin genes have been reported at the cDNA and genomic levels (Takaiwa et al. 1986, 1987a, 1987b, 1989, 1991b; Takaiwa and Oono 1991; Okita et al. 1989; Masumura et al. 1989a). DNA sequences and Southern blot analysis revealed that glutelin is coded for by about 10 copies per haploid genome. Based on their sequences, glutelin genes are clearly classified into two subfamilies, designated as *GluA* and *GluB* (Takaiwa et al. 1991b). The coding sequences between the *GluA* and *GluB* members share 60%-65% homology at the amino acid level, whereas members within the same subfamily are more than 80% homologous to each other. The *GluA* subfamily is composed of at least four members, *GluA-1* (Gt2), *GluA-2* (Gt1), *GluA-3* (Gt3), and *GluA-4* genes (Okita et al. 1989; Takaiwa et al. 1987a; Takaiwa and Oono 1991). The coding sequences between *GluA-1* and *GluA-2* share 95% homology to each other and they are 81% homologous to G*luA-3*. The GluA-4 gene shows 73% and 81% homologies to the GluA-1/GluA-2 and GluA-3 genes. The *GluB* subfamily consists of at least of four members, *GluB-1,GluB-2, GluB-3*, and *GluB-4* genes (Takaiwa et al.

1989; Takaiwa et al. 1991b; Masumura et al. 1989a). Their coding regions share 80-88% homologies to each other. When the amino acid sequences deduced from these glutelin genes are compared with those of dicotyledonous 11-12S globulin, they show 32-35% homologies to them.

All the glutelin genes sequenced so far are interrupted by three short introns wiith less than 130 bp (Takaiwa et al. 1987; Okita et al. 1989; Takaiwa and Oono 1991; Takaiwa et al. 1991b). Two introns occur in the region coding for the acidic subunit and the remaining one resides in the region coding for the basic subunit. Locations of three introns are relatively the same sites observed in 11S globulins genes of dicotyledonous plants. It should be noted that introns found in glutelin genes are characteristic in that they are very short in length, ranging from 77 to 129 bp, and rich in A and T content (68.5%-79.8%), contrasting with the coding region (54 to 57%) (Takaiwa and Oono 1991).

There is little sequence homology in the 5' flanking regions of glutelin genes between the *GluA* and *GluB* subfamilies, although most glutelin genes except for *GluA-3* (Gt3) are coordinately expressed during seed maturation. Conserved motifs found in the 5' flanking regions of all the glutelin genes are limited to the 13-bp AACA motif (AACAAACTCTATC), the 7-bp GCN4 motif (TGAGTCA), GCAA motif (GCAAAA-/ATGA), and prolamin box or -300 element (TGCAAAA) in addition to the TATA box (Takaiwa et al. 1996). The first three motifs occur at relatively the same sites of all the glutelin gene promoters. Therefore, these motifs are proposed as candidates for the *cis*-regulatory elements for endosperm-specific expression of glutelin genes. The prolamin box is repeated many times in the 5' flanking regions of glutelin genes (Okita et al. 1989). RY repeat (CATGCATG motif), which has been shown to be essential for high levels of expression in leguminous seeds, is observed only in *GluA-1* gene. Zhao and Okita (1995) found that the CAACACA sequence is present in all members of the *GluA* subfamily. This motif is conserved in not only most cereal seed storage protein genes (zein, gliadin, and hordein), but also in soybean seed storage protein genes (11S glycinin, lectin, and kunitz trypsis inhibitor) (Goldberg 1986). It is not clear that this motif is functionally similar to the AACA motif conserved in all the glutelin genes.

Several glutelin genes have been mapped on chromosomes by RFLP mapping and *in situ* hybridization. ***In situ*** hybridization using *GluA-1* cDNA showed that it was located on the middle of the short arm of chromosome 2 (Suzuki et al. 1991). Nakamura et al. (1995) have reported that *GluA-1* gene, which linked with a semidwarf gene, was mapped on chromosome 1. Iida et al. (1997) have recently showed that a mutant lacking the largest subunits (type1) among four glutelin acidic subunits was mapped on chromosome 2, a mutant lacking the second largest ones (type 2) was on chromosome 10, and a mutant lacking the third largest one (type 3) was on chromosome 1. It was also shown that type 1 links to a glutelin gene in the *GluB* subfamily and type 2 is a mutated gene of the *GluA* subfamily. The type 3 was mapped at the site of similar distance from a semidwarf gene, indicating that the type 3 mutant is linked to the mutated gene of *GluA-1* gene. Glutelin genes are distributed on many chromosomes. Iida et al. (1993) also isolated a mutant (LGC-1) in which the glutelin content is decreased to half and the prolamin content is doubled. RFLP mapping showed that LGC-1 is located in chromosome 2.

2.2 Prolamins

Alcohol-soluble proteins, prolamins comprise about 18%-20% of the total endosperm protein in rice (Ogawa et al. 1987; Li and Okita 1993), and are classified into 10-, 13-, and 16-kDa groups. The major type of prolamin is a 13-kDa species and the remaining two are minor species. The 13-kDa group consists of at least 10 polypeptides with isoelectric points between pH 6.0 and 8.1 when resolved by two-dimensional polyacrylamide gel electrophoresis (Hibino et al. 1989). Amino acid composition of the prolamins is rich in Gln, Pro, and Leu, but lacking in Lyn and low in Met, Cys, Thr, and Tyr. Southern blotting analysis revealed that the 13-kDa group is encoded by a multigene family of 80 to 100 copies per haploid genome (Kim and Okita 1988). Many of them are pseudogenes. Shyer et al. (1992) estimated that there were 30 to 50 genes encoding the class 1, 2, and 3 prolamins. Some of these genes are tightly clustered in the rice genome as reported by Kim and Okita (1988).

To date at least 10 genes have been cloned and sequenced (Feng et al. 1990; Kim and Okita 1988; Masumura et al. 1989b, 1990; Sha et al. 1996; Shyur and Chen 1991; Shyur et al. 1992; Wen et al. 1993; Yamagata et al. 1992). The amino acid sequences deduced from these nucleotide sequences of 13-kDa prolamin can be divided into three major groups based upon their amino acid sequences (types 1, 2, and 3) and their content of sulfur-containing amino acids. The type 1 prolamin has an open reading frame of 156 amino acids. The types 2 and 3 have an open reading frame coding for the same size of 150 amino acids. The type1 species shares 70% and 72% homologies to those of type 2 and type 3 at the amino acid sequence level, respectively. The deduced amino acid of the type 2 is 58% homologous to that of type3. The number of cysteine residues included in these three types of 13-kDa species differs. The type 2 completely deletes in cysteine, whereas type1 and 3 include four and eight cysteine residues, respectively. It should be noted that there are no repetitive sequences rich in Glu characteristic of other cereal prolamins.

The 10-kDa prolamin consists of about 110 residues with a 24-residue signal peptide (Masumura et al. 1989a). The deduced amino acid sequence is rich in sulfur-containing amino acids, methionine (20%) and cysteine (10%), and it shares little homology with those of other cereal prolamins except for the signal peptide sequence. Although there is no repetitive sequence rich in Gln and Pro is observed in most prolamins of cereal prolamins, a single octapeptide, Gln-Gln-Gln-Cys-Cys-Gln-Gln-Leu sequence, is found in the rice 10-kDa and 13-kDa prolamins. It should be noted that signal peptide sequences of the rice 10-kDa and13-kDa prolamins are highly homologous to those of other cereal prolamins, suggesting the targeting system to protein bodies is common to all the cereal prolamins.

Rice prolamin genes reported so far have no intron in the coding region and 5' untranslated region. In the 5' flanking region, the TG(T/A/C)AAA(G/A/C) resembling the -300 element or prolamin box and GCN4 motif (TGAGTCAT) are found (Sha et al. 1996). It should be noted that these different motifs are linked together. Such a combination, which is named an endosperm box, is also found in other cereal prolamin genes. Functional analysis of the bifacorial endosperm box has been performed in wheat LMW glutelin gene (Albani et al. 1977). Taken together, such a combination may be involved in the endosperm-specific expression of the prolamin gene.

2.3 Globulins

The rice salt-soluble globulins comprise 2%-8% of total endosperm protein and migrate as a single abundant spot with a molecular weight of 26 kDa (α-globulin) and a minor spot with a molecular weight of 16 kDa by two-dimensional electrophoresis (Pan and Reeck 1988). Southern blotting analysis revealed that α-globulin is encoded by a single copy gene (Shorrosh et al. 1992). The molecular weight estimated from the deduced amino acid sequence is 19 KD, thus called the 19 KD globulin.

α-Globulins are synthesized as precursors having signal sequences of about 2000 daltons. The amino acid composition of the deduced amino acid sequence is similar to that of purified globulin, being rich in glutamine/glutamate and arginine residues (Shorrosh et al. 1992). It shows 55% and 58% homologies to those of wheat HMW glutenin and barley D hordein, respectively (Krishnan and Pueppke 1993). This result supports the immunological evidence showing that rice globulin antisera cross-reacted with several proteins from wheat, rye, and triticale (Komatsu and Hirano 1992). The nucleotide sequence of the 5' flanking region between -201 and -71 is also highly homologous (68% and 61%) to those of the wheat HMW glutelin subunit and the barley hordein genes (Nakase et al. 1996a).

The consensus prolamin box and an A+T-rich sequence conserved in promoters of cereal prolamin genes are also found in the 5' flanking region. Such motifs may be involved in the endosperm-specific expression. There is no intron in the cording region of the globulin gene in a similar way as other prolamin genes. This evidence suggests that rice globulin originated from a common ancestral gene and is regulated by a similar or common regulatory mechanisms to achieve endosperm-specific expression. The globulin 26-kDa-lacking mutant gene was found to be a single recessive gene and to be located in chromosome 5 by RFLP analysis (Iida et al. 1998).

2.4 Albumins

Rice seed albumin comprises about 5% of the total endosperm protein. The major albumins exhibits heterogeneity in their molecular size (14 to 16 kDa) and isoelectric point (pI 6 to 8), which shows immunological cross-reactivity (Matsuda et al. 1991). The rice 16-kDa protein was identified as a major allergen in rice seeds based on the results of radioallergosorbent test. Rice albumins (RA) show limited sequence homology (20%-40%), in the number and distribution of cysteine residues, with amylase/trypsin inhibitor family in cereal seeds, showing that RA is a member of this inhibitor family (Izumi et al. 1992). The homologous sequence with those conserved among several cereal prolamines as well as rice 13-kDa prolamin, 10-kDa prolamine, and 19-kDa globulin is found in these RAs. A COOH-terminal region containing cysteine residues is also partially homologous with 10-kDa prolamin.

RAs are encoded by a multigene family, but their number is relatively small compared with that of the prolamine genes (Adachi et al. 1993; Alvarez et al. 1995). Comparison of the coding region of three RA genes showed high homologies (70%-95%) to each other at the nucleotide and amino acid levels. RAs are classified into at least four subfamilies based on the amino acid and nucleotide sequences. A region conserved among RA subfamilies is homologous to that of lipid transfer proteins. The coding region of RA genes is not interrupted by an intron.

Expression of RA genes is limited to the maturing seed. Levels of RA mRNA increased from 10 DAF and reached a maximum level at 15 to 20 DAF and then dropped off. In the 5' flanking region the TGCAAAA prolamin box, which is observed in the rice glutelin promoters, occurs at several sites. Furthermore, CTTTGGTCTT and CTTTAGTCTT sequences similar to the nuclear protein-binding site (CTTTCGTGTA) observed in the glutelin gene exist in RA promoters.

3 Expression of the glutelin and prolamin genes during seed development

The levels of prolamines and glutelins were quantified during seed development by immunoblot analysis (Li and Okita 1993). The molar ratio of glutelins to prolamines was 1.7 in seeds 10 days after flowering, and this ratio steadily decreased to 1.2 in 25 DAF seeds because of the increased synthesis and accumulation of prolamines, specifically during the latter stages of seed development.

Results from nuclear run-on transcript assays and RNA blot studies showed that the expression of glutelin and prolamin genes comprising each multigene family was not coordinately but instead differentially regulated at both the transcriptional and posttranscriptional levels during seed development (Kim et al. 1993). Quantification of their mRNA levels revealed that there are equimolar amounts of glutelin and prolamin transcripts in 5 and 10 DAF seeds and a 40% excess of prolamin transcripts, relative to glutelin transcripts, in seeds of 15 days and older. The mRNAs of *GluA-1* (Gt2) and *GluA-2* (Gt1) are present in 5-day-old developing seeds, and their levels continue to increase throughout seed development. In contrast, the A-3 gene mRNAs reach maximum levels at 10 days after flowering, and gradually decline during seed development. Transcripts of the *GluB* classes accumulate in a coordinate fashion with the A1 and A-2 genes. The run-on transcriptional activities of A-3 (Gt3) and prolamins were relatively high, whereas that for *GluA-1*(Gt2)/*GluA-2* (Gt1) genes was significantly lower in seeds at 5 DAF. At 15 DAF, the transcriptional activity of *GluA-1* (Gt2) /*GluA-2* (Gt1) increased about threefold, while the activity of *GluA-3* (Gt3) remained nearly constant. The transcriptional activity of the prolamin genes did not change substantially during seed development. These data indicate that glutelin and prolamin mRNA levels are controlled at both the transcriptional and posttranscriptional levels.

4 Analysis of promoter sequences of seed storage protein genes

4.1 Endosperm-specific expression of seed storage protein genes

Seed storage protein genes are strictly expressed in the maturing endosperm tissue. When the 5' flanking regions of rice glutelin and 13-kDa prolamin genes were transcriptinally fused to the GUS reporter gene and then these chimeric genes were introduced in tobacco, it was shown that the expression site of the GUS reporter gene was restricted to the endosperm tissue (Zhou and Fan 1993; Takaiwa et al. 1991), irrespective of the heterologous tobacco system. Tissue-specific expression has recently been examined by homologous system (Wu et al. 1998a). When the 5' flanking regions of the *GluA-1, GluA-2, GluA-3*, and *GluB-1* glutelin genes, 13-kDa prolamin gene, α-globulin gene, and allergenic protein genes linked to the GUS reporter gene were introduced into rice, all of these promoters directed expression of the GUS gene in the endosperm tissue, although their expression patterns slightly differed one from another (Fig. 1). In the glutelin promoters, blue staining by the GUS gene was mainly observed in the aleurone and subaleurone cells of the endosperm, being less active in the inner portions of the endosperm. The α-globulin and allergenic protein gene promoters directed expression of the GUS gene in inner starchy endosperm cells rather than the aleurone and subaleurone layers, with stronger activity in the region close to the embryo. The 13-kDa prolamin promoter displayed an evident activity in all endosperm cells with the stronger expression in aleurone and subaleurone cells.

4.2 Characterization of *cis*-regulatory elements

To identify the *cis*-regulatory elements required for the endosperm-specific expression of glutelin genes, three types of glutelin genes (*GluA-2, GluA-3,* and *GluB-1*) have been used for characterization. The *GluA-2* and *GluB-1* genes are coordinately expressed during seed maturation, although promoter sequences are quite divergent besides showing several motifs. On the other hand, the *GluA-3* gene is differentially expressed from other genes in a temporal-specific manner (Okita et

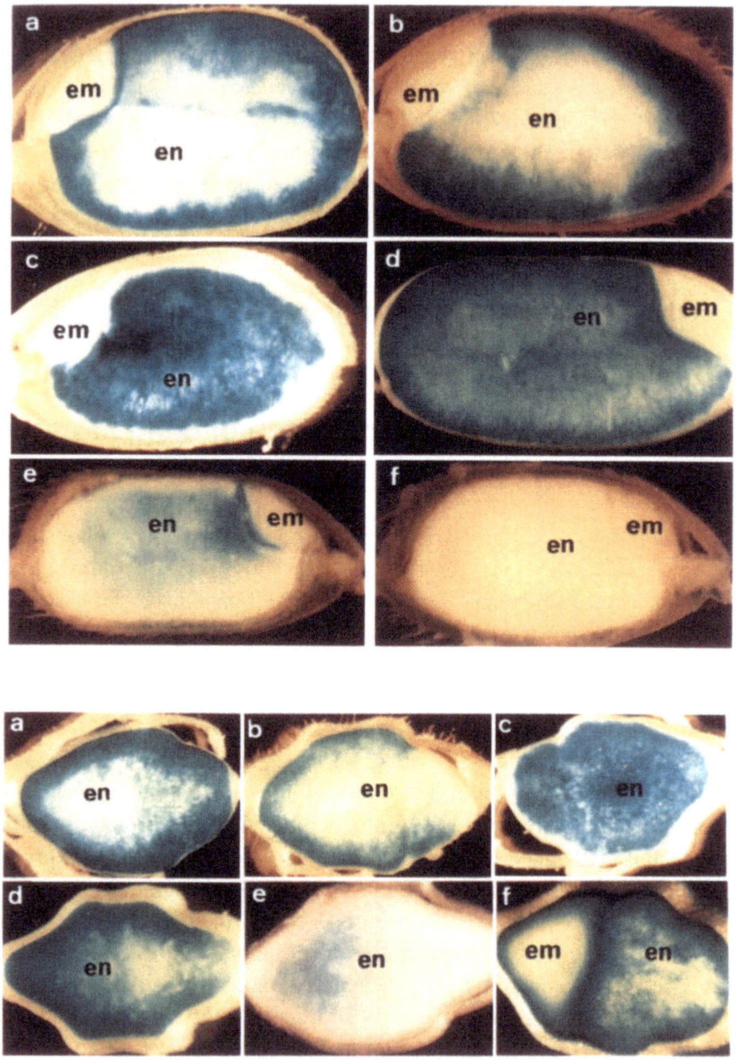

Fig. 1. Histochemical analysis of GUS expression directed by various rice seed storage protein gene promoters. The 5' flanking regions of different rice storage protein genes were fused either transcriptionally (GluA-3, GluB-1, and NRP33) or translationally (Glb-1 and RAG-1) to the GUS reporter gene followed by the Nos terminator. The regions of individual promoter used for this analysis are -897 to +11 for the glutelin GluA-3 gene, -1302 to +18 for the glutelin GluB-1 gene, -652 to -13 from the translation start site of the prolamin NRP33 gene, -980 to the translation start site of the globulin Glb-1 gene, and -498 to the translation start site of the alubumin RAG-1 gene, respectively. A Longitudinal sections: a, glutelin GluB-1 promoter; b, glutelin GluA-3 promoter; c, globulin Gbl-1 promoter; d, prolamin NRP33 promoter; e, alubumin RAG-1 promoter. B Cross sections: a-e, as described in A. A nontransformed seed (f in A) and a seed transformed with a CaMV 35 promoter/GUS construct (f in B) are also shown as negative and positive controls, respectively. Em, embryo; en, endosperm. Taken from Wu et al. (1998a)

al. 1989; Takaiwa and Oono 1991). Although members of the *GluA* subfamily are highly conserved, the 5' flanking region of the *GluA-3* gene was significantly diverged in the upstream region beyond position -150, suggesting that the upstream region is implicated in temporal regulation. On the other hand, when the 5' flanking sequences of *GluA-1* and *GluA-2* genes were compared, there is a large insertion sequence of about 150 to 200 bp in the *GluA-1* gene around position -250 of the *GluA-2* gene (Okita et al. 1989; Kim and Wu 1990). Because the *GluA-1* gene is expressed in embryo and leaf (Hirano et al. 1991; Krishnan and White 1997) as well as in endosperm, it seems likely that such an insertion sequence is related to the alternation of tissue specificity.

4.2.1 *GluA-2* (Gt1) promoter

Analysis of these promoter regions has been mainly performed using the transgenic dicotyledonous tobacco plant. Despite the use of a heterologous dicot system, the rice glutelin promoters of the *GluA-2* and *GluA-3* genes were observed to be sufficient to direct expression in the endosperm of developing seeds in a manner similar to that of rice seed (Leisy et al. 1989; Takaiwa et al. 1991).

5'-Deletion analysis of the *GluA-2* gene first revealed that the region between positions -441 and -237 is required for seed-specific expression, in which there are the GCAA motif, the GCN4 motif, and the endosperm motif (Takaiwa et al. 1991b). The chimeric gene consisting of the progressive 5'-deleted promoters of the corresponding gene and the GUS reporter gene has been introduced into homologous rice (Zheng et al. 1993). Despite the use of a homologous system, when the promoter was deleted to -507 or -154, endosperm specificity was altered and GUS expression was observed in vegetative tissues (phloem of leaves, stems, sheaths, and roots). Further deletion to -214 resulted in complete loss of GUS activity. It is not clear why a longer promoter sequence was required for endosperm-specific expression even in the homologous rice. Substitution mutations (single mutation and simultaneous mutation) in the context of the 1.8-kb promoter, which correspond to the nuclear protein-binding sites (-103 to -86,-124 to -110, -175 to -158, -200 to -217), the conserved AACA motif, and the TATA box, had dramatic effects on the activities of the 1.8-kb promoter. Temporal control during endosperm development was retained in plants of constructs from -5.1 to -155.

4.2.2 *GluA-3* (Gt3) promoter

The *cis*-regulatory elements of the glutelin Gt3 gene were analyzed by a series of 5'-deletions. The results showed that two regions between -945 and -726 and between -346 and -263 are required for temporal regulation and tissue specificity, respectively (Zhao et al. 1994). Removal of the 5'-sequence upstream of -726 resulted in a decrease in quantitative level of promoter activity and a shift in temporal expression from a maximum of 16-20 DAF to about 24 DAF. Similar analysis of the corresponding gene (*GluA-3*) was carried out using a sensitive firefly luciferase (LUC) gene as a reporter gene (Yoshihara and Takaiwa 1996). Progressive deletion fragments between -897 and -437 showed a similar level of promoter activity. Further deletion to -437 resulted in a remarkable reduction of promoter activity, thus indicating that a 121-bp sequence between -437 and -317 from the transcription start site is implicated for quantitative regulation. Furthermore, it was also shown that this region contained at least two functionally different *cis*-regulatory elements. One element acts as a general positive element, and the other is involved in a specific negative regulation in nonseed tissue when fused to the truncated -90 promoter of the CaMV 35S gene. It seems that the AACA motif between -433 and -421 may be responsible for suppressing the expression in vegetative tissue, and either the GCAA motif or endosperm motif acts as a general enhancer. However, the TGACGTGG sequence between -386 and -379, similar to the Opaque 2 binding site, was shown not to be involved in seed-specific expression.

A 3'-deletion analysis of the Gt3 (*GluA-3*) gene between -94 and +7 was also performed by fusing to the truncated promoter (-90) of CaMV35S and analyzed in transgenic tobacco (Croissant-Sych and Okita 1996). These deletion analyses showed that two regions between -278 and -482 and between -181 and +7 were required for maximum expression. The distal region identified here is consistent with the region observed in the 5'-deletion analysis. The proximal region contains the AACA motif, the GCN4 motif, and the sequence similar to the pyrimidine box conserved in the α-amylase gene.

The function of the proximal region between -104 and -60 containing the conserved GCN4 and AACA motifs was examined by fusing to the truncated promoter of the CaMV35S gene (Yoshihara et al. 1996a). Histochemical and quantitative analyses showed that this 45-bp sequence specifically enhanced GUS activity in endosperm tissue. When the base substitutions were introduced into the GCN4 motif or the AACA motif, promoter activity was abolished by mutation into GCN4 and alternation of tissue specificity was observed by the mutation of the AACA motif, respectively. These results suggest that GCN4 acts as a positive element and that the GCN4 motif functions as a specific negative element, suppressing expression in all tissues except the endosperm.

4.2.3 *GluB-1* promoter

The *cis*-regulatory elements required for the endosperm-specific expression of the glutelin *GluB-1* gene have been also characterized by using transgenic tobacco to elucidate the common regulatory mechanisms underlying the endosperm-specific expression observed in all the glutelin genes (Takaiwa et al. 1996). 5'-Deletion analysis of the region beginning at -1302 and ending at +18 showed that promoter activity gradually decreased according to the length of the 5' flanking region. The 245 bp of the promoter region, which is necessary to confer endosperm-specific expression, contains the two AACA motifs, the GCN4 motif, and the G-box-like element. However, further deletion to position -197 resulted in abolishment of promoter activity, indicating that the sequence between -245 and -197 might contain the essential element required for seed-specific expression. The distal AACA motif is located between -212 and -200 in this region, suggesting that this motif acts as a positive element.

3'-Deletion analysis of the flanking region between -573 and +18 was carried out by fusing them to the -90 truncated promoter of the CaMV 35S promoter. The results indicated that the regions between -40 and -93 and between -145 and -227 are necessary for a high level of seed-specific expression. Each of these regions contains the AACA motif. Taken together, the AACA motif may be responsible for quantitative regulation of transcription of the *GluB-1* gene. To define the functional significance of the AACA and GCN4 motifs or other elements involved in seed-specific expression, several internal deletions were introduced into the 5' flanking region between -573 and +18. The elimination of the regions containing the AACA motif reduced the promoter activity to three- to tenfold. The removal of the GCN4 motif resulted in about fivefold reduction of GUS activity. The deletion of the longer region containing the AACA and GCN4 motifs led to the complete loss of GUS activity, suggesting that the combinatorial or synergistic interaction of the AACA motif and the GCN4 motif is required for the seed-specific expression. Furthermore, *cis* elements involved in endosperm-specific expression were characterized by introducing mutagenesis into several candidate motifs within the minimal 245-bp proximal region (Washida et al., submitted). Base substitution analysis showed that the proximal AACA motif between -73 and -61 and the GCN4 motif between -165 and -158 acted as a critical element, since such mutations completely disrupted the promoter activities. The G-box-like motif between -80 and -75 and Skn 1, like the element between -173 and -169, also participated greatly in controlling seed-specific expression. The prolamin box between -136 and -130 functioned as a negative element by itself.

Internal deletion analysis using this minimal promoter suggested that a combinatorial or synergistic interaction of the distal AACA and GCN4 motifs between -245 and -145 and the G-box-like motif and the proximal AACA motif between -90 and +18 was responsible for seed-specific expression. Such cooperative interaction of several motifs was confirmed by gain-of-function experiments. When the distal region between -245 and -145 containing the AACA and GCN4 motifs and the proximal region between -113 and -45 containing the G-box like motif and AACA motifs were fused to the truncated promoter (-90 to +9) of CaMV35S gene fused to the GUS reporter gene, they were sufficient to confer a high level of seed-specific expression. However, when substituted for the core promoter of CaMV35S gene, seed-specific expression was abolished, suggesting that a putative G-box-like motif in the -90 truncated promoter is require for such specific expression in addition to the above two motifs.

Similar analysis using the *GluB-1* promoter/GUS reporter gene has been done using homologous transgenic rice. At first, to determine the minimum length of the *GluB-1* gene required for seed-specific expression, progressive 5'-deletions were generated using -573 construct between -573 and +18 relative to the transcription start site. Contrary to the results obtained by heterologous transgenic tobacco, the region between -197 and +18 was sufficient to confer seed specific GUS activity. Further deletions to -145 and -113 resulted in background level of promoter activ-

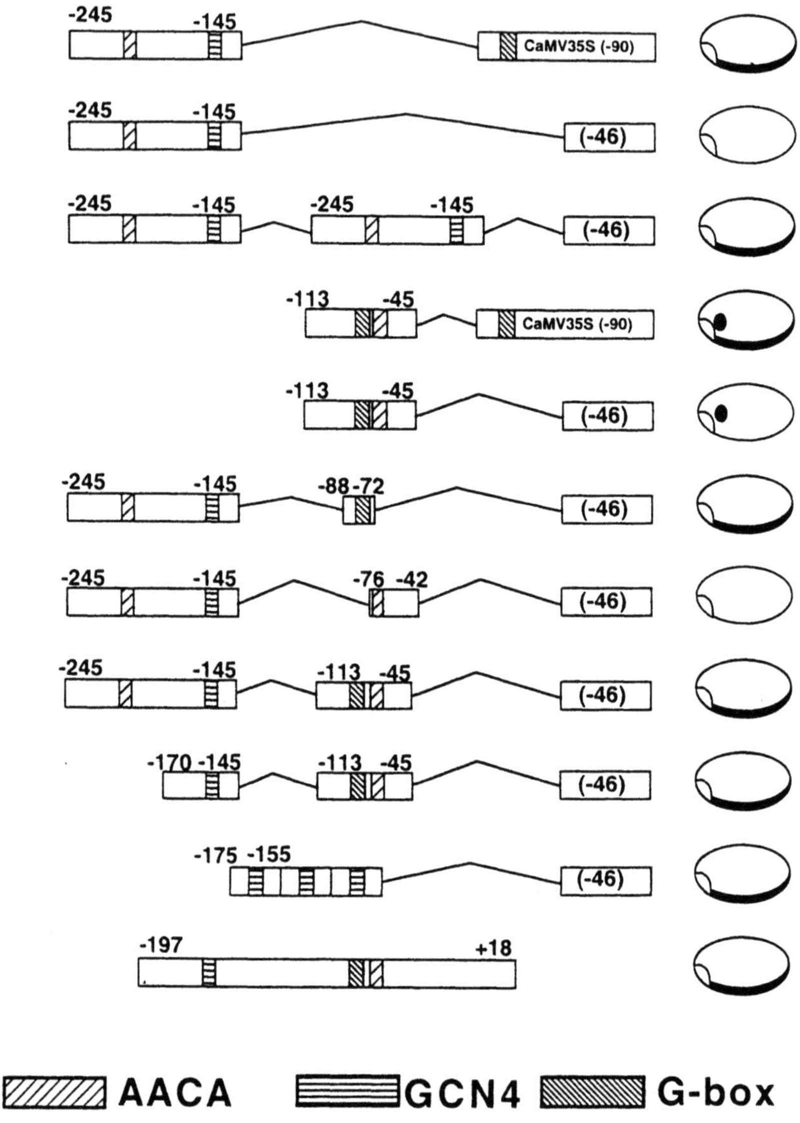

Fig. 2. Identification of *cis*-regulatory elements required for endosperm-specific expression of glutelin GluB-1 gene in transgenic rice. The distal region between -245 and -145 and the proximal region between -113 and -45 were fused to a truncated (-90 to +9) or core promoter (-46 to +1) of CaMV 35S followed by the GUS reporter gene. The G-box-like element, AACA motif, or GCN4 motif was fused to the distal or proximal region, which were then introduced upstream of the -46 core promoter. The trimer of the 21-bp fragment containing the GCN4 motif was also fused upstream of the -46 core promoter. These chimeric genes were introduced into rice by electroporation. GUS staining in maturing seed was examined in stable transgenic rices.

ity. The -145 promoter lacks the distal GCN4 motif, while the -113 promoter construct lacks both the GCN4 and the endosperm motif. Taken together, the results of deletions suggested that the distal GCN4 motif located between -197 and -145 is a critical *cis*-regulatory element for seed-specific glutelin gene expression (Wu et al. 1998b).

The minimum region required for endosperm-specific expression of the *GluB-1* gene was defined by gain-of-function experiments (Fig. 2). The region between -245 and -145 containing the AACA and GCN4 motifs and the region between -113 and -45 containing the G-box-like element and the AACA motif were fused upstream of the truncated (-90) or core (-46) promoter of CaMV35S/GUS reporter. When fused to the truncated promoter, both regions were sufficient to produce high levels of GUS activity in maturing seeds. However, when changed to the core promoter, GUS activities dropped to the background level. Histochemical analysis revealed that blue staining was mainly observed in the aleurone and subaleurone layer, as shown in the native promoter (-243 and -197 constructs). It is notable that the ectopic expression in inner starchy endosperm close to the embryo in addition to the aleurone and subaleurone layer was observed by the region between -113 and -45. This novel expression in the inner starchy endosperm was observed irrespective of background level in GUS activity, when fused to the core promoter. These results indicate that glutelin expression in aleurone and subaleurone seed layers is determined by at least three different motifs, either the AACA and GCN4 motifs in combination with the G-box-like element present naturally in the *GluB-1* promoter or in combination with the G-box-like element (as-1 element) found between -83 and -63 of the truncated promoter of CaMV 35S. The expression in inner starchy endosperm may be due to the effect of the G-box-like element.

The requirement of at least three motifs for endosperm-specific expression was further examined by combining the distal region containing the AACA motif and the GCN4 motif and the proximal regions containing the G-box-like element and the AACA motif, respectively, to several fragments containing the G-box-like element, the AACA motif, or the GCN4 motif. The combinations of the distal region and the G-box-like element and the proximal region and the GCN4 motif conferred endosperm-specific expression, even when fused to the core promoter of CaMV35S promoter/GUS gene.

To elucidate common regulatory mechanisms responsible for coordinated expression between members of *GluA* and *GluB* subfmilies, hybrid promoters among the 5' distal and proximal sequences of *GluA-3* and *GluB-1* genes were constructed and their promoter activities were examined in maturing seeds of transgenic tobacco (Yoshihara et al. 1996b). When the chimeric genes are introduced into tobacco, it has been shown that the AACA motif and GCN4 or G-box-like motif act as *cis* elements required for endosperm-specific expression, and the differences in quantitative regulation between the glutelin genes may be attributed to the differences in the combination and/or the synergistic interaction between these motifs and the additional motifs in the distal region.

5 Transcription factors involved in endosperm-specific expression

5.1 Analysis of binding sites of seed nuclear *trans*-acting factors

The binding sites of nuclear protein factors from maturing seeds have been analyzed in the 5' flanking region of the Gt2 (*GluA-1*) by gel retardation and DNase I footprinting assays (Kim and Wu 1990). At least six protein–DNA complexes are formed. The regions showing the protection are located between -103 and -86 (Box I), between -164 and -146 (Box II), between -122 and -108 (Box III), between -206 and -189 (Box IV), and between -595 and -575 (Box V). Box I and Box III contain the GCN4 motif, which is highly conserved in all the glutelin genes. It has been demonstrated by methylation interference experiments that maturing seed-specific *trans*-acting factor binds to the region between -130 and -120 (Box II) of the *GluA-2* gene (Takaiwa and Oono 1990). It should be noted that the binding activity was closely correlated to the expression level of the glutelin gene. This region covers the target sequence (AAAG) of the Dof class of plant zinc-finger DNA-binding factor (Vicente-Carbajosa et al. 1997).

Three binding sequences interacting with the 5' flanking region of the glutelin Gt3 were identified (Zhao and Okita 1995). One region between -272 and -259 contained the CAACACA sequence conserved among several seed protein genes. The sequence between -861 and -838 was suggested to be implicated in temporal regulation.

Several genes of albumin, prolamin, and glutelin storage proteins are coordinately expressed during seed maturation. Nakase et al. (1996b) showed by gel retardation assays that the TGCAAAA prolamin box and the CTTTC/GGTC/GT sequence, which are found in common in their 5' flanking regions, are recognized by the same transcription factors, indicating that some of the transcriptional machinery is shared among them.

5.2 Characterization of transcription factors

The sequence of the AACA motif is highly conserved in glutelin genes and shares strong homology to that of the gibberellin-responsive element (TAACAa/gANTCt/cGG) conserved in cereal gene promoter (Rogers and Rogers 1992) and the target sequence (TAACAAACT) of the *Petunia* PHMYB3 MYB-related protein (Solano et al. 1995). It has been reported that the GA-regulated Myb transcription factor, GAMYB, from barley can specifically bind to the gibberellin-responsive element (GARE), and it has been implicated in the regulation of α-amylase gene expression (Gubler et al. 1995). Therefore, Suzuki et al. (1997) screened Myb cDNA clones from a rice seed cDNA library. One of them, OSMYB5, shares about 80% homology to the MYB domains of GAMYB and PHMYB3 at the amino acid sequence level, which can bind to the AACA motif. DNase I footprinting analysis demonstrated that OSMYB5 specifically recognized the proximal and distal AACA motifs of the *GluB-1* gene (Suzuki et al. 1998). Analysis of the recognition sequence by scanning mutations in the AACA motif revealed that the AAC-A is the target site of the OSMYB5 protein.

It has been reported that maize Opaque2 (O2) bZIP transcriptional activator binds to the GCN4 motif of the pea lectin promoter (de Pater et al. 1994) and the wheat glutelin promoter (Holdsworth et al. 1995), and can activate the transcription through the motif fused to the core promoter of CaMV 35S in transient assays in protoplasts. When the binding site of the maize O2 in the *GluB-1* promoter was examined by methylation interference analysis, it was shown that the GCN4 motif responsible for endosperm-specific expression is specifically recognized by the O2 (Wu et al. 1998b). When the trimer of a 21-bp fragment containing the GCN4 motif was fused upstream of the core promoter of the CaMV35S and introduced together with the 35S/O2 as an effector into rice protoplasts, a remarkably high level of activation (more than 100 fold) was observed when compared with that obtained by the absence of 35S/O2. Such an O2-mediated transactivation by the GCN4 motif was also observed in transgenic rice plants containing the trimer of GCN4 motif reporter and the 35S/O2 effector. GUS expression in phoem tissue of leaf was detected under the control of the CaMV35S promoter through binding to the GCN4 motif. Furthermore, Wu et al. (1998) showed that the trimer of the 21-bp fragment containing the GCN4 motif, which fused to the core promoter of CaMV35S, gave rise to endosperm-specific expression in transgenic rice. Therefore, it was

	basic region / leucine repeat
RISBZ1	KVKKR KESNRESARRSR SRKAA RLKDLEEQVSLL RVENSSL LRRL ADANQKYSAAA IDNR VLMADIEAL
RISBZ2 (REB)	RLQRR KQSNRESARRSR SRKAA HLNELEAQVSQL RVENSSL LRRL ADVNQKYNDAAVDNR VLKADVETL
blz-1	RLRRR KQSNRESARRSR SRKAA HLNELEAQVSQL RVENSSL LRRL ADVNQKYNGAAVDNR VLKADVETL
OHP2	RQQRR KQSNRESARRSR SRKAA HLNELEAQVAQL RVENSSL LRRL ADVNQKFNEAAVDNR VLKADVETL
OHP1	RLQRR KQSNRESARRSR SRKAA HLNELEAQVAQL RVENSSL LRRL ADVNQKFNEAAVDNR VLKADVETL
OPAQUE2	RVRKR KESNRESARRSR YRKAA HLKELEDQVAQL KAENSCL LRRI AALNQKYNDANVDNR VLRADMETL
O2-sorghum	RVRKR KESNRESARRSR YRKAA HLKDLEDQVDKL KAENSCL LRRL AALNQKYNHATVDNR VLKADMETL
O2-coix	RVRKR KESNRESARRSR YRKAA HLKELEDQVEQL KAENSCL LRRL AALNQKYNEANVDNR VLRADMETL
SPA	KANKR KESNRDSARRSR SRKAA HAKELEEQVSLL RVANNSL MRHL ADVSHRYVNIS IDNR VLKANVETL
RISBZ3 (RITA1)	RMR-RMVSNRESARRSR KRKQA HLADLETQVDQL RGENASL FKQL TDANQQF TTAVTDNR ILKSDVEAL
RISBZ4	RVR-RMVSNRESARRSR KRKQA HLADLESQVDQL RGENASL FKQL TDANQQF TTSVTDNR ILKSDVEAL
RISBZ5	RIR-RMVSNRESARRSR RRKQA QLSELESQVEQL KGENSSL FKQL TESSQQFNTAVTDNR ILKSDVEAL
G/HBF-1	RVR-RMLSNRESARRSR RRKQA HLTELETQVSQL RVENSSL LKRL TDISQKYNEAAVDNR VLKADVETL
CPRF-2	RVR-RMLSNRESARRSR RRKQA HMTELETQVSQL RVENSSL LKRL TDISQRYNDAAVDNR VLKADIETM

Fig. 3. Sequence alignment of leucine-zipper domain in five rice O2 homologues to the corresponding domains from O2, OHP1, OHP2, O2-sorgham, O2-coix, SPA, G/HBF-1, and CPRF-2.

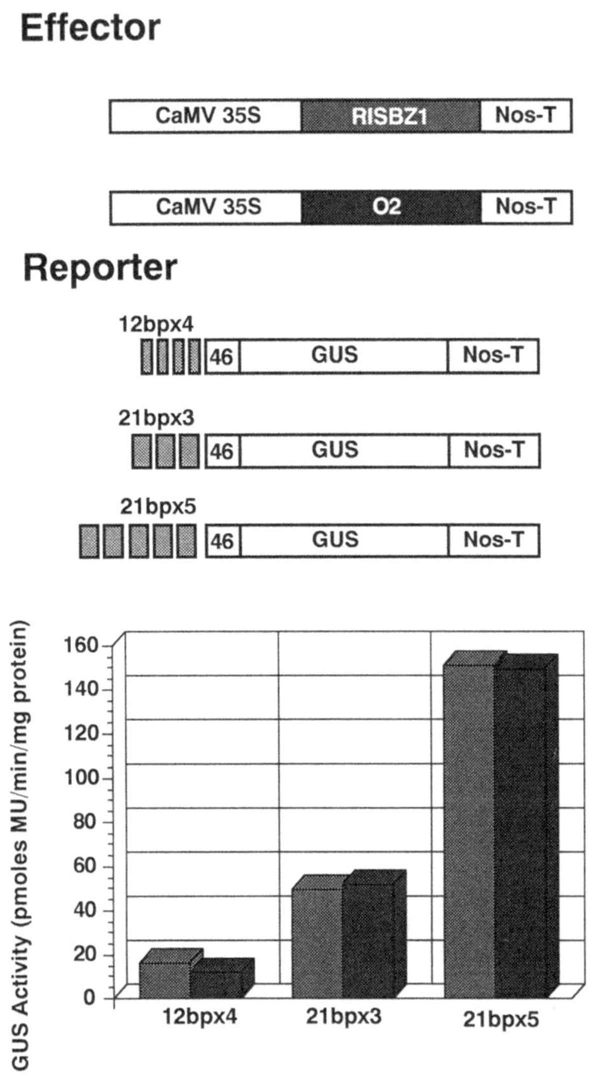

Fig. 4. Transactivation from 4x12 bp sequence, 3x21 bp sequence, and 5x21 bp sequence containing the GCN4 motif by rice RISABZ1 and maize O2 in rice protoplasts. RISBZ1 and O2 effector plasmids were used in the transactivation assays.

suggested that the GCN4 motif acts as a critical *cis*-regulatory element.

The maize O2 homologues were screened from a cDNA library from maturing rice seeds. Five types of cDNA clones were isolated, two of which corresponded to the RITA-1 and REB bZIP transcription factors (Fig. 3). RITA-1 is a basic leucine zipper transcription factor that is mainly expressed in aleurone and endosperm cells of developing seeds (Izawa et al. 1994). It was shown that RITA-1 displays a broad binding specificity for palindromic ACGT elements. We showed that the

GCN4 motif was also recognized by RITA-1, but the transactivation level was remarkably low compared to that of maize O2. The REB gene has been isolated as a novel bZIP protein which shows high homology to the O2 heterodimerizing protein (OHP) of maize (Pysh et al. 1993; Nakase et al. 1997). Nakase et al. (1997) showed that the target sequence of REB is the GCCACGT(A/C)AG sequence, which is different from that of O2. However, irrespective of high homology to REB, RISBZ1 isolated by our group specifically bound to the GCN4 motif in a similar way as the O2. When the RISBZ1 gene was expressed under the control of the CaMV35S promoter, a high level of activation (more than 100 fold) of the 3xGCN4 motif/GUS reporter gene was observed through interaction with the GCN4 motif in a transient assay (Fig. 4).

It was also confirmed by methylation interference experiments that the GCN4 motif found in glutelin genes is specifically recognized by the RISBZ1. It has recently been reported that the bZIP transcription factors wheat SPA1 (Albani et al. 1997) and barley BLZ1 (Vicente-Carbajosa et al. 1998) activate expression of prolamin genes (LMW glutelin and hordein genes) through binding to the GCN4 motif. It is noteworthy that these genes are also closely related to the maize OHP and rice REB in a similar way as the RISBZ1. On the other hand, the remaining two genes, which are highly homologous to the RITA1, specifically bind to the GCN4 motif sequence, but they could not transactivate the expression of 3xGCN4 motif/GUS in a manner similar to that of the RITA-1. Thus, the RISBZ1 is the real *trans*-acting factor involved in the expression of glutelin gene, and it is suggested that binding of the RISBZ1 to the GCN4 motif is the critical step of endosperm-specific expression.

Acknowledgments. F.T. is supported by research grants from the Enhancement of Center-of-Excellence, the Special Coordination Funds for Promoting Science and Technology of Japan, the Bio-oriented Technology Research Advancement Institution (PROBRAIN), and the Ministry of Agriculture, Forestry and Fisheries of Japan.

References

Adachi T, Izumi H, Yamada T, Tanaka K, Takeuchi S, Nakamura R, Matsuda T (1993) Gene structure and expression of rice seed allergenic proteins belonging to the α-amylase/trypsin inhibitor family. Plant Mol Biol 21:239-248

Albani D, Hammond-Kosack MCU, Smith C, Conlan S, Colot V, Holdsworth M, Bevan M (1997) The wheat transcriptional activator APA: a seed-specific bZIP protein that recognizes the GCN4-like motif in the bifactorial endosperm box of prolamin genes. Plant Cell 9:171-184

Alvarez AM, Adachi T, Nakase M, Aoki N, Nakamura R, Matsuda T (1995) Classification of rice allergenic protein cDNAs belonging to the α-amylase/trypsin inhibitor gene family. Biochim Biophys Acta 1251:201-204

Croissant-Sych Y, Okita TW (1996) Identification of positive and negative regulatory cis-elements of the rice glutelin *Gt3* promoter. Plant Sci 116:27-35.

De Pater S, Katagiri F, Kijne F, Chua NH (1994) bZIP proteins bind in a palindromic sequence without an ACGT core located in a seed-specific element of the pea lectin promoter. Plant J 6:133-140.

Feng G, Wen L, Huang J-K, Shorrosh BS, Muthukrishnan S, Reeck GR (1990) Nucleotide sequence of a cloned rice genomic DNA fragment that encodes a 10 kDA prolamin polypeptide. Nucleic Acids Res 18:683

Goldberg RB (1986) Regulation of plant gene expression. Philos Trans R Soc Lond B 314:343-353

Gubler F, Kalla R, Roberts JK, Jacobsen JV (1995) Gibberellin-regulated expression of a myb gene in barley aleurone cells: evidence for Myb transactivation of a high-pI α-amylase gene promoter. Plant Cell 7:1879-1891.

Hibino K, Kizu T, Masumura T, Otsuki K, Kawabata K, Fujii S (1989) Amino acid composition of rice prolamin polypeptides. Agric Biol Chem 53:513-518

Higuchi W, Fukazawa C (1987) A rice glutelin and a soybean glycinin have evolved from a common ancestral gene. Gene 55:245-253

Hirano H, Komatsu S, Nakamura A, Kikuchi F, Kajiwara H, Tsunasawa S, Sakiyama F (1991) Structural homology between semidwarfism-related proteins and glutelin seed protein in rice (*Oryza sativa* L.). Theor Appl Genet 83:153-158

Holdsworth MJ, Munoz-Blanco J, Hammond-Kosack M, Colot V, Schuch W, Bevan MW (1995) The maize transcription factor Opaque-2 activates a wheat glutelin promoter in plant and yeast cells. Plant Mol Biol 29:711-720

Iida S, Amano E, Nishio T (1993) A rice (*Oryza sativa* L.) mutant having a low content of glutelin and a high content of prolamin. Theor Appl Genet 87:374-378

Iida S, Kusaba M, Nishio T (1997) Mutants lacking glutelin subunits in rice: mapping and combination of mutated glutelin genes. Theor Appl Genet 94:177-181

Iida S, Miyahara K, Nishio T (1998) Rice mutant lines lacking α-globulin. Breed Sci 48:45-49.

Izawa T, Foster R, NakajimaM, Shimamoto K, Chua N-H (1994) Plant Cell 6:1277-1287

Izumi H, Adachi T, Fujii N, Matsuda T, Nakamura R, Tanaka K, Urisu A, Kurosawa Y (1992) Nucleotide sequence of a cDNA clone encoding a major allergenic protein in rice seeds. FEBS Lett 302:213-216

Kim SY, Wu R (1990) Multiple protein factors bind to a rice glutelin promoter region. Nucleic Acids Res 18:6845-6852

Kim WT, Li X, Okita TW (1993) Expression of storage protein multigene families in developing rice endosperm. Plant Cell Physiol 34:595-603

Kim WT, Li X, Okita TW (1993) Expression of storage protein multigene families in developing rice endosperm. Plant Cell Physiol 34:593-603

Komatsu S, Hirano H (1992) Rice seed globulin: a protein similar to wheat seed glutenin. Phytochemistry 31:3455-3459

Krishnan HB, Pueppke SG (1993) Nucleotide sequence of an abundant rice seed globulin: homology with the high molecular weight glutelins of wheat, rye and triticale. Biochem Biophys Res Commun193:460-466

Krishnan HB, White JA (1997) Protein body formation and immunocytochemical localization of globulins and glutelins in developing rice (*Oryza sativa* L.) embryos. Crop Sci 37:932-939

Leisy DJ, Hnilo J, Zhao Y, Okita TW (1989) Expression of a rice glutelin promoter in transgenic tobacco. Plant Mol Biol 14:41-50

Li X, Okita TW (1993) Accumulation of prolamines and glutelins during rice seed development: a quantitative evaluation. Plant Cell Physiol 34:385-390

Limas GG, Salinas M, Moneo I, Fischer S, Wittman-Liebold B, Hendez E (1990) Purification and characterization of ten new rice NaCl-soluble proteins: identification of four protein-synthesis inhibitors and two immunoglobulin-binding proteins. Planta 181:1-9

Luthe DS (1983) Storage protein accumulation in developing rice (*Oryza sativa* L.) seeds. Plant Science Lett 32:147-158

Masumura T, Kidzu K, Sugiyama Y, Mitsukawa N, Hibino T, Tanaka K, Fiji S (1989a) Nucleotide sequence of a cDNA encoding a major rice glutelin. Plant Mol Biol 12:723-725

Masumura T, Shibata D, Hibino T, Kato T, Kawabe K, Takeba Go, Tanaka K, Fujii S (1989b) cDNA cloning of an mRNA encoding a r sulfur-rich 10 kD prolamin polypeptide in rice seeds. Plant Mol Biol 12:123-130

Masumura T, Hibino T, Kidzu K, Mitsukawa N, Tanaka K, Fujii S (1990) Cloning and characterization of a cDNA encoding a rice 13 kDa prolamin. Mol Gen Genet 221:1-7

Matsuda T, Nomura R, Sugiyama M, Nakamura R (1991) Immunochemical studies on rice allergenic proteins. Agric Biol Chem 55:509-513

Nakamura A, Komatsu S, Xia BS, Hirano H (1995) Linkage analysis of the gene loci for seed glutelin (Glu-1), semidwarfism (sd-1) and shatering habit (sh-2) in rice (*Oryza sativa* L). Breed Sci 45:185-188

Nakase M, Hoot H, Adachi T, Aoki N, Nakamura R, Masumura T, Tanaka K, Matsuda T (1996a) Cloning of the rice seed α-globulin-encoding gene: sequence similarity of the 5'-flanking region to those of the genes encoding wheat high-molecular-weight glutamin and barley D hordein. Gene 170:223-226

Nakase M, Yamada T, Kira T, Yamaguchi J, Aoki N, Nakamura R, Matsuda T, Adachi T (1996b) The same nuclear proteins bind to the 5'-flanking regions for the rice seed storge protein: 16 kDa alubumin, 13 kDa prolamin and type II glutelin. Plant Mol Biol 32:621-630

Nakase M, Aoki N, Matsuda T, Adach T (1997) Characterization of a novel rice bZIP protein which binds to the α-globulin promoter. Plant Mol Biol 33:513-522

Okita TW, Hwang YS, Hnilo J, Kim WT, Aryan AP, Larson R, Krishnan HB (1989) Structure and expression of rice glutelin multigene family. J Biol Chem 264:12573-12581

Pan SJ, Reeck GR (1988) Isolation and characterization of rice α-globulin. Cereal Chem 65:316-319

Pysh LO, Aukerman MJ, Schmidt RJ (1993) OHP1: a maize basic domain/leucine zipper protein that interacts with Opaque2. Plant Cell 5:227-236

Rogers JC, Rogers SW (1992) Definition and functional implications of gibberellin and abscisic acid *cis*-acting hormone response complexes. Plant Cell 4:1443-1451

Sha S, Sugiyama Y,Mitsukawa N, Masumura, T, Tanaka K (1996) Cloning and sequencing of a rice gene encoding the 13-kD prolamin polypeptide. Biosci Biotechnol Biochem 60:335-337

Shibasaki M, Suzuki S, Nemoto H, Kuroume T (1977) Allergenicity and lymphocyte-stimulating property of rice protein. J Allergy Clin Immunol 64:259-265

Shorrosh BS, Wen L, Zen K-C, Huang J-Kl, Pan JS, Hermodson MA, Tanaka K, Muthukrishnan S, Reek GR (1992) A novel cereal storage protein: molecular genetics of the 19 kDa globulin of rice. Plant Mol Biol 18:151-154

Shyur L-F, Chen C-S (1991) Nucleotide sequence of two rice prolamin cDNA. Nucleic Acids Res 18:6683

Shyur L-F, Wen T-N, Chen C-S (1992) cDNA cloning and gene expression of the major prolamins of rice. Plant Mol Biol 20:323-326

Solano R, Nieto C, Avila J, Canas L, Diaz I, Paz-Ares J (1995) Dual DNA binding specifity of a petal epidermis-specific MYB transcription factor (MYB.Ph3) from *Petunia hybrida*. EMBO J 14:1773-1784

Suzuki H, Futsuhara Y, Takaiwa F, Kurata N (1991) Localization of glutelin gene in rice chromosome by in situ hybridization. Jpn J Genet 66:305-312

Suzuki A, Suzuki T, Tanabe F, Toki S, Washida H, Wu C-Y, Takaiwa F (1997) Cloning and expression of five myb-related genes from rice seed. Gene 198:393-398

Suzuki A, Wu C-Y, Washida H, Takaiwa F (1998) Rice MYB protein OSMYB5 specifically binds to the AACA motif conserved among promoters of genes for storage protein glutelin. Plant Cell Physiol 39:555-559

Takaiwa F, Kikuchi S, Oono K (1986) The structure of rice storage protein glutelin precursor deduced from cDNA. FEBS Lett 206:33-35

Takaiwa F, Kikuchi S, Oono K (1987a) A rice glutelin gene family - a major type of glutelin mRNAs can be divided into two classes. Mol Gen Genet 208:15-22

Takaiwa F, Ebinuma H, Kikuchi S, Oono K (1987b) The complete nucleotide sequence of a rice glutelin gene. FEBS Lett 221:43-47

Takaiwa F, Kikuchi S, Oono K (1989) The complete nucleotide sequence of new type cDNA coding for rice storage protein glutelin. Nucleic Acids Res 17:3289

Takaiwa F, Oono K (1990) Interaction of an immature seed-specific trans-acting factor with the 5' upstream region of a rice glutelin gene. Mol Gen Genet 224:289-293

Takaiwa F, Oono K (1991) Genomic DNA sequences of two new genes for new storage protein glutelin in rice. Jpn J Genet 66:161-171

Takaiwa F, Oono K, Wing D, Kato A (1991a) Sequence of three members and expression of a new major subfamily of glutelin genes from rice. Plant Mol Biol 17:875-885

Takaiwa F, Oono K, Kato A (1991b) Analysis of the 5' flanking region responsible for the endosperm-specific expression of a rice glutelin chimeric gene in transgenic tobacco. Plant Mol Biol 16:49-58

Takaiwa F, Yamanouchi U, Yoshihara H, Washida H, Tanabe F, Kato A, Yamada K (1996) Characterization of common *cis*-regulatory elements responsible for the endosperm-specific expression of members of a rice glutelin multigene family. Plant Mol Biol 30:1207-1221

Tanaka Y, Hayashida S, Hongo M (1975) The relationship of the feces protein particles of rice protein bodies. Agric Biol Chem 39:515-518

Vicente-Carbajosa J, Moose SP, Parsons RL, Schmidt RJ (1997) A maize zinc-finger protein binds the prolamin box in zein gene promoters and interacts with the basic leucine zipper transcriptional activator Opaque2. Proc Natl Acad Sci USA 94:7685-7690

Vicente-Carbojosa J, Onate L, Lara P, Diaz I, Carbonero P (1998) Barley BLZ1: a bZIP transcriptional activator that interacts with endosperm-specific gene promoters. Plant J 13:629-640

Wen T-J, Luthe DS (1985) Biochemical characterization of rice glutelin. Plant Physiol 78:172-177

Wen T-N, Shyer L-F, Su J-C, Chen C-S (1993) Nucleotide sequence of a rice *(Oryza sativa)* storage protein gene, RP6. Plant Physiol 101:1115-1116

Wu C-Y, Adach T, Hatano T, Washida H, Suzuki A, Takaiwa F (1998a) Promoters of rice seed protein genes direct endosperm-spcific expression in transgenic rice. Plant Cell Physiol 39:885-889

Wu C-Y, Suzuki A, Washida H, Takaiwa F (1998b) The GCN4 motif in a rice glutelin gene is essential for endosperm-specific gene expression and is activated by Opaque-2 in transgenic rice plants. Plant J 14:673-683

Yamagata H, Sugimoto T, Tanaka K, Kasai Z (1982) Biosynthesis of storage proteins in developing rice seeds. Plant Physiol 70:1094-1100

Yamagata H, Nomura T, Arai S, Tanaka K, Iwasaki T (1992) Nucleotide squence of a cDNA that encodes a rice prolamin. Biosci Biotechnol Biochem 56:537

Yoshihara T, Takaiwa F (1996) *cis*-regulatory elements responsible for quantitative regulation of the rice seed storage protein glutelin *GluA-3* gene. Plant Cell Physiol 37:107-111

Yoshihara T, Washida H, Takaiwa F (1996a) A 45-bp proximal region containing AACA and GCN4 motif is sufficient to confer endosperm-specific expression of the rice storage protein glutelin gene, *GluA-3.* FEBS Lett 383:213-218

Yoshihara T, Washida H, Takaiwa F (1996b) Assessment of common regulatory regions required for the endosperm-specific expression of rice storage protein glutelin genes by hybrid promoters. Plant Sci 121:63-73

Zhao W-M, Gatehouse JA, Boulter D (1983) The purification and partial amino acid sequence of a polypeptide from the glutelin fraction of rice grains; homology to pea legumin. FEBS Lett 162:96-102

Zhao Y, Leisy DJ, Okita, TW (1994) Tissue-specific expression and temporal regulation of the rice glutelin Gt3 gene are conferred by at least two spatially separated cis-regulatory elements. Plant Mol Biol. 25:429-436

Zhao Y, Okita TW (1995) Interactions of the glutelin Gt3 5' flanking regulatory regions with rice nuclear proteins. Plant Cell Physiol 36:1657-1667

Zheng Z, Kawagoe Y, Xiao S, Okita T, Hau TL, Lin A, Murai N (1993) 5' distal and proximal cis-acting regulatory elements are required for developmental control of a rice seed storage protein glutelin gene. Plant J 4:357-366

Zhou X, Fan Y-L (1993) The endosperm-specfic expression of a rice prolamin chimaeric gene in transgenic tobacco plants. Transgenic Res 2:141-146

11
Molecular biology of host-pathogen interactions in rice diseases

Guo-Liang Wang[1] *and Hei Leung*[2]

[1] Institute of Molecular Agrobiology, 1 Research Link, National University of Singapore, Singapore 117604
[2] Division of Entomology and Plant Pathology, International Rice Research Institute, P.O. Box 933, 1099 Manila, Philippines

1 Introduction

In the classic book on rice diseases, Ou (1985) cited more than 70 diseases caused by fungi, bacteria, viruses, nematodes, and physiological disorders. While most of the recorded diseases are of minor economic importance, several major diseases represent serious constraints to high productivity throughout the history of rice production. In temperate and tropical Asia, where over 90% of rice is grown and consumed, the most serious diseases are bacterial blight, blast, sheath blight, and tungro. Taking into account the frequency of occurrence and disease severity across the region, a conservative estimate of 5% yield loss could be attributed to diseases (Evenson et al. 1996). Because of the large area planted to rice, this translates into billions of dollars in losses. Solving disease problems therefore becomes a prerequisite for sustainable rice production.

Host plant resistance has been the cornerstone of disease control throughout the rice-producing countries. Considering the large acreage of rice planted over time, remarkable success has been achieved through the use of host resistance in combating diseases. This is largely attributed to the steady progress in various aspects of rice disease research that has lead to better strategies for overall disease control. However, some diseases, particularly those caused by highly adaptive pathogens, continue to be major constraints for rice production. Thus, a continuing understanding of the epidemiology, physiology, molecular biology, genetics, and popula-

Key words. avirulence gene, blast, bacterial blight, durable resistance, defense gene, *Oryza sativa*, QTL

tion biology of rice is necessary to sustain rice productivity.

During the past decade, one of the most significant achievements has been the systematic approach taken in developing two rice diseases—bacterial blight and blast—as genetic models for understanding host-pathogen interactions. This, together with the rapid ascent of rice as a model crop plant for molecular studies has accelerated the work on molecular biology of rice diseases (Izawa and Shimamoto 1996). The genetically defined pathosystems make it possible to probe the fascinating biology of host-pathogen coevolution. From a practical standpoint, the delineation of the molecular mechanisms in host response and pathogenesis can lead to better design of control strategies.

Because of the variability of pathogens, a long-standing challenge in disease control has been the identification of durable resistance and the understanding of its basis. Johnson (1983) defined durable resistance as a form of host resistance that provides protection to a crop cultivar for a long time and over a broad geographic area. Durable resistance is therefore a retroactive assessment of the quality of resistance. Although durable resistance has been generally associated with nonspecific and quantitative resistance, there is no clear understanding of the biochemical nature of such resistance. Through genetic analysis, we have progressed from a phenotypic description of durable resistance to its genetic interpretation (Wang et al. 1994). The recent successful cloning of disease resistance genes in rice and other plant species will undoubtedly shed light on the molecular structure and biochemical basis of resistance genes. The future challenge is to build upon this expanding knowledge to begin to understand the molecular basis of durable resistance.

The purpose of this chapter is to summarize the recent advances in understanding the molecular biology of rice diseases and to synthesize a strategy to apply this emerging knowledge toward disease control. We will first discuss the genetic and molecular basis of host resistance in rice. This is followed by a discussion of microbial strategies in attacking the rice plant. We conclude by discussing the opportunities ahead, especially when a vast DNA database of rice and other allied species is available. Our emphasis will be on bacterial blight and blast where most information has been accumulated. Viral diseases are treated elsewhere (see chapter 13) because of the unique strategies in engineering virus resistance.

2 Genetic basis for host-pathogen interactions

Most plants are resistant to microbial attack in the natural environment. This type of resistance is generally referred to as nonhost resistance, which may involve passive mechanical barriers or preformed chemicals that prevent pathogen ingress or establishment. To become a plant pathogen, the microbes need to develop mechanisms to overcome these basic barriers, which are often unique to a plant species or genera. Once a basic compatible interaction is established, an elaborate interplay in genetics and biochemistry between the host and the pathogen begins. This results in a range of host-pathogen interactions that can span from opportunistic parasitism to highly coevolved genetic systems. The level of intimacy of such coevolution further determines the underlying genetic control of resistant phenotypes.

In a highly coevolved host-pathogen relationship, disease resistance can be generally classified into two types based on the effect of resistance to a pathogen and its subsequent reproduction. Qualitative resistance confers strong inhibition against specific pathogen strains, preventing reproduction of the pathogen, while quantitative resistance reduces the extent of pathogen reproduction within the context of a basic compatible interaction. Race-specific and nonrace-specific resistance are used to describe whether a resistance reaction is against a specific pathogen genotype or across diverse genotypes, respectively. Of the major rice diseases, bacterial blight (caused by *Xanthomonas oryzae* pv. *oryzae* or *Xoo*) and blast (caused by *Magnaporthe grisea*) and possibly all viral diseases represent highly coevolved host-pathogen systems. In these interactions, one would expect the evolution of resistance with high levels of specificity. In contrast, specific resistance is less likely to develop in a disease such as sheath blight caused by *Rhizoctonia solani*, an opportunistic parasite with a broad host range.

Based on the observed pattern of genetic interactions between flax (*Linum usitatissmum*) and rust (*Melampsora lini*), Flor (1971) first formulated the gene-for-gene hypothesis which states that resistance is expressed when a plant containing a specific resistance gene recognizes a pathogen with a corresponding avirulence (*avr*) gene. This genetic complementarity has provided the theoretical basis for understanding host-pathogen interactions and eventually led to the molecular cloning of pathogen *avr* genes and their corresponding plant resistance genes. Molecular studies have in most part confirmed the prediction of the gene-for-gene model that can now be further refined in molecular terms. Induction of the plant defense response is initiated by the recognition of specific signal molecules (elicitors) encoded directly or indirectly by *avr* genes. Resistance genes are thought to encode receptors for these elicitors (Staskawicz et al. 1995). This specific recognition activates a cascade of host genes that lead to the hypersensitive response and inhibition of pathogen growth (Keen 1990).

While the gene-for-gene model concerns genes involved in the initial host-pathogen recognition, many genetic factors contributing to incremental resistance do not entail a high level of specificity. Such genes are often referred as quantitative trait loci (QTLs). Although quantitative resistance is generally considered nonrace specific, it is important to note that not all quantitative resistance has been tested for its nonspecificity. It is possible that some quantitative resistance may reflect a reduced expression of qualitative resistance and hence exhibits a pattern of specificity similar to that of qualitative resistance.

Because of their small and quantitative effects, QTLs are difficult to define genetically. The development of molecular tools, e.g., RFLP, RAPD, STS, AFLP, and microsatellites, makes it possible to investigate the inheritance of these complex traits and to locate individual genetic factors associated with these traits. As we are approaching saturation mapping and the beginning of complete genome sequencing in several crop species, QTLs for resistance are likely to be cloned in the near future.

3 Structure and functions of major resistance genes

3.1 Categories of cloned disease resistance genes in plants

Over the past 5 years, the successful cloning of several disease resistance genes has dramatically advanced our understanding of the molecular basis of disease resistance in plants (Staskawicz et al. 1995; Baker et al. 1997). Sequence analysis of the predicted proteins reveals that resistance genes of diverse origin and pathogen specificity share similar structural motifs, indicative of the common surveillance strategy adopted by plant species to detect invading pathogens. Based on the similarity in predicted protein structures, the cloned resistance genes can be grouped into four main classes. The first class includes the tomato gene *Pto*, which confers resistance to *Pseudomonas syringae* pv. *tomato* containing the avirulence gene *avrPto* (Martin et al. 1993; Ronald et al. 1992a). *Pto* encodes a serine/threonine kinase that interacts with the *avrPto* gene product of the pathogen (Scofield et al. 1996; Tang et al. 1996b). Eight resistance genes against diverse pathogens from *Arabidopsis* (*RPS2, RPP5* and *Rpm1*), tobacco (*N*), flax (*L6* and *M*), and tomato (*Prf* and *I2*) constitute the second class of plant resistance genes. All of them contain a nucleotide-binding site (NBS) and leucine-rich repeats (LRR) (Bent et al. 1994; Mindrinos et al. 1994; Whitham et al. 1994; Grant et al. 1995; Lawrence et al. 1995; Salmeron et al. 1996; Parker et al. 1997; Anderson et al. 1997; Ori et al. 1997). Three subclasses can be further divided according to the presence of extra putative domains containing a putative leucine zipper (LZ) (i.e., *RPS2, RPM1,* and *Prf*), the *Drosophila* developmental gene *Toll,* and the mammalian immune response gene encoding the interleukin-1 receptor (TIR) (e.g., *N, M, L6,* and *RPP5*). The *I2* gene only contains NBS and LRR (Ori et al. 1997). The *Cf-9, Cf-2,* and *Cf-4* genes, which mediate resistance to the fungal pathogen *Cladosporium fulvum* in tomato (Jones et al. 1994; Dixon et al. 1996; Thomas et al. 1997), form a third class of resistance genes. These *Cf* genes encode putative membrane-anchored proteins with the LRR motif in the extracellular domain and a short C-terminal tail in the intracellular domain. The fourth class of disease resistance gene is represented by a single member *Xa21*, a rice gene conferring resistance to *Xoo* (Song et al. 1995). Detailed comparison of the function and structure of these resistance genes is provided in a number of excellent reviews (Bent 1996; Hammond-Kosack and Jones 1997; Jones and Jones 1997; Baker et al. 1997).

Compared to the study in dicot species, research on rice disease resistance genes is relatively limited, but results in the last few years on bacterial blight resistance genes have provided new insights into the structure and function of disease resistance genes. We focus here on recent studies on bacterial blight and blast resistance genes in rice.

3.2 Bacterial blight resistance genes in rice

Genetic studies on the inheritance of resistance genes to bacterial blight began about three decades ago in Japan and at the International Rice Research Institute (Mew 1987). Major genes from different sources of resistant donors have been identified. As of now, 25 bacterial blight resistance genes have been found on different chromosomal loci (Kinoshita 1995; Lin et al. 1998; Zhang et al. 1997). Some of these have been widely used in rice breeding programs, and some of them are targets of cloning efforts in several laboratories. Thus far, two resistance genes, *Xa21* and *Xa1*, have been isolated using a map-based cloning strategy (Song et al. 1995; Yoshimura et al. 1998).

3.2.1 *Xa21* encodes a receptor-like kinase

Xa21 was originally identified in the wild species *Oryza longistaminata*, and subsequently transferred to the cultivated rice IR24, generating the near-isogenic line IRBB21 (Ikeda et al. 1990; Khush et al. 1990). In the initial screening, IRBB21 conferred resistance to many known *Xoo* races in India and the Philippines (Khush et al. 1990) and was mapped onto chromosome 11 (Ronald et al. 1992b). Using a map-based cloning strategy, Song et al. (1995) cloned a gene at the *Xa21* locus and discovered that *Xa21* is a member of a small multigene family containing eight members. The deduced amino acid sequence of *Xa21* encodes a receptor kinase-like protein carrying LRRs in the putative extracellular domain, a single pass transmembrane domain, and a serine/threonine kinase intracellular domain. Compared with other cloned plant resistance genes, the structure of *Xa21* is unique in carrying both the receptor domain LRR, presumably for recognition, and the kinase domain for subsequent signal transduction (Ronald 1997).

To determine whether the multi-isolate resistance observed for line IRBB21 was due to a single gene or multiple genes at the *Xa21* locus, isolates from diverse locations were used to inoculate *Xa21*- transgenic plants. The transgenic plants expressing the cloned *Xa21* gene conferred resistance to 29 of 32 diverse isolates collected from eight different countries, but were susceptible to three isolates from Korea. The resistance spectrum of the engineered line was identical to that of the donor line indicating that the single cloned gene is sufficient to confer multi-isolate resistance (Wang et al. 1996).

To investigate the evolution of the *Xa21* gene family, seven *Xa21* gene family members, designated A1, A2, B, C, D, E, and F were cloned and sequenced (Song et al. 1997). They are grouped into two classes based on sequence similarity. The *Xa21* class contains *Xa21*, as well as members D and F. The A2 class contains members A1, A2, C, and E. While nucleotide sequences within each class are nearly identical (98.0% average identity for the members of the *Xa21* class; 95.2%

average identity for the members of the A2 class), sequence identity between members of the two classes is low (63.5% identity between *Xa21* and A2) (Song et al. 1997). A highly conserved (HC) 233-bp sequence was identified among the seven family members such that recombination at the HC region between family members resulted in a precise swapping of the promoter regions. Large sequence duplications were generated by a presumed unequal crossover event in the intergenic regions. Furthermore, 15 transposon-like elements were found within the gene cluster and 2 of them were located in the open reading frames (ORFs) of members D and E, suggesting that these elements may play a role in the diversification of the *Xa21* family members (Song et al. 1997). Based on these results, it is proposed that duplication, recombination, and transposition contribute to the amplification and diversification of the *Xa21* gene family (Song et al. 1997).

Recent work on the *M* locus of flax and *Cf* genes of tomato demonstrated that alterations in the LRR domain play a significant role in the evolution of resistance genes and production of new recognition specificities (Anderson et al. 1997; Thomas et al. 1997). Within the LRR domain, amino acid differences distinguishing *Cf-4* and *Cf-9* are confined to their N-termini, delimiting a region that determines the recognition specificity of ligand binding. The majority of these differences are in residues interstitial to those of the LRR consensus motif (Thomas et al. 1997). Similarly, distinct amino acid differences are found in the LRRs of the *Xa21* gene family members, suggesting that the LRR region is responsible for different specificities. To test this hypothesis, Wang et al. (1998a) investigated the inheritance of resistant patterns in transgenic rice plants carrying family members A1, A2, C, D, E, and F to eight *Xoo* races of the Philippines. It was found that members A1, A2, C, E, and F conferred no observable resistance in transgenic plants whereas the *Xa21* class member D, designated *Xa21D,* conferred the same resistance spectrum as *Xa21* (Wang et al. 1998a). However, the resistance level in the *Xa21D* transgenic plants was intermediate to that observed for *Xa21*. The presumed ORF of *Xa21D* encodes a receptor-like molecule lacking the transmembrane and kinase domains due to insertion of a retrotransposon element at the 3'-region of the LRR domain, supporting the hypothesis that the *Xa21D* LRR domain is involved in pathogen recognition.

Nucleotide substitution patterns in members of the *Xa21* gene family were analyzed to gain insight into the function and evolution of particular coding domains. The type of selection acting on a gene family can be inferred through a comparison of synonymous and nonsynonymous substitution rates per synonymous/nonsynonymous site (Hughes and Nei 1988). In the absence of selection pressure, the nonsynonymous/synonymous ratio is predicted to be 1. The nonsynonymous/synonymous ratio in the *Cf-4* and *Cf-9* LRR region was found to be greater than 1, indicating positive selection for diversification of the gene family (Thomas et al. 1997). Comparison of nucleotide substitutions in the LRR coding regions of *Xa21* and *Xa21D* revealed that, although *Xa21* and *Xa21D* share 99.1% sequence identity, nonsynonymous substitutions occur significantly more frequently than synonymous substitutions in the LRR. Thus, the results further support the hypothesis that LRRs play a role in ligand binding and that the LRR domain is subject to adaptive evolution (Wang et al. 1998a).

3.2.2 *Xa1* gene encodes NBS and LRR motifs and its expression is inducible

Xa1 confers a high level of specific resistance to race 1 strains of *Xoo* in Japan. The gene was identified from the cultivar Kogyoku and mapped onto chromosome 4. *Xa1* has been extensively used in the Japanese breeding programs because *Xoo* race 1 is the predominant race in Japan. Using a map-based strategy, a 340-kb YAC clone spanning the *Xa1* locus was isolated (Yoshimura et al. 1996). Using this YAC clone as a probe, 33 positive clones were identified from a cDNA library constructed from the *Xa1* donor line. Among them, seven candidate cDNAs were fully sequenced. One of these cDNAs, c42, has significant homology to NBS/LRR-type resistance genes. A genomic cosmid clone was subsequently identified and used to transform the susceptible plants. The complementation test confirmed that the *Xa1* gene was confined to a 35-kb region of the cosmid clone. The specificity of resistance shown by the transgenic plants was identical with that of the *Xa1* donor line, IRBB1 (Yoshimura et al. 1998).

The derived 1802-amino-acid sequence for the *Xa1* gene contains several regions with similarity to the deduced polypeptide domains of other resistance genes (e.g., *RPS2*, *RPM1*, *N*, and *L6*); it has two motifs of NBS and an LRR domain. The *Xa1* LRR is comprised of six almost perfect repeats, each 93 amino acids long. In each repeat unit, there are six occurrences of a consensus sequence [LXXLXL/IXXN/CXX]. Of the six direct repeat units, the first to the fifth are almost identical not only at the amino acid level (93% homology), but also at the nucleotide sequence level (98%). The sixth unit is less similar to the other five units with 62.7% similarity to the amino acid sequence and 73.8% at the nucleotide sequence level (Yoshimura et al. 1998). Finally, in contrast to the constitutive expression seen in all cloned resistance genes, *Xa1* expression is induced upon inoculation with *Xoo* or by wounding. It is not clear whether the inducibility of *Xa1* is related to the enhanced resistance observed phenotypically.

3.3 Blast resistance genes

Genetic studies of resistance to rice blast began when Goto established the differential system for races of the blast fungus in Japan in the early 1960s (Ou 1985). Since then, the inheritance of resistance has been extensively studied, and more than 25 major resistance genes have now been identified (Kinoshita 1995; Naqvi and Chattoo 1996). Chromosomal positions have been reported for at least 18 loci conferring qualitative resistance (Kinoshita 1995; Yu et al. 1991; Mackill and Bonman 1992; Zhu et al. 1993; Wang et al. 1994; Naqvi and Chattoo 1996). Thus far, successful cloning of blast resistance genes has not been reported; however, a number of genes (*Pib*, *Pita*, *Pita*^{2}, *Pi5*, and *Pi7*) are being actively pursued by several laboratories (Tsunoda et al. 1998; Wang et al. 1998b; Chen et al. 1998).

3.3.1 Map-based cloning of blast resistance genes

Pib was introgressed independently from two Indonesian and two Malaysian cultivars into various *japonica* cultivars (Yokoo et al. 1978). This gene is highly resistant to most blast pathogen races in Japan. The gene was fine-mapped near the terminal region of chromosome 2 using RFLP and RAPD markers (Miyamoto et al. 1996). *Pib* co-segregated with RAPD marker b-1 and RFLP marker RZ123, and it was bracketed between flanking RFLP markers at 0.5 cM on the centromeric side and 1.9 cM on the telomeric side. A BAC library was constructed from the *Pib* donor line, and a contig of 400 kb spanning 2.4 cM around the gene was established. Recently, *Pib* was mapped onto a single BAC clone (180 kb) based upon F_2 analysis (Kawasaki et al. 1997). Using this BAC clone as a probe, 17 cDNAs were recovered. A cDNA clone with a protein kinase motif that differed between the resistance and susceptible lines was identified. A subgenomic library spanning the *Pib* region was constructed (average, 30 kb/clone), and the complementation tests are in progress (Tsunoda et al. 1998; Jwa et al. 1998). However, at the same locus, Wang et al. (1998b) found partial DNA sequences within the BAC contigs that have high homology to NBS identified in *RPM1* and *RPS2* of *Arabidopsis*. With the rapid progress being made by different laboratories, the true nature of the *Pib* locus should be known very shortly.

Pita^2 and *Pita,* two closely linked genes or different alleles at the same locus, have been mapped within the centromeric region of chromosome 12 (Wu et al. 1996; Nakamura et al. 1997). Wu et al. (1996) constructed a high-resolution genetic and physical map using identified BACs around the *Pita* region with three molecular markers within 0.5 cM of the gene. Using a similar approach, Nakamura et al. (1997) have constructed a contig map spanning a 300-kb region between markers within 0.3 cM from *Pita*^{2}. However, this centromeric region has a low recombination rate and a high frequency of repetitive sequences, posing difficulties in approaching the gene by chromosome walking. Recently, Rybka et al. (1997) defined the *Pita*^{2} with flanking markers at 0.3 cM on both sides. *Pita* was mapped to a position overlapping *Pita*^{2} by graphical genotype analysis of a near-isogenic line with a very narrow introgressed region. The result suggested that the two functionally related genes are allelic or at least very closely linked and that they may be derived from a common ancestral gene (Rybka et al. 1997).

3.3.2 Transposon tagging

In addition to the map-based cloning approach, Shimamoto and co-workers have been applying transposon tagging in an attempt to isolate blast resistance genes. Since an endogenous and well-characterized transposon system is not yet available in rice, heterologous transposon tagging systems have been engineered into rice (Izawa et al. 1991). The details of designing a *Ac-Ds* transposon tagging system are described by K. Shimamoto (see chapter 4). With respect to tagging blast resistance genes, some mutant lines with a loss of *Piz*[5] resistance have been identified. DNA fragments flanking *Ds* elements have been isolated by inverse PCR and genetic and molecular analyses of mutant lines are in progress.

Insertional mutagenesis has been highly successful in the cloning of other resistance genes (*Cf-9*, *N*). Thus, we expect that as soon as an efficient gene tagging system in rice is achieved, this approach will be adopted widely to clone not only disease resistance genes but also other agronomically important genes.

4 Disease response genes and quantitative resistance

The successful cloning of major genes provides the starting point to delineate the downstream genes involved in host-pathogen interactions and defense responses. Although there is emerging molecular and biochemical evidence that at least some resistance gene products physically interact with *avr* genes from the pathogen (Tang et al. 1996b; Scofield et al. 1996), how the defense pathway is activated in infected plants after initial recognition is still mostly unknown. Lessons learned from other pathosystems may provide useful hints in discovering new genes controlling the defense pathway in rice.

4.1 Genetic approach to define genes downstream from initial recognition

Several genes involved in events downstream from initial recognition have been cloned in *Arabidopsis*, barley, and tomato. For example, *Prf*, a gene required for both *Pto*-mediated resistance and fenthion sensitivity in tomato, was isolated by a mutational approach (Salmeron et al. 1996). Using the *avrPto* avirulence gene product as a bait in a yeast two-hybrid system, a protein kinase gene (*Pti1*, for *Pto*-interacting), and three genes encoding novel proteins with characteristics of transcriptional factors (*Pti4*, *Pti5*, and *Pti6*) were identified (Zhou et al. 1995; Zhou et al. 1997). The *Pti1* kinase is involved in the hypersensitive response, and *Pti4*, *Pti5*, and *Pti6* are believed to bind to *cis*-elements that are conserved among pathogenesis-related (PR) genes. Thus, analysis of downstream events reveals a wealth of genetic factors regulating the expression of resistance.

Another set of mutants has been isolated from *Arabidopsis* based on their inability to initiate systemic acquired resistance (SAR). Mutants with lesions in two genes *NPR1* and *NIM1* in *Arabidopsis* are defective in their response to salicyclic acid, and hence become more susceptible to pathogens. *NPR1* encodes a novel protein containing ankyrin repeats that are presumably involved in protein–protein interactions (Cao et al. 1997). Transformation of the *NPR1* gene into *npr1* mutants not only complemented the mutation, but also rendered the transgenic plants more resistant to infection by *Pseudomonas syringae* in the absence of SAR induction.

The *NIM1* gene cloned by Ryals et al. (1997) has the same sequence and is therefore considered allelic to *NPR1*. Using a differential display technique, a rice gene with enhanced expression in blast-resistant plants has been isolated (He et al. 1997). This gene, designated *RANK1* (for rice ankyrin repeats), encodes a protein containing ankyrin repeats with significant homology to the NPR1/NIM1 protein and the mammalian inhibitor *I-kB*. RT-PCR and northern analysis showed that the *RANK1* message rapidly accumulated in resistant plants as early as 4 h after inoculation with the blast fungus. Since the *NPR1/NIM1* gene product is involved in the signal transduction pathway leading to SAR in *Arabidopsis*, it is envisaged that *RANK1* may play a similar role in mediating defense to blast infection.

Century et al. (1997) isolated an *Arabidopsis* gene *NDR1* that is essential for conferring resistance against bacterial and fungal pathogens (*Pseudomonas syringae* pv. *tomato* and *Peronospora parasitica*). This gene is required for the expression of resistance triggered by race-specific host resistance genes. Mutation at *NDR1* results in susceptibility to multiple races of both fungal and bacterial pathogens. The gene encodes a predicted product of 219 amino acids with two transmembrane domains. It is postulated that *NDR1* is downstream from the avirulence-resistance gene interactions and may function in the transmission of signals generated from initial recognition.

Hypersensitive response (HR) triggered by pathogen invasion is the most common feature associated with active host resistance (Dangl et al. 1996). Because the HR appears to be so intimately involved in resistance, considerable efforts have been made to define the events that trigger HR and to define the steps culminating in cell death. Using a mutational approach, several mutants have been identified in *Arabidopsis* that exhibit a disease or HR-like phenotype in the absence of pathogens (Dietrich et al. 1994). These mutants expressed histochemical and molecular markers associated with disease resistance response and showed enhanced resistance to pathogens when lesions were present. It appears that these cell-death phenotypes can trigger pathogen-nonspecific resistance resembling SAR (Dietrich et al. 1994). One of the identified mutants, *lsd1*, was cloned by a map-based cloning strategy (Dietrich et al. 1997). The predicted LSD1 protein contains three zinc finger domains. It is possible that LSD1 regulates transcription via either repression of a pro-death pathway or activation of an "anti-death" pathway, in response to signals generated from cells undergoing pathogen-induced hypersensitive cell death.

The powdery mildew resistance gene *mlo* in barley also exhibits a spontaneous cell-death phenotype under pathogen-free conditions (Buschges et al. 1997). This recessive gene confers broad-spectrum resistance to almost all known isolates of *Erysiphe graminis* f. sp. *hordei*. The cloned *Mlo* gene encodes a novel 60-kDa protein, which is predicted to be membrane-anchored by at least six membrane-spanning helices. One model suggests that the *Mlo* wild-type sequence is a negative regulator of the cell-death or defense response; thus, defects in this protein may lead to a deregulation of the defense response, resulting in broad-spectrum resistance. Buschges et al. (1997) also reported that highly homologous sequences to *Mlo* have been found in rice and *Arabidopsis*, suggesting that the MLO protein may be a common regulatory element in both monocots and dicots. A better understanding of genes involved in lesion formation may therefore open new opportunities for enhancing plant defense functions.

In rice, Takahashi et al. (1997) isolated 170 mutant lines that spontaneously formed necrotic lesions on leaves in the absence of pathogen. Among them, three mutants exhibited a nonrace-specific resistance response against the blast pathogen. Northern blot analysis showed that two defense-related genes, *LOX* (lipoxygenase gene) and *PBZ1* (probenazole-inducible gene), are constitutively expressed in these mutants. Enhanced expression of these two genes is correlated with lesion formation, indicating that these mutations may be involved in the signal transduction pathway. We have analyzed nine spotted leaf (*spl*) mutants (kindly provided by Dr. G.S. Khush at IRRI) and found at least two mutants expressing enhanced resistance to blast infection. The lesion formation in these two mutants can be triggered by blast inoculation (Wang et al., unpublished data, 1998). Further characterization of these two mutants will shed light on how the defense responses are activated and the relationship between cell death and blast resistance in rice.

4.2 Induction of host defense response genes

In addition to the genetic approach, a direct assessment of the biochemical and physiological changes during disease development can be used to identify genes involved in the disease response pathway. Enhanced expression of genes during infection could be interpreted as an involvement of these genes in host defense. This approach has been used in a number of studies to determine what genes are turned on during *Xoo* and blast infection.

Zhu and Lamb (1991) isolated the first defense-related gene encoding a rice basic chitinase induced in suspension culture cells by a fungal cell wall elicitor. As of now, ten family members of the rice class I chitinases have been cloned (Xu et al. 1996). There is increasing evidence that class I chitinases are important in defending the plant against fungal pathogens. Enhanced chitinase levels in transgenic rice plants appears to be responsible for the observed resistance against blast infection (Zhu et al. 1994). Exogenous chitinase genes and other constructs encoding PR proteins have been introduced into rice to enhance sheath blight resistance (Lin et al. 1995; S. Datta, IRRI, personal communication, 1998). Promising levels of resistance have been achieved based on greenhouse tests; however, due to the strong environmental influence on sheath blight disease, accurate assessment of the resistance awaits results from field experiments (T.W. Mew, personal communication, 1998). Since different chitinase genes hydrolyze different substrates, combining a variety of chitinase genes with other antimicrobial genes may have potential for broad-spectrum protection against pathogens.

For bacterial blight resistance, the defense response in rice is characterized by an increase in peroxidase activities, the deposition of lignin into the plant cell wall, host cell death, and limitation of bacterial multiplication (Reimers et al. 1992).

Two peroxidase genes, *POX8.1* and *POX22.3*, are predominantly expressed during resistant interactions (Chittoor et al. 1997). Mapping analysis showed that *POX22.3*, *POX8.1*, and another wound-inducible peroxidase, *POX5.1*, are located near the telomeric region of chromosome 7 (Chittoor et al. 1998). To determine the role of peroxidases in the defense to bacterial blight pathogen, transgenic plants overexpressing sense and antisense constructs of the pathogen-inducible peroxidases are currently being developed (J. Leach, personal communication, 1998).

In mammalian systems, phospholipase D (PLD) is a critical component in cellular signal transduction (Divecha and Irvine 1995). Recent evidence indicates that PLD may also play a role in plant signaling or metabolism. Young et al. (1996) investigated the location and distribution of PLD during the defense response in rice. It was found that PLD clustered on the plasma membranes at regions where bacteria are immediately adjacent to the cell wall. This phenomenon only occurs in the resistant interactions (Young et al. 1996). PLD may be involved in the regulation of increased localized secretion of defense response compounds, and may therefore regulate the magnitude of the defense response.

As more defense genes are being discovered, a number of practical questions arise. What is the contribution of these defense genes at the phenotypic level? How many genes are involved in the process, and what are their effects on pathogen growth? Are there sufficient functional variations in these defense genes to allow for effective selection? Overexpression and underexpression (antisense) approaches in transgenic plants may provide information on gene function. However, gene expression studies may be complicated by the fact that many defense genes are functionally redundant. Nonetheless, the rapid identification of defense-related genes provides an expanding pool of candidate genes that can be genetically correlated to phenotypic effects, a topic explored further below.

5 Pathogenesis: microbial strategies for attacking plants

Because disease resistance is a dynamic process defined by the genotypes of both the host and the pathogen, a parallel analysis of the genetic and biochemical determinants of pathogenesis is crucial to fully understand host resistance. The analysis of pathogenesis serves two practical purposes. First, identification of essential genes for pathogenesis may provide novel targets for intervention. This information is particularly relevant for the biorational design of chemical or biological agents specifically against the pathogen without having negative effects on nontarget organisms. Second, pathogenic factors that "cross-talk" with the host are instruments to detect resistance genes. Over the past several years, a number of pathogen genes involved in recognition and subsequent parasitic growth in planta have been cloned and characterized. We describe here the nature of some of these genes and the insight gained toward understanding disease resistance.

5.1 Genes involved in penetration and entry of host tissues

To establish parasitic growth in the host tissue, a plant pathogen must first overcome the physical barrier of the cell wall. This can be achieved through mechanical penetration, enzymatic dissolution of the cell wall, or entry through natural openings. In the case of bacterial blight, the pathogen enters primarily through hydathodes and wounds (Huang and de Cleene 1989). For the blast fungus, cutinase was originally thought to be involved in weakening the cell wall, but gene disruption experiments suggested that the inactivated cutinase was not essential for penetration (Sweigard et al. 1992). Instead, mechanical penetration is the most important, if not the only, means of the infection process.

Central to the penetration process of the blast fungus is the formation of an appressorium, a specialized infection structure formed by a number of phytopathogenic fungi (Mendgen et al. 1996). After a conidium lands on the plant surface, a germ tube is produced. Within 8-12 h, the tip of the germ swells and forms a hook. The swollen germ tube attains a spherical structure and is firmly attached onto the plant surface. Within the fully developed appressorium, enormous turgor pressure is built up enabling the fungus to puncture the host epidermis with a penetration peg (Howard et al. 1991). Proper melanization of appressorium is essential to maintain this high turgor pressure. Genetic lesions in the melanin biosynthetic pathway lead to loss of penetration and pathogenicity (Chumley and Valent 1990).

Because of the importance of appressorial formation in penetration, studies on blast infection have focused on genetic and biochemical signals that affect the development and function of appressoria. Lee and Dean (1993) found that hydrophobic surface (such as the leaf epidermis) is inducive to appressorium formation and that cAMP serves to induce appressorium formation in the absence of physical stimuli. Mutation in a cAMP-dependent protein kinase (*CPKA*) gene was originally thought to abolish appressorium formation and pathogenicity (Mitchell and Dean 1995); however, recent work by Xu et al. (1997) showed that *cpkA* mutants still retain the ability to form appressoria after 24-h incubation. The mutant appressoria, however, are smaller relative to the wild type and defective in penetration. To add support to the importance of the cAMP pathway in appressorial development, Choi and Dean (1997) cloned the adenylate cyclase gene (*MAC1*), which catalyzes the conversion of ATP to cAMP. Disruption of *MAC1* resulted in not only a loss of appressorium formation but also reduced somatic growth, sporulation, and spore germination.

In addition to cAMP, other signaling pathways are also involved in appressorium formation. Xu and Hamer (1996) showed that a MAP kinase gene, designated *PMK1*, is essential for appressorium formation and invasive growth in planta. The gene is homologous to the MAP kinases *FUS3* and *KSS1* in *Saccharomyces cerevisiae*. *Pmk* mutants cannot form appressoria but still retain the ability to respond to exogenous cAMP by forming a swollen and hooked tip cell. This suggests that the *PMK*-mediated MAP kinase signaling pathway is downstream from the cAMP pathway (Xu and Hamer 1996). Shi and Leung (1995) identified a series of spore morphogenetic mutants that express a full range of defects in appressorium formation. For example, mutations at the *CON7* and *CON1* produce not only abnormal spore shape but also completely abolish formation of appressoria. Thus, appressorium formation seems to be at the "intersection" of multiple pathways such that mutations in any one pathway would lead to reduced or total loss of appressorium formation.

5.2 Genes involved in invasive growth in planta

Relative to the process of penetration and early infection, much less is known about the genetic control of in planta colonization. However, some clues are emerging from several recent studies. In *M. grisea*, the *PMK1* and *CON7* genes are particularly interesting as they represent new classes of genes not conventionally considered pathogenicity determinants yet are essential for pathogenesis. Both *PMK1* and *CON7* are dispensable for saprophytic growth but absolutely required for infectious growth in rice. These mutants cannot colonize host tissue even though the penetration step is bypassed by wound inoculation. Xu and Hamer (1996) suggest that the *PMK1*-encoded MAP kinase may activate appressorial-specific genes that lead to new cell wall biosynthesis and buildup of turgor pressure. Upon entry into the host cell, *PMK1* may be required for signal activation of a new set of genes required for parasitic growth.

Different from *PKM1*, the *CON7* gene is a putative transcriptional regulator that appears to control the shift from saprophytic to parasitic growth of the pathogen (Shi 1995). Mutation at *CON7* has strong pleiotropic effects on spore morphology, appressorium formation, and pathogenesis. The predicted product of *CON7* is a 441-amino-acid protein containing a single zinc finger motif with a C-terminal putative nuclear localization signal. Interestingly, the predicted CON7 protein has strong functional and structural similarity with StuA, a DNA-binding protein involved in the regulation of spore formation in *Asperigillus nidulans*. *CON7* may transcriptionally control multiple target genes essential for invasive growth in planta. Another developmental gene essential for in planta growth is *CON1*, which was first discovered as a mutation that causes a drastic morphological change from the pyriform to needle-shaped conidia (Shi and Leung 1994). Preliminary sequence analysis of the *CON1* locus showed no sequence similarity in the database (Z. Shi and H. Leung, unpublished data, 1998). It is not known whether *CON1* is a regulatory gene or simply codes for a structural component determining cell morphology. Liu and Dean (1997) showed that disruption of a G-protein α-subunit gene (*magB*) reduced the infection and colonizing ability of the fungus, but the defects seem to be generalized to both the vegetative and parasitic phases of the pathogen. Together, these studies indicate that genes involved in developmental regulation are essential for regulating parasitic growth. Discovery of such genetic factors may open up new avenues for intervention.

Relatively little is known about the genes required for bacterial growth after the

Xoo enters the plant tissues. In *Xanthomonas* species, genes responsible for the biosynthesis of lipopolysaccharide (LPS) and extracellular polysaccharides (EPS) are known to play a role in the survival of the pathogen in planta (Kingsley et al. 1993; Dow et al. 1995). In *X. campestris pv. campestris*, the black rot pathogen of cabbage, mutations in genes producing LPS and EPS reduce bacterial numbers by 100- to 1000 fold relative to the wild type (Newman et al. 1994). Tang et al. (1996a) identified in *Xoo* a gene homologous to the gene cluster *rpfC*, which regulates synthesis of extracellular enzymes and exopolysaccharides in *X. campestris pv. campestris*. Mutation in this gene reduced exopolysaccharides production as well as disease symptoms on rice. These genes apparently are important in establishing a favorable environment for bacterial growth.

5.3 Genes involved in recognition

An avirulence gene is genetically defined as a pathogen gene that produces, directly or indirectly, a product recognized by the host to trigger the resistance response. Avirulence genes can be highly diverse in structure or function. Changes in avirulence genes are directly related to the varietal "breakdown" commonly seen in crop cultivars. Therefore, avirulence genes can be perceived as instruments through which diverse disease resistance genes can be identified. An understanding of the molecular basis of avirulence genes may also be a key to the identification of useful and durable forms of host resistance.

5.3.1 Avirulence genes in *Xoo*

In general, avirulence genes from multiple genera and species of bacteria do not share sequence similarity (Leach and White 1996). The unique case where avirulence genes share an extensive homologous region is found in *Xanthomonas* spp. (Vivian and Gibbon 1997). The three avirulence genes (*avrXa10*, *avrXa7*, and *avrxa5*) identified in *Xoo* belong to a gene family typified by the *avrBs3* gene from *X. campestris pv. vesicatoria* (Hopkins et al. 1992). In *Xoo*, these avirulence genes are present in 12 to 14 copies based on Southern hybridization, but not all of these copies are necessarily active (Leach and White 1996).

The *avrBs3*-type gene family is widespread in several *Xanthomonas* pathovars and has two important structural features. First, the gene includes a central domain containing a series of 102-bp direct repeats, each coding for a 34-amino-acid repeat domain. The number of repeats varies among different *avrBs3* homologs. For example, *avrXa10* of *Xoo* contains 15.5 copies of the repeats whereas *avrXa7* contains 25.5 copies (Hopkins et al. 1992). It has been shown in *avrBs3* that the repetitive region has a physical role for specificity. Deletion analyses of the central domain suggest that the organization (i.e., number and arrangement) of the repeats are responsible for specificity. New specificity can be created through reorganization of these repetitive motifs (Herbers et al. 1992; Bonas et al. 1993).

The second important feature of the *avr* gene is the carboxy terminal containing one to three regions that resemble nuclear localization signals (NLS). Yang and Gabriel (1995) made a fusion between the NLS of *pthA* from *X. ctiri* with the GUS reporter gene and showed that the *pthA*-GUS fusion protein moved into onion nuclei. Mutation of the NLS resulted in a loss of nuclear targeting function. Since then, additional nuclear localization experiments with other *Xanthomonas avr* genes have confirmed that the NLS is responsible for nuclear targeting in eukaryotic cells (van der Ackerveken et al. 1996). It is important, however, to note that some *avr* genes in *Xanthomonas* do not have NLS but still retain the avirulence gene function (Leach and White 1996). Whether NLS is dispensable in some cases or being compensated by other functional domains of the gene remains to be investigated.

5.3.2 Avirulence genes in *M. grisea*

Relative to bacterial pathogens, only a few avirulence genes in phytopathogenic fungi have been cloned and characterized. Avirulence genes of *C. fulvum* code for preproteins that are processed by a combination of fungal and plant proteases into short peptides (e.g., 86 and 28 aa for *Avr4* and *Avr9*, respectively; Joosten et al. 1997; De Wit 1997). These peptides serve directly as elicitors in the apoplastic fluid of the tomato plant (De Wit 1995). The avirulence gene *NIP1* of *R. secalis* also codes for an extracellular peptide of 60 amino acids that serves as an elicitor against the specific *Rrs1* resistance genes in barley (Rohe et al. 1995).

In *M. grisea*, two types of specificity genes have been cloned using a map-based cloning strategy (Valent and Chumley 1994). A cultivar specificity gene, *AVR2-YAMO*, controls the incompatible interaction with rice cultivar Yashiro-mochi. The predicted protein of *AVR2-YAMO* consists of 223 amino acids and contains a zinc-protease motif. It is postulated that AVR2-YAMO is either the elicitor itself, or that its protease function may produce the final elicitor. The second class of specificity gene *PWL* (*p*athogenicity on *w*eeping *l*ovegrass) controls recognition on the host weeping lovegrass (*Eragrostis curvula*) (Sweigard et al. 1995). A series of homologs of *PWL* genes have been cloned from *M. grisea* isolates which parasitize a variety of grass hosts (Kang et al. 1995). The "prototype" *PWL2* gene encodes a glycine-rich, hydrophilic protein that contains motifs of eukaryotic secretory signal sequences. Changes from avirulence to virulence in *PWL* can result from single-base mutations in the coding sequence or alteration of the promoter region, leading to loss of expression (Kang et al. 1995). Recently, Leong et al. (1997) reported the cloning of another avirulence gene conferring specificity on cultivar Co39; sequence analysis showed no similarity to other known avirulence genes. *AvrCo39* has three open reading frames; site-directed mutagenesis showed that ORF1 and ORF3 are required for avirulence function.

5.3.3 Delivery system of avirulence gene products

For almost all avirulence genes in bacteria, the sequence structure has not provided any hint as to the site of interaction between the avirulence gene product and the host resistance gene. Evidence provided by immunoelectron microscopy suggests that the AVRXa10 protein is present in the cytoplasm of *Xoo* (Young et al. 1994). With the elucidation of the interaction between *hrp* (hypersensitive response and pathogenicity) gene cluster and *avr* genes, a model emerges that could account for the trafficking of AVR proteins into plant cells. Excellent reviews are available on the functional relationship between *hrp* gene and the *avr* genes (He 1996; Alfano and Collmer, 1997).

The *hrp* gene clusters, of approximately 25-30 kb, have been found in the four major genera of phytopathogenic bacteria: *Pseudomonas*, *Erwinia*, *Ralstonia*, and *Xanthomonas*. Sequence analysis of the *hrp* genes showed homologies to genes responsible for the secretion of virulence proteins in *Yersinia* species. Some of the genes in the *hrp* cluster encode the "type III" protein secretion apparatus. Roine et al. (1997) showed that the *hrpA* gene of *P. syringae pv. tomato* encodes a novel pili protein which can be formed on the cell surface when the bacteria are grown on pili-inducing medium. Thus, there is physical evidence that the secretion apparatus is encoded by the *hrp* gene cluster in *P. syringae* pv. *tomato* and that these *Hrp* pili are responsible for the delivery of avirulence gene products or other pathogenicity factors into the host.

For some *Xanthomonas avr* genes, the delivery of AVR protein does not end in the cytoplasm but in the nucleus. As explained earlier, NLSs are present at the C-terminus of several *avrBs3*-type genes. Van der Ackerveken et al. (1996) introduced AvrBs3 into pepper plants carrying resistance gene *Bs3*, and showed that AvrBs3 required at least two of the three NLSs to elicit the hypersensitive response in the plant. The fact that NLS from SV40 T-antigen can be used to substitute the NLS in AvrBs3 and restore elicitor function suggest that the NLS is solely needed for nuclear targeting (van der Ackerveken et al. 1996). The *avrXa10* gene of *Xoo* has three putative NLSs in the C-terminus. Work is in progress to define the functions of these NLSs (J. Bai and J. Leach, personal communication, 1998). An interesting question is whether quantitative variation in nuclear targeting may lead to a different level of resistance.

Delivery of fungal avirulence genes is more straightforward since most of the gene products known so far appear to function directly as extracellular elicitors. For example, the *Avr4* and *Avr9* products of *C. fulvum* have secretion signal sequences and are known to be active in the apoplastic fluid of tomato plants. The *PWL* gene families of *M. grisea* have putative signal peptides but the actual function of these sequences has not been proven. On the other hand, the product of *AVR2-YAMO* does not have any signal peptide that would suggest an extracellular function. In this case, the avirulence gene may function to produce other elicitors, as in the case of *avrD* gene of *Pseudomonas syringae* pv *tomato* (Keen et al. 1990).

5.3.4 Functions of avirulence genes

It has long been an enigma that avirulence genes are maintained in the population despite the seeming disadvantage of producing targets recognized by resistance genes. While the sequence structure reveals little about their function, marker-exchange experiments have demonstrated that some avirulence genes indeed code for functions that are important for either parasitic or saprophytic competitiveness of the pathogen. The first indication that the AVR gene product may carry an additional function besides being a recognition factor comes from *Xanthomonas campestris* pv. *vesicatoria* in which insertional inactivation of the *avrBs2* gene (matching the pepper disease resistance *Bs2* gene) resulted in reduced pathogenic fitness of the mutant (Kearney and Staskawicz 1990). Nearly all field strains carry the a*vrBs2* gene, suggesting that the avirulence gene may play an essential role for the pathogen's survival in nature. Sword et al. (1996) showed that *avrBs2*, which is not related to the *avrBs3* family or *avrBs1*, codes for a predicted protein with similarity to *Agrobacterium tumefaciens* agrocinopine synthase and *Escherichia coli UgpQ* enzyme. Both enzymes are involved in the synthesis and hydrolysis of phosphodiester linkages. It is postulated that the *avrBs2*-encoded enzyme modifies plant glucan to promote intercellular adaptation of the pathogen. While the modified glucan may provide a more favorable environment for the pathogen, it may also serve as the signal molecule triggering resistance response. Hence, *avrBs2* has the dual phenotype of conferring avirulence and pathogenic fitness.

Lorang et al. (1994) demonstrated that *avrA* and *avrE in Pseudomonas syringae* pv. *tomato* are required together to confer the full pathogenicity function. Yang et al. (1996) used marker eviction to demonstrate in *Xanthomonas campestris* pv. *malvacearum* that the avirulence genes contribute additively to pathogenicity (water soaking) on cotton. Elimination of seven members of the ten avirulence genes leads to loss of symptom development and more than 1000-fold reduction in the release of bacteria onto the leaf surface. They concluded that multiple copies of the *avr* gene family are needed for pathogenicity in cotton although they are not required for in planta growth.

The biological functions of most fungal avirulence genes are not known. One exception may be *NIP1*, which codes for a necrosis-inducing protein causing cell death, hence providing the food base for the necrotrophic growth of *R. secales*. In *C. fulvum*, inactivation of *Avr9* and *Avr4* does not appear to affect saprophytic or parasitic characteristics. In *M. grisea*, a detailed assessment of the functions of the *AVR2-YAMO* and *PWL* gene products has not been made (Valent and Chumley 1994; Sweigard et al. 1995). In the initial detection of *PWL2* mutations in the laboratory, both single-base and deletion mutants were found but they showed no apparent negative defects other than a change from avirulence to virulence on weeping lovegrass. Based on Southern blot analysis, *PWL2* homologs are detected in nearly all field isolates even though these isolates have different host specificity (Kang et al. 1995). This is taken as evidence that these genes are probably needed for pathogen survival in the field.

Leung et al. (1997) reported preliminary data on the identification of *M. grisea* mutants adapted to the resistance gene *Pi2*. In this experiment, avirulent isolate Po6-6 was screened for a change from avirulence to virulence on a series of near-isogenic lines carrying different resistance genes. Two independently derived mutants showed compatibility to *Pi2* and at the same time showed a significant reduction in vitro growth (90% of wild type) and sporulation (60% of wild type). The parasitic fitness of these mutants was tested on plants by measuring the number of lesions formed and in planta sporulation, but no significant differences were observed. It has not yet been determined whether the mutation is a single gene mutation or results from loss of a larger piece of DNA (as in a deletion or chromosome loss event) that is responsible for the apparent pleiotropic effects. Although the results are far from conclusive, fitness measurement of genetically defined virulent mutants may be the key toward understanding the nature of pathogen adaptation.

Some promising data have been accumulated in *Xoo* to suggest that a fitness penalty is associated with changes in specific avirulence genes. Inactivation of one copy of *avrXa7* in the multigene family in Pxo86 results in a loss of fitness in terms of the ability to cause lesion and bacterial multiplication on IR24, a cultivar without *Xa7* (S.H. Choi and J. Leach, personal communication, 1997). Introducing a cosmid carrying the *avrXa7* gene can restore this loss of fitness. To address the fitness question in the natural environment, field experiments were conducted to monitor the evolution of bacterial strains adapted to *Xa7*. Despite the presence of strains virulent on *Xa7*, there were no severe epidemics in the field over six cropping seasons (C.M. Vera Cruz, unpublished data, 1997). The result supports the hypothesis that pathogen adaptation results in a lowering of the competitiveness of the pathogen in nature.

6 Integrating strategies to achieve durable resistance

6.1 Identification and diversification of resistance genes

The discovery of structural similarity among resistance genes from different plant taxa has greatly expanded the capacity to use the large repertoire of resistance genes present in germplasm (Michelmore 1996). In theory, resistance genes with known conserved sequences could be readily isolated using PCR technology. Yu et al. (1996) designed degenerate oligonucleotides primers based on the NBS region of *N* and *RPS2* to amplify resistance gene orthologs in soybean. Eleven classes of sequences were isolated representing a superfamily of NBS-containing soybean. Among them, five have been mapped to the vicinity of known soybean genes for resistance to polyviruses. Using a similar approach, Kanazin et al. (1996) identified nine classes of resistance gene analogs and mapped several of them near known resistance genes. Using primers derived from the LRR of *RPS2* in *Arabidopsis* and *N* gene of tobacco, Leister et al. (1996a) identified resistance gene loci corresponding to the *Gro1* (nematode resistance) and *R7* (late blight resistance) in potato. These studies have established the validity of the approach to identify candidate gene loci in diverse plant taxa. More recently, Chen et al. (1998) developed a genome scanning technique to survey a large number of conserved motifs in the genomes of rice, barley, and wheat. This technique vastly improves the power of finding putative candidate genes, and provides an efficient way to determine the functional diversity of resistance genes in rice germplasm.

Obviously, the PCR approach to identify or clone resistance depends on similarity in gene sequences. At present, there is no realistic estimate on the proportion of resistance genes that have common conserved motifs. Will we miss a lot of genes using this approach alone? Are there fundamental differences between recessive and dominant resistance genes? Another limitation of this approach is that it is difficult to identify the function of the isolated candidate genes even though they are mapped to a region containing known resistance genes. This is because genes with similar conserved domains may have completely different biological functions in plant cells. As pointed out by Leister et al. (1996), the unusually high levels of genetic diversity of NBS-LRR homologs and their clustering organization in genomes tend to complicate the isolation of specific resistance genes by homology. Refined strategies for gene isolation with the PCR approach are still needed.

Despite some technical limitations, the prognosis seems good in isolating a large number of resistance genes based on conserved motifs. With the identity of resistance genes clearly defined, it is possible to incorporate diverse resistance sources into rice cultivars. This in turn will provide the needed diversity to buffer against epidemics as a result of rapid pathogen changes.

6.2 Linking pathogen genetics to developing durable resistance

With the PCR approach to detect major resistance genes, many more resistance genes can be rapidly identified and molecularly tagged. It appears that the future challenge is not necessarily the identification of resistance genes but rather the determination of the agronomic usefulness of the mapped or cloned genes. As discussed, there is emerging evidence that some avirulence genes encode fitness function. It is logical therefore to suggest that resistance gene combinations that collectively impose a high fitness penalty should be more durable than those that have trivial fitness impact on the pathogen.

With the excellent genetic systems available in *M. grisea* and *Xoo*, it is possible to examine the cost of adaptation of the pathogens on specific resistance genes. In the laboratory, this can be done by performing precise gene replacement studies in which a specific *avr* mutation can be substituted into the wild type and the cost of the mutation estimated by direct comparison between the mutant and the wild type strain. In cases where genetic manipulation of the pathogen is not possible, an alternative approach is to use DNA fingerprinting to track the evolution of strains *before* and *after* adaptation to a particular resistance gene combination. With DNA fingerprinting, we can identify pairs of isolates that only differ in their virulence on a particular resistance gene. Fitness changes due to avirulence mutation can then be measured in an essentially isogenic background. Provided a set of near-isogenic hosts with single or multiple resistance genes are available, it is possible to systematically determine the agronomic value of resistance genes. Information from this type of analysis should provide a rational basis in constructing effective gene pyramids.

6.3 Mapping QTL with candidate defense genes

Quantitative resistance has long been considered a desirable form of resistance presumably because of its nonrace-specific nature. Due to the masking effect of major gene resistance, it is however difficult to select for quantitative resistance in the presence of major genes. Through marker-aided genetic analysis, Wang et al. (1994) showed that durable resistance in a rice cultivar Moroberekan is contributed by both qualitative resistance and QTL. Ten chromosomal segments were found to be associated with effects on lesion number. Three of the markers associated with QTL for partial resistance were linked to complete blast resistance reported in previous studies. QTL identified in greenhouse tests were good predictors of blast resistance at two field sites (Wang et al. 1994). Thus, in theory, marker-aided selection could be used to combine high-quality resistance genes in a background of nonrace-specific quantitative resistance. However, as a general problem in QTL mapping, most markers associated with QTL are too far apart to have sufficient predictive power in a selection program (Tanksley and Nelson 1995). Thus, there is not yet a successful program for implementing marker-aided selection for quantitative resistance in rice.

This situation may change as our knowledge of defense response increases. In *Arabidopsis*, expressed sequence tags (ESTs) presumably involved in plant stress response are being discovered at a rapid rate. Botella et al. (1997) have identified 94 resistance gene-like ESTs and four *Pto*-like sequences from the *Arabidopsis* database. A total of 47 loci were identified and mapped, and several ESTs mapped to chromosomal regions containing genetically defined disease resistance genes. A similar collection of ESTs in rice is accumulating as more DNA sequences are available. It will be possible to shift from the use of random markers to candidate genes in QTL mapping. This candidate gene approach has been used successfully to map insect resistance against corn earworm (Byrne et al. 1996) and should be equally applicable to rice. By using candidate genes as probes, it is possible to map known genes onto the phenotypically determined QTL conferring resistance. Once the phenotypic contribution of a candidate gene is confirmed by co-segregation analysis, the gene can be used to identify alleles of higher phenotypic values in the germplasm pool.

7 Perspectives

7.1 Toward understanding the molecular basis of durable resistance

Although disease resistance genes have been purposefully incorporated in rice cultivars for disease control for over half a century, the number of genes used in commercial rice cultivars is still small compared to the large reserve of genetic resources in the rice germplasm and its wild relatives. There are ample opportunities to continue to deploy genetic variations in the field so long as we have the means to

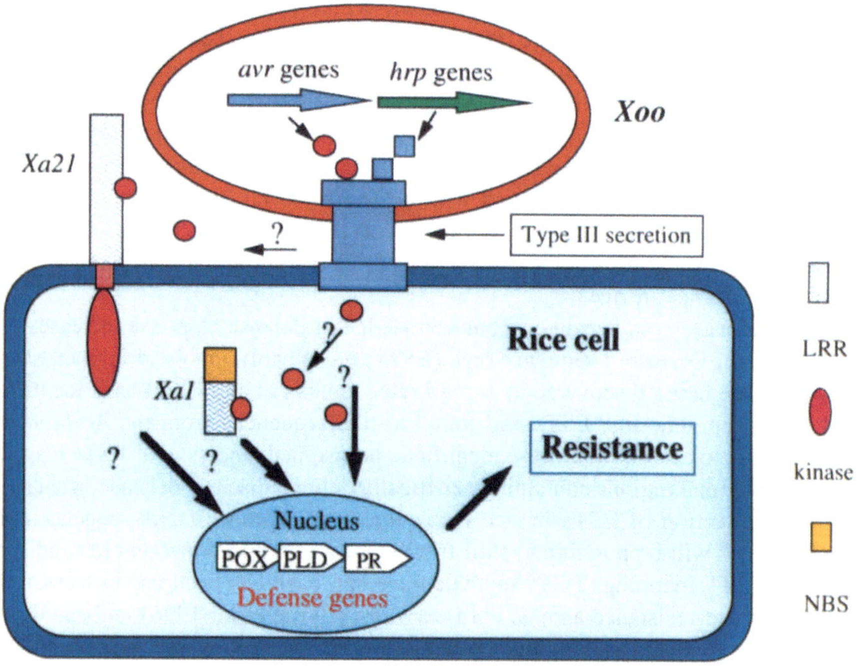

Fig. 1. Hypothetical model for the interactions between *Xanthomonas oryzae* pv. *oryzae* (*Xoo*) and rice cells. The bacterial protein is delivered into the rice cell via a type III secretion system or it is secreted outside of the bacterium. The square (*blue*) is the *hrp* gene product, and the circle (*red*) is the *avr* gene product. The predicted products of known *Xoo avr* genes have a nuclear localization signal, suggesting nuclear targeting and a potential role in gene regulation. The products of yet undiscovered avirulence genes may interact with resistance gene products on the cell surface or in the cytoplasm. The sites of interaction between products of *Xa21* and *Xa1* genes with corresponding *avr* genes are predicted mainly based on their gene structures (extracellular and intracellular). Some plant defense genes known to be involved in this interaction are shown: *POX* (peroxidase), *PLD* (phospholipase), and *PR* (pathogenesis-related proteins).

detect useful variation. The use of conserved DNA primers corresponding to disease resistance genes has greatly enhanced our ability to find more resistance genes. It will definitely hasten identifying novel and untapped resistance sources in the rice gene pool.

For the first time, we know the DNA sequence and predicted molecular structure of disease resistance gene products in rice. Some of the pathogen counterparts in *Xoo* and *M. grisea* are also known. Based on the available data on bacterial blight, the interactions between rice and the pathogen are shown in Fig. 1. Despite the progress made, are we closer to understanding why some resistance genes are more durable than others?

The answer to this question may come from both the host and pathogen. First, the durability of resistance genes can be viewed as a pathogen adaptation. Thus, the ability to determine the molecular changes and the associated fitness cost involved in adaptation will be key to determining the relative merits of resistance genes. At the molecular level, discovering the function and structure of the pathogen molecules interacting with a resistance gene may help us determine the domains responsible for recognition. For example, *Xa21* encodes a receptor-like kinase that confers resistance to multiple *Xoo* races. It is possible that the *Xa21* gene product may recognize a single determinant present in all races or unique determinants produced by each race. Cloning of the *avrXa21* gene is being pursued to address this question (P. Ronald, personal communication, 1998). With the AVR gene product available, it may be possible to identify key amino acids responsible for the interaction between the AVR and resistance gene products. This in turn may lead to design of novel and durable resistance through protein engineering.

Second, structural information of resistance genes with broad-spectrum resistance may shed light on the molecular basis of durability. A relevant example is the recessive gene *mlo* that confers broad-spectrum resistance to many races of the powdery mildew fungus. It is speculated that *Mlo* may be a negative regulator of the defense response, and thus a recessive mutation will lead to enhanced defense. Is it possible to search for *Mlo*-like gene families as has been done with the other conserved motifs in resistance genes? Finally, structural analysis of resistance gene clusters has provided a glimpse of the mechanisms by which new specificities against different pathogen races can be created (Hulbert 1998). It commonly appears that resistance genes are organized in a complex locus in which multiple copies of homologous resistance genes are physically next to each other. In the *Rp1* locus in sorghum, unequal crossing-over and gene conversion are important mechanisms creating new specificities (Richter et al. 1995). The *Xa21* locus is another example of a complex locus where intragenic recombination is suspected to give rise to divergence in specificity (Song et al. 1995; Wang et al. 1996). Can multiple specificities be engineered into a single gene cassette (gene pyramiding in the fine scale) to achieve multiple resistance?

7.2 Engineering nonpathogen-specific resistance: focusing on downstream pathways

The avirulence resistance gene recognition represents only the first step of the signal transduction pathway. The recent surge in discovery of novel genes downstream from the initial recognition suggests that many genes outside the realm of the perceived defense pathway await discovery. We have initiated mutant screening experiments to identify downstream genes that are involved in common signal transduction pathways shared by different pathogens. Thus far, we have recovered fast neutron-induced mutants that have changed from resistant to susceptible to both blast and bacterial blight. Thus, as in the case with the *NDR1* locus in *Arabidopsis*, it is possible to identify rice genes required for resistance against both fungal and bacterial pathogens. Once these genes are cloned, a variety of approaches can be explored to enhance resistance.

One potential drawback with the genetic engineering of defense genes or pathways is the genetic load associated with the unnecessary expression of defense genes. On the one extreme, we may create plants that are hypersensitive to any environmental perturbation, similar to that which we observe in the lesion that mimics mutants in nature. Even a subtle loss of vigor could be magnified in the field, leading to unacceptable yield penalty. One possible approach is to engineer a tight control of gene expression through manipulation of promoter specificity. A more conservative approach is to search for natural allelic variation at these loci. It is possible that some of these loci are phenotypically expressed as QTL. Despite these technical challenges, the potential benefit of producing broad-spectrum resistance in plants warrants more exploration of this approach.

7.3 Disease resistance research in the postgenomics era

The International Rice Genome Sequencing Project was launched in February 1998. The goal of this project is to completely sequence the rice genome within 10 years. Approximately 170 Mb of the 430-Mb rice genome is expected to be completed in the first 5 years. From this project, abundant ESTs and genomic sequences will be available in the next 5 to 10 years. Assigning biological functions to DNA sequences will be a major task in the postgenomics era. As shown in human and yeast, genome-wide analysis using gene-expression chips can dramatically improve our ability to determine the coordinated response of the host to either biotic or abiotic stresses (Schena 1996; Schena et al. 1996; DeRisi et al. 1997). Thus, with DNA chip technology, it is feasible to discover more candidate genes through comparative analysis of mutants and parental types. Production of mutants with defects in disease defense will be a prerequisite to taking full advantage of the expanding DNA database.

While traits with prominent phenotypes can be readily associated with DNA sequences , those that have small effects need further genetic dissection. There will be a flood of candidate genes, cloned or as molecular markers, available to aid selection. It will soon be necessary to consider the mechanisms and logistics by which newly found variation can be efficiently channeled to production. There will be a continuing need for precise biological characterization, particularly in different genetic backgrounds and under different environments. Ultimately the performance of a particular genotype, or gene combinations, will depend upon the interacting biotic and abiotic environments.

Acknowledgments. We thank Jan Leach and Pam Ronald for critical reading of the manuscript.

References

Alfano JR, Collmer A (1997) The type III (*Hrp*) secretion pathway of plant pathogenic bacteria: trafficking harpins, Avr proteins, and death. J Bacteriol 179:5655-5662.

Anderson PA, Lawrence GJ, Morrish BC, et al (1997) Inactivation of the flax rust resistance gene *M* associated with loss of a repeated unit within the leucine-rich repeat coding region. Plant Cell 9:641-651.

Baker B, Zambryski P, Staskwicz B, et al (1997) Signaling in plant-microbe interactions. Science 276:726-732.

Bent AF, Kunkel BN, Dahlbeck D, et al (1994) *RPS2* of *Arabidopsis thaliana*: a leucine-rich repeat class of plant disease resistance genes. Science 265:1856-1859.

Bent AF (1996) Plant disease resistance genes: function meets structure. Plant Cell 8:1757-1771.

Bonas U, Conrads-Strauch J, Balbo I (1993) Resistance in tomato to *Xanthomonas campestris* pv. *vesicatoria* is determined by alleles of the pepper-specific avirulence gene *avrBs3*. Mol Gen Genet 238:261-269.

Botella MA, Coleman MJ, Hughes DE, et al (1997) Map positions of 47 *Arabidopsis* sequences with sequence similarity to disease resistance genes. Plant J 12:1197-1211.

Buschges R, Hollricher K, Panstruga R, et al (1997) The barley *Mlo* gene: a novel control element of plant pathogen resistance. Cell 88:695-705.

Byrne PF, McMullen MD, Snook ME, et al (1996) Quantitative trait loci and metabolic pathways: genetic control of the concentration of maysin, a corn earworm resistance factor, in maize silks. Proc Natl Acad Sci USA 93:8820-8825.

Cao H, Glazebrook J, Clarke JD, et al (1997) The *Arabidopsis NPR1* gene that controls systemic acquired resistance encodes a novel protein containing ankyrin repeats. Cell 88:57-63.

Century KS, Shapiro AD, Repetti PP, et al (1997) *NDR1*, a pathogen-induced component required for *Arabidopsis* disease resistance. Science 278:1963-1965.

Chen D, Wang GL, Ronald PC (1997) Location of the rice blast resistance locus Pi5(t) in Moroberekan by AFLP bulk analysis. Rice Genetics Newsletter 14:95-97.

Chen XM, Line RF, Leung H (1998) Genome scanning for conserve motifs of disease resistance genes in rice barley, and wheat using high resolution electrophoresis. Theor Appl Genet (in press).

Chittoor JM, Leach JE, White FF (1997) Differential induction of a peroxidase gene family during infection of rice by *Xanthomonas oryzae* pv. *oryzae*. Mol Plant-Microbe Interact 10:861-871.

Chittoor JM, White FF, Leach JE, et al (1998) A family of inducible peroxidase genes are clustered in the rice. Abstract presented at Plant and Animal Genome, San Diego, CA, USA, 18-22 Jan, 1998, p 381.

Choi W, Dean RA (1997) The adenylate cyclase gene *MAC1* of *Magnaporthe grisea* controls appressorium formation and other aspects of growth and development. Plant Cell 9:1973-1983.

Chumley FG, Valent B (1990) Genetic analysis of melanin-deficient, nonpathogenic mutants of *Magnaporthe grisea*. Mol Plant-Microbe Interact 3:135-143.

Dangl JL, Dietrich RA, Richberg MH (1996) Death don't have no mercy: cell death programs in plant-microbe interactions. Plant Cell 8:1793-1807.

De Wit PJGM (1995) Fungal avirulence genes and plant resistance genes: unraveling the molecular basis of gene-for-gene interactions. Adv Bot Res 21:147-185.

De Wit PJGM (1997) Pathogen avirulence and plant resistance: a key role for recognition. Trends Plant Sci 2:452-458.

DeRisi JL, Iyer VR, Brown PO (1997) Exploring the metabolic and genetic control of gene expression on a genomic scale. Science 278:680-686.

Dietrich RA, Delaney TP, Uknes SJ, et al (1994) *Arabidopsis* mutants simulating disease resistance response. Cell 77:565-577.

Dietrich RA, Richberg MH, Schmidt R, et al (1997) A novel zinc finger protein is encoded by the *Arabidopsis LSD1* gene and functions as a negative regulator of plant cell death. Cell 88:685-694.

Divecha N, Irvine RF (1995) Phospholipid signaling. Cell 80:269-278.

Dixon M, Jones D, Keddie J, et al (1996) The tomato *Cf-2* disease resistance locus comprises two functional genes encoding leucine rich repeat proteins. Cell 84:451-459.

Dow JM, Osbourn AE, Wilson TJG, et al (1995) A locus determining pathogenicity of *Xanthomonas campestris* is involved in lipopolysaccharide biosynthesis. Mol Plant-Microbe Interact 8:768-777.

Evenson RE, Herdt RW, Hossain M (1996) Rice research in Asia: progress and priorities, CAB International, Wallingford, UK, p 418.

Flor HH (1971) The current status of the gene-for-gene concept. Annu Rev Phytopathol 9:275-296.

Grant MR, Godiard L, Straube E, et al (1995) Structure of the *Arabidopsis Rpm1* gene enabling dual specificity disease resistances. Science 269:843-846.

Hammond-Kosack KE, Jones JDG (1997) Plant disease resistance genes. Annu Rev Plant Physiol Plant Mol Biol 48:575-607.

He SY (1996) Elicitation of plant hypersensitive response by bacteria. Plant Physiol 112:865-869.

He C, Fong SHT, Ding Y et al (1997) Isolation of a differentially expressed cDNA from blast resistant rice showing homology to genes containing ankyrin repeats. Abstract presented at the Fifth International Congress of Plant Molecular Biology, Singapore, 21-27 September 1997, p. 775.

Herbers K, Conrads-Strauch J, Bonas U (1992) Race-specificity of plant resistance to bacterial spot disease determined by repetitive motifs in a bacterial avirulence protein. Nature 356:172-174.

Hopkins CM, White FF, Choi SS, et al (1992) A family of avirulence genes from *Xanthomonas oryzae* pv. *oryzae*. Mol Plant-Microbe Interact 5:451-459.

Howard RJ, Ferrari MA, Roach DH, et al (1991) Penetration of hard substrates by a fungus employing enormous turgor pressures. Proc Natl Acad Sci USA 88:11281-11284.

Huang JS, De Cleene M (1989) How rice plants are infected by *Xanthomonas campestris* pv. *oryzae*. In: Bacterial Blight of Rice. Proceedings of the International Workshop on Bacterial Blight of Rice, International Rice Research Institute, 14-18 March 1988, pp. 31-42.

Hughes AL, Nei M (1988) Pattern of nucleotide substitution at major histocompatibility complex class I loci reveals overdominant selection. Nature 335:167-170.

Hulbert SH (1998) Arming plants against pathogens - recombination in resistance loci. Trend Plant Sci (in press).

Ikeda R, Khush GS, Tabien RE, et al (1990) A new resistance gene to bacterial blight from *Oryza longistaminata*. Jpn J Breed 40:280-281.

Izawa T, Miyazaki C, Yamamoto M, et al (1991) Introduction and transposition of the maize transposable element *Ac* in rice (*Oryza sativa* L.). Mol Gen Genet 227:391-396.

Izawa T, Shimamoto K (1996) Becoming a model plant: the importance of rice to plant science. Trends Plant Sci 1:95-99.

Johnson R (1983) Genetic background of durable resistance In: Lamberti F, Waller JM, van der Graff NA (eds) Durable resistance in crops. Plenum Press, New York, pp 5-26.

Jones DA, Thomas CM, Hammond-Kosack KE, et al (1994) Isolation of the tomato *Cf-9* gene for resistance to *Cladosporium fulvum* by transposon tagging. Science 266:789-793.

Jones DA, Jones JDG (1997) The role of leucine-rich repeat proteins in plant defense. Adv Bot Res 24:89-167.

Joosten MH, Vogelsang R, Cozijnsen TJ, et al (1997) The biotrophic fungus *Cladosporium fulvum* circumvents *Cf-4*-mediated resistance by producing unstable AVR4 elicitors. Plant Cell 9:367-379.

Jwa NS, Tsuonda Y, Nakamura S, et al.(1998) Characterization of all cDNA and protein kinase in a BAC clone region spanning rice blast resistance gene *Pi-b*. Abstract presented at Plant and Animal Genome, San Diego, CA, USA, Jan 18-22, 1998, p 148.

Kanazin V, Marek LF, Shoemaker RC (1996) Resistance gene analogs are conserved and cluster in soybean. Proc Natl Acad Sci USA 93:11746-11750.

Kang S, Sweigard JA, Valent B (1995) The *PWL* host specificity gene family in the blast fungus *Magnaporthe grisea*. Mol Plant-Microbe Interact 8:939-948.

Kawasaki S, Tsunoda Y, Akiyama K, et al (1997) Characterization of the region close to a rice blast resistance gene *Pi-b* and spanning of it by high capacity binary vector RAC (Ri-ori driven artificial chromosome) contig. Abstract presented at the Fifth International Congress of Plant Molecular Biology, Singapore, 21-27 Sept. 1997, p 1491.

Kearney B, Staskawicz BJ (1990) Widespread distribution and fitness contribution of *Xanthomonas campestris* avirulence gene *avrBs2*. Nature 346:385-386.

Keen NT (1990) Gene-for-gene complementarity in plant-pathogen interaction. Annu Rev Genet 24:447-463.

Keen NT, Tamaki S, Kobayashi D, et al (1990) Bacteria expressing avirulence gene D produce a specific elicitor of the soybean hypersensitive reaction. Mol Plant-Microbe Interact 3:112-121.

Khush GS, Bacalangco E, Ogawa T (1990) A new gene for resistance to bacterial blight from *O. longistaminata*. Rice Genet Newsl 7:121-122.

Kingsley MT, Gabriel DW, Marlow GC, et al (1993) The *opsX* locus of *Xanthomonas campestris* affects host range and biosynthesis of lipopolysaccharide and extracellular polysaccharide. Bacteriol 175:5839-5850.

Kinoshita T (1995) Report of committee on gene symbolization, nomenclature and linkage groups. Rice Genet Newsl 12:9-115.

Lawrence GJ, Finnegan EJ, Ayliffe MA, et al (1995) The *L6* gene for flax rust resistance is related to the *Arabidopsis* bacterial resistance gene *RPS2* and the tobacco viral resistance gene *N*. Plant Cell 7:1195-1206.

Leach JE, White FF (1996) Bacterial avirulence genes. Annu Rev Phytopathol 34:153-179.

Lee YH, Dean RA (1993) cAMP regulates infection structure formation in the plant pathogenic fungus *Magnaporthe grisea*. Plant Cell 5:693-700.

Leister D, Ballvora A, Salamini F, et al (1996) A PCR-based approach for isolating pathogen resistance genes from potato with potential for wide application in plants. Nat Genet 14:421-429.

Leong SA, Farman M, Punekar N, et al (1997) Molecular characterization of *avr1-CO39* from *Magnaporthe grisea*. Abstract presented at the general meeting of the International Program on Rice Biotechnology, Malacca, Malaysia, Sept 15-19, 1997, p. 53.

Leung H, Baraoidan M, George ML, et al (1997) Linking pathogen genetics to development of durable resistance. Abstract presented at the general meeting of the International Program on Rice Biotechnology, Malacca, Malaysia, Sept 15-19, 1997, p. 57.

Lin W, Anuratha CS, Datta K, et al (1995) Genetic engineering of rice for resistance to sheath blight. Bio/Technology 13:686-691.

Lin X, Wang C, Wen G, et al (1998) Progress in map-based cloning of *Xa22(t)*, a new gene for bacterial blight resistance in rice. Paper presented at Plant and Animal Genome, San Diego, CA, USA, Jan 18-22, 1998.

Liu S, Dean RA (1997) G protein a subunit genes control growth, development, and pathogenicity of *Magnaporthe grisea*. Mol Plant-Microbe Interact 10:1075-1086.

Lorang JM, Shen H, Kobayashi D, et al (1994) *avr*A and *avr*E in *Pseudomonas syringae* pv. tomato PT23 play a role in virulence on tomato plants. Mol Plant-Microbe Interact 7:508-515.

Mackill DJ, Bonman JM (1992) Inheritance of blast resistance in near-isogenic lines of rice. Phytopathology 82:746-749.

Martin GB, Brommonschenkel SH, Chunwongse J, et al (1993) Map-based cloning of a protein kinase gene conferring disease resistance in tomato. Science 262:1432-1436.

Mendgen K, Hahn M, Deising H (1996) Morphogenesis and mechanisms of penetration by plant pathogenic fungi. Annu Rev Phytopathol 34:367-386.

Mew TW (1987) Current status and future prospects of research on bacterial blight of rice. Annu Rev Phytopathol 25:359-382.

Michelmore R (1996) Flood warning-resistance genes unleashed. Nat Genet 14:376-378.

Mindrinos M, Katagiri F, Yu GL, et al (1994) The *A. thaliana* disease resistance gene *RPS2* encodes a protein containing a nucleotide-binding site and leucine-rich repeats. Cell 78:1089-1099.

Mitchell TK, Dean RA (1995) The cAMP-dependent protein kinase catalytic subunit is required for appressorium formation and pathogenesis by the rice blast pathogen *Magnaporthe grisea*. Plant Cell 7:1869-1878.

Miyamoto M, Ando I, Rybka K, et al (1996) High resolution mapping of the *indica*-derived rice blast resistance genes.I. *Pi-b*. Mol Plant-Microbe Interact 9:6-13.

Nakamura S, Asakawa S, Ohmido N, et al (1997) Construction of an 800-kb contig in the near-centromeric region of the rice blast resistance gene *Pi-ta*^{2} using a highly representative rice BAC library. Mol Gen Genet 6:611-620.

Naqvi NI, Chattoo BB (1996) Development of a sequence characterized amplified region (SCAR) based indirect selection method for a dominant blast-resistance gene in rice. Genome 39:26-30.

Newman MA, Conrads-Strauch J, Scofield G, et al (1994) Defense-related gene induction in *Brassica campestris* in response to defined mutants of *Xanthomonas campestris* with altered pathogenicity. Mol Plant-Microbe Interact 7:553-563.

Ori N, Eshed Y, Paran I (1997) The wilt disease resistance gene locus belongs to the nucleotide binding, leucine-rich repeat superfamily of plant resistance genes. Plant Cell 9:521-532.

Ou SH (1985) Rice diseases. 2nd edn. Commonwealth Mycology Institute, Surrey, England, p 380.

Parker JE, Coleman MJ, Szabo V, et al (1997) The *Arabidopsis* downy mildew resistance gene RPP5 shares similarity to the Toll and Interleukin-1 receptors with N and L6. Plant Cell 9:879-894.

Reimers PJ, Guo A, Leach JE (1992) Increased activity of a cationic peroxidase associated with an incompatible interaction between *Xanthomonas oryzae* pv. *oryzae* and rice (*Oryza sativa*). Plant Physiol 99:1044-1050.

Richter TE, Pryor TJ, Bennetzen JL, et al (1995) New rust resistance specificities associated with recombination in the *rpl*. Genetics 141:373-381.

Rohe M, Gierlich A, Hermann H, et al (1995) The race-specific elicitor, NIP1, from the barley pathogen, *Rhynchosporium secalis*, determines avirulence on host plants of the *Rrs1* resistance genotype. EMBO J 14:4168-4177.

Roine E, Wei W, Yuan JY, et al (1997) Hrp pilus: an *hrp*-dependent bacterial surface appendage produced by *Pseudomonas syringae* pv. *tomato* DC3000. Proc Natl Acad Sci USA 94:3459-3464.

Ronald PC, Carland F, Salmeron J, et al (1992a) Cloned avirulence gene *avrPto* induces disease resistance on *Pto* tomato cultivars. J Bacteriol 174:1604-1611.

Ronald PC, Albano B, Tabien R, et al (1992b) Genetic and physical analysis of the rice bacterial blight disease resistance locus, *Xa21*. Mol Gen Genet 236:113-120.

Ronald PC (1997) The molecular basis of disease resistance in rice. Plant Mol Biol 35:179-186.

Ryals J, Weymann K, Lawton K, et al (1997) The *Arabidopsis* NIM1 protein shows homology to the mammalian transcription factor inhibitor IkB. Plant Cell 9:425-439.

Rybka K, Miyamoto M, Ando I, et al (1997) High resolution mapping of the indica-derived rice blast resistance genes *Pi-ta*^{2} and *Pi-ta* and a consideration of their origin. Mol Plant-Microbe Interact 10:517-524.

Salmeron JM, Oldroyd GED, Rommens CMT, et al (1996) Tomato *Prf* is a member of the leucine-rich repeat class of plant disease resistance genes and lies embedded within the *Pto* kinase gene cluster. Cell 86:123-133.

Schena M, Shalon D, Heller R, et al. (1996) Parallel human genome analysis: microarray-based expression monitoring of 1000 genes. Proc Natl Acad Sci USA 93:10614-10619.

Schena M (1996) Genome analysis with gene expression microarrays. BioEssays 18:427-431.

Scofield SR, Tobias CM, Rathjen JP, et al (1996) Molecular basis of gene-for-gene specificity in bacterial speck disease of tomato. Science 274:2063-2063.

Shi Z, Leung H (1994) Genetic analysis and rapid mapping of a sporulation mutation in *Magnaporthe grisea*. Mol Plant-Microbe Interact 7:113-120.

Shi Z, Leung H (1995) Genetic analysis of sporulation in *Magnaporthe grisea* by chemical and insertional mutagenesis. Mol Plant-Microbe Interact 8:949-959.

Shi Z (1995) Genetic and molecular analysis of sporulation in the rice blast fungus. Ph.D. thesis. Washington State University.

Song WY, Wang GL, Chen L, et al (1995) A receptor kinase-like protein encoded by the rice disease resistance gene *Xa21*. Science 270:1804-1806.

Song WY, Pi LY, Wang GL, et al (1997) Evolution of the rice *Xa21* disease resistance gene family. Plant Cell 9:1279-1287.

Staskawicz BJ, Ausubel FM, Baker B, et al (1995) Molecular genetics of plant disease resistance. Science 268:661-667.

Sweigard JA, Chumley FG, Valent B (1992) Disruption of a *Magnaporthe grisea* cutinase gene. Mol Gen Genet 232:183-190.

Sweigard JA, Carroll AM, Kang S, et al (1995) Identification, cloning and characterization of *PWL2*, a gene for host-species specificity in the rice blast fungus. Plant Cell 7:1221-1233.

Sword KMM, Dajbeck D, Keraney B, et al (1996) Spontaneous and induced mutations in a single open reading frame alter both virulence and avirulence in *Xanthomonas campestris* pv. *vesicatoria avrBs2*. J Bacteriol 178:4661-4669.

Takahashi A, Kawasaki T, Satoh H, et al (1997) Lesion mimic mutants of rice exhibiting resistance to rice blast fungus. Abstract presented at the Fifth International Congress of Plant Molecular Biology, Singapore, Sep 21-27, 1997, pp 603.

Tang JL, Feng JX, Li QQ, et al (1996a) Cloning and characterization of the *rpfC* gene of *Xanthomonas oryzae* pv. *oryzae*: involvement in exopolysaccharide production and virulence to rice. Mol Plant-Microbe Interact 9:664-666.

Tang X, Frederick RD, Zhou J, et al (1996b) Initiation of plant disease resistance by physical interaction of *avrPto* and *Pto* kinase. Science 274:2060-2063.

Tanksley SD, Nelson JC (1995) Advanced backcross QTL analysis: a method for the simultaneous discovery and transfer of valueable QTLs from unadapted germplasm into elite breeding lines. Theor Appl Genet 92:191-203.

Thomas CM, Jones DA, Parniṣke M, et al (1997) Characterization of the tomato *Cf-4* gene for resistance to *Cladosporium fulvum* identifies sequences that determine recognitional specificity in Cf-4 and Cf-9. Plant Cell 9:2209-2224.

Tsunoda Y, Akiyama K, Sasaki N, et al (1998) Construction of a contig spanning 180 kb of blast resistance gene *Pi-b* region from an average 42 kb subgenome library of a high capacity binary vector RAC; Ri-ori driven artificial chromosome. Abstract presented at Plant and Animal Genome, San Diego, CA, USA, Jan 18-22, 1998, p. 114.

Valent B, Chumley FG (1994) Avirulence genes and mechanisms of genetic instability in the rice blast fungus In: Zeigler RS, Leong S, Teng P (eds) The rice blast disease. International Rice Research Institute, Los Banos and Commonwealth Agricultural Bureaux, Cambridge, pp 111-134.

van den Ackerveken G, Marois E, Bonas U (1996) Recognition of the bacterial avirulence protein AvrBs3 occrus inside the host plant cell. Cell 87:1307-1316.

Vivian A, Gibbon MJ (1997) Avirulence genes in plant-pathogen bacteria: signal or weapons. Microbiology 144:693-704.

Wang GL, Mackill DJ, Bonman JM, et al (1994) RFLP mapping of genes conferring complete and partial resistance to blast in a durably resistant rice cultivar. Genetics 136:1421-1434.

Wang GL, Song WY, Ruan DL, et al (1996) The cloned gene *Xa21*, confers resistance to multiple *Xanthomonas oryzae* pv. *oryzae* isolates in transgenic plants. Mol Plant-Microbe Interact 9:850-855.

Wang GL, Ruan DL, Song WY, et al (1998a) The rice disease resistance gene, *Xa21D*, encodes a receptor-like molecule with a LRR domain that determines race specific recognition and is subject to adaptive evolution. Plant Cell 10:765-779.

Wang ZX, Yamanouchi U, Yano M, et al (1998b) Construction of a high-resolution genetic map and a cosmid contig of the genome region of a blast resistance gene, *Pi-b*, in rice. Abstract presented at Plant and Animal Genome, San Diego, CA, USA, Jan 18-22, 1998, p 84.

Whitmam S, Dinesh-Kumar SP, Choi D, et al (1994) The product of the tobacco mosaic virus resistance gene *N*: similarity to Toll and the interleukin-1 receptor. Cell 78:1101-1115.

Wu K, Martinez C, Lentini Z, et al (1996) Progress in cloning the rice blast resistance gene, Pi-ta(t). 8th International Congress on Plant Microbe Interactions, Knoxville, TN, July 14-19, 1996, p B103.

Xu JR, Hamer JE (1996) MAP kinase and cAMP signaling regulate infection structure formation and pathogenic growth in the rice blast fungus *Magnaporthe grisea*. Genes Dev 10:2696-2706.

Xu Y, Zhu Q, Panbangred W, et al (1996) Regulation, expression and function of a new basic chitinase gene in rice (*Oryza sativa* L.). Plant Mol Biol 30:387-401.

Xu JR, Urban M, Sweigard JA, et al (1997) The *CPKA* gene of *Magnaporthe grisea* is essential for appressorial penetration. Mol Plant-Microbe Interact 10:187-194.

Yang Y, Gabriel DW (1995) *Xanthomonas* avirulence/pathogenicity gene family encodes functional plant nuclear targeting signals. Mol Plant-Microbe Interact 8:627-631.

Yang Y, Yuan Q, Gabriel DW (1996) Watersoaking function(s) of XcmH1005 are redundantly encoded by members of the *Xanthomonas avr/pth* gene family. Mol Plant-Microbe Interact 9:105-113.

Yokoo M, Kikuchi F, Fujimaki H, et al (1978) Breeding of blast resistance lines (BL1-7) from indica-japonica crosses of rice. Jpn J Breed 28:359-385.

Yoshimura S, Yamanouchi U, Toki S, et al (1996) Characterization of candidate clones of *Xa-1*, the bacterial blight resistance gene in rice, isolated by map-based cloning. Theor Appl Genet 93:117-122.

Yoshimura S, Yamanouchi U, Katayose Y, et al (1998) Expression of *Xa1*, a bacterial blight resistance gene in rice, is induced by bacterial inoculation. Proc Natl Acad Sci USA 95:1663-1668.

Young SA, White FF, Hopkins CM, et al (1994) AVRXa10 protein is in the cytoplasm of *Xanthomonas oryzae* pv. *oryzae*. Mol Plant-Microbe Interact 7:799-804.

Young SA, Wang X, Leach J (1996) Changes in the plasma membrane distribution of rice phospholipase D during resistant interactions with *Xanthomonas oryzae* pv. *oryzae*. Plant Cell 8:1079-1090.

Yu ZH, Mackill DJ, Bonman JM, et al (1991) Tagging genes for blast resistance in rice via linkage to RFLP markers. Theor Appl Genet 81:471-476.

Yu YG, Buss GR, Saghai Maroof MA (1996) Isolation of a superfamily of candidate disease resistance genes in soybean based on a conserved nucleotide-binding site. Proc Natl Acad Sci USA 93:11751-11756.

Zhang Z, Naqvi N, Sim S, et al (1997) Fine mapping of blast and bacterial blight resistance genes introgressed from wild species *Oryza minuta*. Abstract presented at the general meeting of the International Program on Rice Biotechnology, Malacca, Malaysia, Sept 15-19, 1997, p 278.

Zhou J, Loh YT, Bressan RA, et al (1995) The tomato gene *Pti1* encodes a serine/threonine kinase that is phosphorylated by *Pto* and is involved in the hypersensitive response. Cell 83:925-935.

Zhou J, Tang X, Martin GB (1997) The *Pto* kinase conferring resistance to tomato bacterial speck disease interacts with proteins that bind a cis-element of pathogenesis-related genes. EMBO J 16:3207-3218.

Zhu Q, Lamb C (1991) Isolation and characterization of a rice gene encoding a basic chitinase. Mol Gen Genet 226:289-296.

Zhu LH, Chen Y, Xu YB, et al (1993) Construction of a molecular map of rice and gene mapping using a double haploid population of cross between *indica* and *japonica* varieties. Rice Genet Newsl 10:132-135.

Zhu Q, Maher EA, Masoud S, et al (1994) Enhanced protection against fungal attack by constitutive co-expression of chitinase and glucanase genes in transgenic tobacco. Bio/Technology 12:807-812.

Part 4

Transformation and rice improvement

12
Agrobacterium-mediated transformation

Yukoh Hiei, Toshihiko Komari, and Tomoaki Kubo

Plant Breeding and Genetics Research Laboratory, Japan Tobacco Inc.,
700 Higashibara, Toyoda, Iwata, Shizuoka 438-0802, Japan

1 Introduction

Agrobacterium has been used routinely for the transfer of genes to dicotyledonous plants. Until recently, however, monocotyledonous plants, including important cereals, were thought to be outside the range of this technology since they are generally not included within the host range of crown gall. Various attempts to infect monocotyledons with *Agrobacterium* were made in the 1970s and 1980s, but conclusive evidence of integrative transformation was not obtained until a few years ago. Therefore, attention was focused primarily on methods that involve the direct uptake of DNA. In many varieties of rice (*Oryza sativa* L.), the production of transgenic plants by bombardment-based methodology has now become routine (Christou 1997). Improvements in methods for transformation of rice using *Agrobacterium* have been considerable in the past few years. It used to be possible to transform rice only by direct gene-delivery methods, but now rice is among the plants that can be transformed routinely by methods that involve *Agrobacterium tumefaciens* (Aldemita and Hodges 1996; Dong et al. 1996; Hiei et al. 1994; Rashid et al. 1996; Toki 1997). In this chapter, details of such methods, as well as key factors involved in the transformation of rice, are discussed.

Key words. *Agrobacterium tumefaciens*, rice, *Oryza sativa* L., transformation, monocotyledon

2 *Agrobacterium tumefaciens* and plant transformation

Agrobacterium tumefaciens can transform plant cells genetically with a segment of DNA (transfer DNA, abbreviated as T-DNA) from a tumor-inducing plasmid (Ti plasmid) with the resultant production of a crown gall, which is a plant tumor. Another species of *Agrobacterium*, *A. rhizogenes*, causes hairy root disease in higher plants via an identical process. *A. tumefaciens* is the most extensively studied species of *Agrobacterium*, and the term "*Agrobacterium*-meditated transformation" usually refers to transformation mediated by *A. tumefaciens*.

2.1 Outline of the mechanism of *A. tumefaciens*-mediated gene transfer

Since details of the mechanisms of the transfer of DNA from *A. tumefaciens* have been reviewed by many authors (e.g., Hooykaas and Beijersbergen 1994; Hooykaas and Schilperoort 1992; Klee et al. 1987; Rathore et al. 1993; Zambryski 1992; Zupan and Zambryski 1995), we shall simply summarize here the process by which crown gall is induced in nature. This process consists of a number of discrete and essential steps. Wounding of the plant allows the entry of bacteria and provides the phenolic compounds that activate the DNA-transfer machinery of *A. tumefaciens*. The sap around a plant wound is an acidic solution that contains, among other compounds, a variety of phenolic compounds, such as derivatives of intermediates in the biosynthetic pathways to lignin and flavonoid compounds. The expression of virulence genes is induced by such sap. Bolton et al. (1986) identified seven phenolic compounds that induce the expression of virulence genes. The most effective inducers are monocyclic phenolic compounds, such as acetosyringone (Stachel et al. 1985). The bacteria multiply in the wound sap and attach to the walls of plant cells. In the Ti plasmid, the genes responsible for the transfer of T-DNA are clustered together in a region known as the virulence region. T-DNA is transferred to plant cells and integrated into the plant genome. Only the 25-bp direct repeats at the ends of the T-DNA are required, in *cis*, for mobilization of the T-DNA. Virtually any DNA sequence flanked by such border repeats can be transferred to plant cells by the virulence system. Tumors develop as a result of cell divisions that are triggered by the continuous production of auxin and cytokinin by enzymes encoded in the T-DNA. In addition, the tumor cells produce and secrete opines that are synthesized by other enzymes encoded in the T-DNA, and these opines are consumed specifically by the infecting bacteria. These findings have allowed the modification of the *Agrobacterium* system to yield various plant-transformation vectors.

2.2 Host range of crown gall

De Cleene and De Ley (1976) reported a list of natural host plants for *Agrobacterium* that contained 42 gymnosperms, 596 dicotyledonous plants, and 5 monocotyledons. The number of monocotyledons in the list was very small and no agronomically important cereals were included. Thus, the general statement that *Agrobacterium* does not infect monocotyledons is not literally true, but it reflects general beliefs over a long period of time.

2.3 Vector systems for *Agrobacterium*-mediated transformation

Various sophisticated plant-transformation vectors have been developed. The first remarkable development was the removal of wild-type T-DNA from Ti plasmids to create "disarmed strains" such as LBA4404 (Hoekema et al. 1983) and C58Cl (pGV3850) (Zambryski et al. 1983). Two strategies were then developed for the introduction of engineered T-DNA into *A. tumefaciens*. In one strategy, *E. coli* vectors are used that can be integrated into a disarmed Ti plasmid to yield what is known as an intermediate vector system (Fraley et al. 1983; Zambryski et al. 1983). The second strategy exploits the fact that the transfer of T-DNA can occur even if the virulence genes and the T-DNA are located on separate replicons in an *Agrobacterium* cell (Hoekema et al. 1983). In this system, small plasmids are used that can be replicated in both *Agrobacterium* and *E. coli*. Artificial T-DNA is engineered within the plasmids and then introduced into disarmed strains, in a so-called binary vector system. Vectors such as pBIN19 (Bevan 1984), pBI121(Jefferson 1987), and pGA482 (An et al. 1988) have been extensively utilized in various attempts to exploit this strategy.

Strain A281 (Watson et al. 1975) is a "supervirulent" strain, and its host range is wider and its transformation efficiency is higher than those of other strains (Hood et al. 1987; Komari 1989). These characteristics are due to the Ti plasmid, pTiBo542, that is harbored by this strain. Two types of system based on pTiBo542 have been developed. The first system involves strain EHA101 (Hood et al. 1986) or EHA105 (Hood et al. 1993), which carries a "disarmed" version of pTiBo542. The second involves a "super-binary" vector. In this system, a DNA fragment that includes *virB*, *virC*, and *virG* from the virulence region of pTiBo542 is introduced into a small T-DNA-carrying plasmid (Komari 1990) in a binary vector system, in which the disarmed Ti plasmid has its own complete set of virulence genes. *virB* and *virG* are the genes responsible for the hypervirulence of A281, and extra copies of these genes might also enhance the virulence.

3 Development of methods for direct delivery of genes in rice

Early success in the production of transgenic rice involved electroporation (Shimamoto et al. 1989; Toriyama et al. 1988; Zhang et al. 1988) or the polyethylene glycol-mediated (PEG-mediated) transfer of genes (Datta et al. 1990; Hayashimoto et al. 1990; Zhang and Wu 1988). Procedures for protoplast-mediated transformation of rice have certain problems. It remains difficult to regenerate plants from protoplasts of elite Japonica and the majority of Indica varieties, although some successful results have been reported (Ayres and Park 1994). Plants regenerated from protoplasts are often sterile and phenotypically abnormal (Abdullah et al. 1989; Battraw and Hall 1992; Datta et al. 1992). Some of these problems are probably related to changes in ploidy and the use of cells that have been cultured for long periods of time (Yamagishi et al. 1996).

Microprojectile bombardment, which is also referred to as the biolistic method or the particle gun method, has been attempted in many laboratories. Christou et al. (1991, 1992) reported that immature embryos are good targets for biolistic transformation of important cultivars of Japonica, Javanica, and Indica rice. Biolistic transformation of Japonica rice has become a routine procedure (Li et al. 1993). In Indica rice, successful transformation was recently reported by several research groups (Jian et al. 1996; Sivamani et al. 1996; Zhang et al. 1996). The molecular and genetic properties and the agronomic performance of transgenic rice plants that had been generated by bombardment methods were analyzed in recent studies (Cooley et al. 1995; Oard et al. 1996). Another method for the direct transfer of genes to intact tissues, developed for use in rice, involves the electroporation of mature embryos, and Xu and Li (1994) reported the production of fertile transgenic Indica rice by this method.

Use of direct gene-transfer systems often results in integration into the genome of fragmented and rearranged multiple copies of transgenes (Battraw and Hall 1992; Cooley et al. 1995; Hayashimoto et al. 1990; Peng et al. 1992; Rathore et al. 1993; Tada et al. 1990). This feature of transgenic plants is undesirable since the number of copies of a transgene influences the co-expression of introduced genes (Finnegan and McElroy 1994; Flavell 1994). Increased copy number that resulted in reduced co-expression was observed in rice after gene transfer by biolistic methods (Christou 1997; Cooley et al. 1995). In spite of concerns that systems of direct gene delivery might yield transgenic plants with multiple copies of transgenes, a recent report (Christou 1997) indicated that a majority of transgenic rice plants generated by bombardment contained only one to three copies of the transgene(s).

4 Attempts at the *Agrobacterium*-mediated transformation of monocotyledons

In early studies, various monocotyledonous plants, including cereals, were infected with wild-type strains of *Agrobacterium* and the formation of tumors or the biosynthesis of opines were examined. For example, Hernalsteens et al. (1984) infected stem segments of *Asparagus officinalis* with *A. tumefaciens* and observed induction of tumors. They also detected nopaline in the tumor cells. Similar observations were made in many monocotyledonous species that included cereals. However, data from Southern hybridization analysis that proved the integration of T-DNA into the genome were presented only for tumors in *Asparagus officinalis* (Bytebier et al. 1987) and yam (Schäfer et al. 1987).

Grimsley et al. (1986) developed a new technique, designated "agroinfection," by which the sequence of a viral genome can be introduced into a higher plant via *Agrobacterium* with the resultant systemic infection of the plant by the virus. They demonstrated that maize could be infected with maize streak virus that had been delivered by agroinfection (Grimsley et al. 1987). These studies provided clear evidence that DNA can be transferred from *Agrobacterium* to cereals, as well as a useful technique for plant virologists. Since infectious viral genomes can be produced from unintegrated T-DNA, the integration of T-DNA in chromosomes was not demonstrated in these studies.

In the early 1990s, attempts were made to introduce a gene for β-glucuronidase (GUS; Jefferson 1987) and drug resistance genes into cereal species. For example, segments of young seedlings of rice (Li et al. 1992) were infected with *Agrobacterium* and transient expression of GUS was detected. The *intron-gus* marker, with an intron in the coding sequence of GUS, has been an especially powerful tool because the presence of this gene does not result in the expression of GUS activity in *Agrobacterium* (Ohta et al. 1990; Vancanneyt et al. 1990).

Kanamycin- or G418-resistant cells that expressed GUS were obtained from rice embryos (Raineri et al. 1990), from wheat embryos that had been enzymatically digested (Mooney et al. 1991), and from segments of rice roots (Chan et al. 1992) that had been cocultivated with *Agrobacterium*. The cited studies included results of Southern hybridization. Some results have generated considerable controversy. Potrykus (1990a, 1990b, 1991) and Langridge et al. (1992) published critical reviews of methods of *Agrobacterium*-mediated transformation. They proposed the following criteria for conclusive proof of integrative transformation. (1) Expression of marker genes in plant cells must be distinguished from expression of the genes in *Agrobacterium* or other contaminating microorganisms. (2) Credible results of Southern hybridization analysis should be presented to confirm the integration of T-DNA in the plant chromosomal DNA. (3) Mendelian inheritance of transgenes should be demonstrated and confirmed by Southern hybridization. (4) The data should be obtained from a number of independent transgenic plants.

Chan et al. (1993) obtained a few transgenic rice plants by inoculating immature embryos with a strain of *A. tumefaciens*. They proved the inheritance of the transferred DNA to progeny plants by Southern hybridization but they only analyzed the progeny of one transformed plant. The efficient transformation of a cereal species, with sufficient data to satisfy all the above-listed criteria, was first reported in rice by Hiei et al. (1994). Their evidence was based on molecular analysis and on genetic studies of a large number of transgenic plants, as well as the analysis of sequences of T-DNA junctions in rice. Similar methods also proved effective in Javanica rice (Dong et al. 1996) and in Indica rice (Rashid et al. 1996). Aldemita and Hodges (1996) reported that transgenic plants were produced from immature embryos of Japonica and Indica rice that had been cocultivated with *Agrobacterium*.

Ishida et al. (1996) developed an efficient system for *Agrobacterium*-mediated transformation of another important cereal, maize. The successful transformation of wheat and barley using *A. tumefaciens* has also been reported (Cheng et al. 1997; Tingay et al. 1997) The frequencies of transformation in the latter studies were somewhat lower than those in studies with rice and maize. However, the controversy over the possible transformation of monocotyledonous plants by *Agrobacterium* has finally been resolved.

5 The recalcitrance of some monocotyledonous plants to infection by *Agrobacterium*

It is useful to consider why monocotyledonous plants remained resistant to infection by *Agrobacterium* for so long and why it suddenly became possible to transform these plants using *Agrobacterium*. In this context, special attention should be paid to wounding.

Early experiments with *Agrobacterium* suggested that wounding of the host plant might be essential for subsequent transformation. Obviously, wounding allows penetration by the bacteria of the host plant. Wounding also results in the production of phenolic compounds that induce the expression of *vir* genes. Binns and Thomashow (1988) pointed out the distinct correlation in reports in the early literature between the wound-induced division of cells and the competence of such cells to be transformed by *Agrobacterium*. They proposed that processes related to the synthesis of DNA and cell division might be required for the incorporation of foreign DNA into a host genome. This hypothesis is consistent with our understanding that integration into a host genome of pieces of foreign DNA is catalyzed by host enzymes, most probably enzymes that are part of the DNA replication/repair machinery.

The wound responses of many monocotyledons differ from those of many dicotyledons. Three main types of wound reaction can be observed in higher plants (Kahl 1982). Some plants, including many monocotyledons, make use of the damaged cells or their debris as a protective layer, which acts physically to prevent access by plant pathogens and also prevent extensive loss of water (type 1). The cells underneath this protective layer usually respond to the wound stimulus by lignification or impregnation with phenol of their cell walls but never by cell division. By contrast, wound responses of many dicotyledons are characterized by the division and proliferation of cells that result in formation of "wound callus" (type 2) or "wound periderm" (type 3) for reestablishment of the normal shape of the wounded organ.

In many routine protocols for the transformation of dicotyledons, explants, such as leaf disks, stem segments, or tuber sections, are cocultivated with *Agrobacterium*. Such explants provide ideal materials for transformation since they include cells that are actively dividing or are about to divide. These processes depend strongly on an active response to wounding (Potrykus 1991). It appears that routine methods of transformation cannot be applied to dicotyledons or monocotyledons that do not exhibit an active response to wounding. Potrykus (1991) stated that "transformation of 'monocots' is of no importance in this context: it is not because they are monocots that cereals are difficult to transform but because they do not have the proper wound response." However, methods for transformation had proved so successful in dicotyledons and so efficient that many efforts were made to exploit similar approaches in monocotyledonous plants. This is one of the reasons why development of efficient methods for *Agrobacterium*-mediated transformation of cereals was delayed for so long.

The main roles of wounding in the transformation process can be summarized as the production of *vir*-inducing molecules, the induction of DNA synthesis, and the induction of the rapid division of cells. Therefore, it would seem likely that actively growing, regenerable tissues from monocotyledons might be transformable in the presence of *vir*-inducing compounds. Reports of the successful transformation of cereals clearly support this possibility (Hiei et al. 1994, Ishida et al. 1996). Tissues consisting of actively dividing, embryonic cells were cocultivated with Agrobacterium in the presence of 100 μM acetosyringone in these successful studies.

6 Key factors involved in *Agrobacterium*-mediated transformation of rice

It is clear that many factors are of critical importance in the *Agrobacterium*-mediated transformation of rice. We shall discuss the key factors, including the type and stage of tissue that is infected, the kind of vector, the genotype of rice, the various conditions of tissue culture, and details of infection.

6.1 Choice of tissue for infection

The type and quality of the starting material is a very important factor, as is always the case in tissue culture. Callus cultures are excellent sources of cells for the production of transgenic rice (Dong et al. 1996; Hiei et al. 1994; Rashid et al. 1996;

Toki 1997). The use of actively growing, embryogenic calli is very important for efficient transformation of rice. Such calli can be obtained from mature or immature embryos. The production of transgenic rice plants from isolated shoot apices was reported by Park et al. (1996), but the frequency of transformation was very low. Long-term culture does not significantly affect the efficiency of transformation, but the risk of so-called somaclonal variations might be enhanced during such long-term culture.

Unlike the embryogenic calli of Japonica rice, the embryogenic calli of many Indica varieties are poor starting materials for *Agrobacterium*-mediated transformation. By contrast, isolated immature rice embryos consist of actively dividing cells and, although immature rice embryos proved initially to be poor starting materials (Chan et al. 1993; Hiei et al. 1994), transformants can now be produced efficiently from immature rice embryos as a result of optimization of procedures (Aldemita and Hodges 1996).

Cocultivation of rice cells that had been grown in suspension culture with *Agrobacterium* yielded low frequencies of transient expression of GUS activity and low frequencies of stable transformation even though such cells were actively proliferating (Hiei et al. 1994). However, preliminary culture of cells on solid medium for several days before cocultivation resulted in the high-frequency transfer of genes.

6.2 Bacterial strains and vectors

The most common strains of *Agrobacterium* used for transformation of higher plants are LBA4404 (Hoekema et al. 1983) and EHA101 (Hood et al. 1986). Both strains efficiently mediate transformation of rice (Dong et al. 1996; Hiei et al. 1994; Rashid et al. 1996). To date, numerous strains of *A. tumefaciens* have been isolated and characterized but only a few of them have been modified for use in the transformation of higher plants. It may be possible to identify strains that are more suitable for transformation of rice by screening large numbers of additional strains.

Various binary vectors can be used for transformation of rice, for example, pIG121Hm (Hiei et al. 1994), which is a derivative of the most commonly used vectors pBI121 (Jefferson 1987) and pBIN19 (Bevan 1984). Other choices are super-binary vectors, which are particularly efficient. Hiei et al. (1994) and Dong et al. (1996) compared a normal vector, pIG121Hm, with a super-binary vector, pTOK233, in rice transformation. In their transformation experiments, pTOK233 was as effective or slightly more effective than pIG121Hm with "easy" genotypes, and it was definitely much more effective with "difficult" genotypes, which included the Koshihikari cultivar and some cultivars of Javanica rice. These data suggest that the choice of vector and strain is important for the transformation of difficult cultivars. Efficient transformation of maize has only been achieved with super-binary vectors (Ishida et al. 1996).

6.3 Infection and cocultivation

Agrobacterium can be grown either in liquid medium or on agar-solidified medium. The concentration of the inoculum affects the efficiency of transformation. Transformation of rice is possible when the concentration of *A. tumefaciens* in the inoculum ranges between 1.0 x 10^6 and 1.0 x 10^{10} cfu/ml, and the optimal concentration is between 1.0 x 10^8 and 1.0 x 10^{10} cfu/ml (Aldemita and Hodges 1996; Dong et al. 1996; Hiei et al. 1994). We normally culture *Agrobacterium* on AB agar medium (Chilton et al. 1974) at 28°C for 3 days and suspend one loopful of bacteria in 1 ml of infection medium. This method for preparation of a uniform inoculum is reasonably reproducible; the concentration of bacteria can vary by as much as twofold but not more in such preparations.

After infection of tissues with a liquid inoculum, plant tissues and bacteria are cultivated together for 3 to 5 days. One technique that is commonly used in the transformation of dicotyledonous plants is the addition of a phenolic compound, such as acetosyringone, to cocultures or bacterial cultures, and the addition of such compounds is indispensable if the plant does not produce sufficient amounts of various signal molecules (Van Wordragen and Dons 1992). Rice cells might be capable of producing a certain amount of signal molecules (Raineri et al. 1990; Vijayachandra et al. 1995) but the levels appear to be very low. Hiei et al. (1994) demonstrated that acetosyringone at 100 μM is the key to the successful transformation of rice. The level of transient expression of GUS after cocultivation was very low when acetosyringone was omitted. Addition of an exogenous *vir*-induction agent is not necessary if the tissue produces sufficient amounts of signal molecules, as in the case, for example, of immature embryos of barley or suspension-cultured cells of wheat (Tingay et al. 1997; Cheng et al. 1997). The transfer of T-DNA is initiated at an early stage of cocultivation in the presence of acetosyringone, and pretreatment of bacteria with acetosyringone is not essential.

The temperature of cocultivation is another important factor. The optimal temperature for the growth of *Agrobacterium* is between 28°C and 30°C, and cereal cells in tissue culture are usually incubated between 25°C and 30°C. Induction of *vir* genes is, however, more efficient at temperatures between 15°C and 25°C (Alt-Moerbe et al. 1988; Alt-Mörbe et al. 1989). Fullner and Nester (1996) found that the mobilization of the plasmid by the T-DNA-transfer machinery is most active at temperatures as low as 19°C. In rice, the efficiency of transformation is high between 22°C and 28°C (Hiei et al. 1994).

Other factors during cocultivation, such as an acidic pH (Alt-Moerbe et al. 1988; Turk et al. 1991) and high osmotic pressure (Usami et al. 1988), have also been reported to be important for the expression of *vir* genes. Veluthambi et al. (1989) found that opines stimulated the induction of *vir* genes. In addition, Shimoda et al. (1990) and Cangelosi et al. (1990) reported that a group of aldoses, including D-glucose and some noncatabolizable sugars, markedly enhanced the expression of *vir* genes. The effects of sugars were particularly clear when levels of phenolic inducers were limited. Hiei et al. (1994) confirmed that the addition of D-glucose,

D-galactose, or L-arabinose and addition of nopaline did not have any significant synergistic effect in the presence of 100 μM acetosyringone.

The composition of the cocultivation medium affects the efficiency of transformation, and media that can support the active division of cells are preferable. Simple modified N6 medium containing acetosyringone, 2,4-dichlorophenoxyacetic acid (2,4-D), and casamino acids is one suitable medium for cocultivation (Hiei et al. 1994). Media solidified with a gelling agent are better for cocultivation than liquid media. Optimization of the medium may be necessary, depending on the genotype of rice or the type of material, and it can be achieved by monitoring the transient expression of marker genes. Conditions that allow high-level expression of markers after cocultivation are generally associated with high-frequency production of transgenic rice calli.

6.4 Selection of transformants

In general, the choice of marker genes and of selective agents has a major influence on the efficiency of transformation. The gene for neomycin phosphotransferase [*nptII*, also called *aph(3')II* or *neo*] was used in many early attempts at the direct transformation of rice. This gene confers resistance to the amino-glycoside antibiotic kanamycin. G418 is a related amino-glycoside antibiotic that is also inactivated by neomycin phosphotransferase. Not only is selection of transformed calli more efficient with G418 than with kanamycin, but calli selected on G418 regenerate into green plants at a higher frequency than those selected on kanamycin (Ayers and Park 1994). In *Agrobacterium*-mediated transformation, stably transformed calli can be produced efficiently with G418, but the absence of regenerants suggests that long-term exposure of cells to G418 might inhibit regeneration (Aldemita and Hodges 1996).

Another widely used and more effective selectable marker is the gene for hygromycin phosphotransferase (*hpt*, *aphIV*, or *hph*), which confers resistance to the amino-glycoside antibiotic hygromycin. Hygromycin allows clear discrimination between transformed and nontransformed rice tissues, and no problems with albinos have been reported (Ayers and Park 1994). The *hpt* gene has been used as an efficient marker gene for selection after *Agrobacterium*-mediated transformation in rice (Aldemita and Hodges 1996; Dong et al. 1996; Hiei et al. 1994; Rashid et al. 1996).

Other selective agents include herbicides. Genes have been isolated that endow plants with resistance to various commercially important herbicides. Among them, the *bar* gene for phosphinothricin acetyltransferase might be the most valuable in rice (Ayres and Park 1994). The *bar* gene confers resistance to L-phosphinothricin (PPT), glyphosinate (an ammonium salt of PPT), and bialaphos (a derivative of PPT). The *bar* gene has been successfully used to select transgenic rice calli and plants in a number of laboratories (Cao et al. 1992; Christou et al. 1991; Datta et al. 1992; Rathore et al. 1993; Toki 1997).

In many successful attempts to transform cereals, genes for selectable markers have been driven by the constitutively expressed 35S promoter of cauliflower mosaic virus (CaMV). This promoter appears to direct sufficiently strong expression of genes for the selection of transformants at high frequency. The level of expression of this promoter in cereals is, however, as much as 100 fold lower than in dicotyledons (Fromm et al. 1985; Hauptmann et al. 1987). Different promoters, such as those of genes for ubiquitin 1 (*Ubi1*) and alcohol dehydrogenase 1 (*Emu*) from maize (Cornejo et al. 1993; Li et al. 1997) and that of a gene for actin (*Act1*) from rice (Zhang et al. 1991), and elements that enhance gene expression, such as introns, have been successfully tested in cereals (Vasil 1994). Use of more powerful promoters and enhancer elements for the expression of marker genes should further improve the efficiency of transformation of rice by *Agrobacterium*.

Transgenic rice plants can be regenerated via somatic embryogenesis after *Agrobacterium*-mediated transformation by the methods described by Hiei et al. (1994). Maintenance of the embryogenic status of cells on the selection medium is important for the efficient recovery of regenerants. Techniques for somatic embryogenesis from callus cultures have been well established for rice of various genotypes. Hence, it seems likely that *Agrobacterium*-mediated methods should be useful for many varieties of rice with the introduction of the media that allow successful application of such techniques. The conditions of tissue culture and the composition of media that support the active growth of plant cells facilitate efficient selection and shorten the time required for the selection and regeneration steps (Toki 1997). Inclusion of a selective agent in the regeneration and rooting media greatly favors the production of transgenic plants that stably express transgenes (Hiei et al. 1994; Komari et al. 1996; Toki 1997).

6.5 Genotypes of rice

Hiei et al. (1994) reported that various rice cultivars of Japonica, including Koshihikari, can be transformed efficiently by *Agrobacterium*. Koshihikari is well known for the high quality of its grain and many of the new, elite Japanese cultivars are related to Koshihikari. Koshihikari and related cultivars are also known for their poor responses in tissue culture and for the difficulties associated with their transformation. The establishment of a routine and efficient method for the transformation of Koshihikari would contribute considerably to efforts aimed at the improvement of Japanese rice cultivars by genetic engineering. Dong et al. (1996) reported the *Agrobacterium*-mediated transformation of two Javanica cultivars, Gulfmont and Jefferson, which are, respectively, widely cultivated or about to be commercially cultivated in the southern United States.

Rashid et al. (1996) obtained transgenic plants from Basmati cultivars of Indica rice. However, most Indica varieties belong to a different group, group I, when rice cultivars are classified on the basis of results of isozyme analysis (Glaszmann 1987). These latter Indica varieties are generally recalcitrant both to growth in tissue culture and to attempts at transformation. Aldemita and Hodges (1996) reported the successful transformation by *Agrobacterium* of two cultivars in group I, TSC10 and IR72. They infected immature embryos with a super-binary vector LBA4404 (pTOK233). We have also succeeded in the efficient production of transformants of group I varieties of Indica rice, including Nan Jin 11, Xin Qing Ai 1, Suewon 258, IR8, IR24, IR26, IR36, IR54, IR64, and IR72, by a similar method, with transformation frequencies above 30% in ten varieties.

Unlike those of Japonica rice, the embryogenic calli of Indica varieties have been poor starting materials for *Agrobacterium*-mediated transformation, probably because Indica calli tend to grow very slowly. By contrast, freshly isolated immature rice embryos that consist of actively dividing cells may be suitable for genotype-independent introduction of genes. Thus, for the transformation of most Japonica and some Indica varieties of rice (Rashid et al. 1996), which produce rapidly growing embryogenic calli, callus cultures are the preferred starting materials. Large numbers of pieces of calli can be prepared more easily than immature embryos. For the transformation of other varieties, such as group I varieties of Indica rice, immature embryos are probably the most suitable materials.

7 Characteristics of transgenic rice plants

In a study in 1994, Hiei et al. found that almost all their transgenic Japonica rice plants had normal morphology and 70% were fully fertile. Similar observations were made with transgenic plants derived from Indica varieties. The genomes of 60 independent, GUS-positive, hygromycin-resistant, transgenic Japonica rice plants that have been transformed with EHA101 (pIG121Hm) or LBA4404 (pTOK233) were analyzed by Southern hybridization (Hiei and Komari 1996). The number of copies of integrated genes ranged from one to six, but the majority of the transformants contained one or two copies of integrated genes. In several plants, the numbers of copies of *hpt* and *gus* genes were different, probably as a result of the rearrangement of DNA during transformation (Hiei and Komari 1996).

Plants transformed with LBA4404 (pTOK233) that gave negative results in a histochemical assay of GUS activity were occasionally found among hygromycin-resistant regenerants. Some other transformants showed scattered expression of GUS. Between 10% and 20% of rice transformants typically had these undesirable characteristics. We analyzed these latter transformants by Southern hybridization. All the plants contained intact copies of the *hpt* gene, but intact copies of the gene for GUS were not found in 13 of 20 plants that were negative for GUS activity. Thus, part of the gene had been lost during transformation. The other 7 negative plants and 30 plants with scattered expression of GUS had one or more copies of the intact gene for GUS. Similar inactive expression of an introduced gene was also observed in studies of the biolistic transformation of rice (Cooley et al. 1995; Kohli et al. 1996; Kumpatla et al. 1997).

The stable inheritance and expression of foreign genes are of critical importance in the application of genetically engineered rice to agriculture. Hiei and Komari (1996) analyzed the inheritance of transgenes as far as the R4 generation for 18 primary transgenic rice plants that had been produced by *Agrobacterium*-mediated methods. They confirmed the Mendelian transmission of the two marker genes, *gus* and *hpt*, and analyzed the DNA from selected offspring by Southern hybridization. A close correlation between phenotype and genotype was demonstrated.

For the transformation of various dicotyledons, the advantages of *Agrobacterium*-mediated transformation include the transfer of pieces of DNA with defined ends with minimal rearrangement, the transfer of relatively large segments of DNA, and the integration of small numbers of copies of genes into plant chromosomes (Klee et al. 1987; Zambryski 1988). Since studies to date have provided strong support for the hypothesis that T-DNA is transferred from *Agrobacterium* to dicotyledons and monocotyledons by an identical molecular mechanism, it is likely that all these advantages also apply to the *Agrobacterium*-mediated transformation of rice. *Agrobacterium*-mediated transformation of rice depends on tissue culture techniques, and problems associated with tissue culture, such as somaclonal variation, remain to be overcome. Somaclonal mutation can occur even when the primary callus culture is used for regeneration (Yamagishi et al. 1996). Therefore, further studies of the characteristics of transgenic plants, including detailed molecular analyses of transferred genes, are necessary for the future genetic engineering of rice.

8 A co-transformation system for generation of marker-free transgenic rice

Selectable marker genes are almost always included in transformation systems, but they are usually not required once transgenic plants have been produced. The safety of each marker gene is another potential problem in the genetic engineering of plants. These issue were reviewed extensively by Yoder and Goldsbrough (1994).

Several techniques for the elimination of marker genes have been developed to cope with this problem (Yoder and Goldsbrough 1994). Co-transformation systems appear to be the simplest among them and have been widely used in direct transformation. A marker gene and the genes of interest are encoded by separate DNA molecules, and the molecules are introduced into plant genomes as unlinked fragments. Consequently, the nonselectable genes can segregate from the marker gene in the progeny. Co-transformation by *A. tumefaciens* has been examined in several laboratories (Yoder and Goldsbrough 1994), but it is not a routine technique. Komari et al. (1996) constructed super-binary vectors that carried two separate T-DNAs for co-transformation. In a study of co-transformation of rice, progeny that contained the nonselectable T-DNA and that were also free of the selectable marker were obtained from more than 25% of regenerants. Since multiple fragments of DNA were integrated at one locus in most cases of co-transformation when direct transformation methods were applied (Christou et al. 1989; Rathore et al. 1993; Spencer et al. 1992), super-binary vectors appear to provide a co-transformation system for the production of marker-free rice transformants in an efficient manner.

Table 1. Summary of studies of *Agrobacterium*-mediated transformation of rice*

Variety	Strain	Gene	Tissue	Selection	Comments	References
Japonica Nipponbare Fujisaka5	A281, A856 LBA4404	nos-*nptII* CaMV35S-*gus* oncogene	Mature embryos from germinated seeds	Kanamycin (200 mg/l)	Kanamycin-resistant calli; no plants; T-DNA detection by Southern analysis	Raineri et al. 1990
Indica Taichung native 1	C58C1 LBA4404	nos-*nptII* CaMV35S-*gus*	Roots from germinated seeds	G418 (20 mg/l)	G418-resistant calli; no plants; T-DNA detection by Southern analysis	Chan et al. 1992
Japonica Tainung 62	A281	nos-*nptII* α-*Amy8*-*gus*	Cultured immature embryos	G418 (40 mg/l)	Four transgenic plants; one plant produced progeny; potato cells in suspension culture required for transformation	Chan et al. 1993
Japonica Tsukinohikari Asanohikari Koshihikari	LBA4404 EHA101	CaMV35S-*hpt* CaMV35S-*int-gus*	Calli derived from scutella, immature embryos, suspension-cultured cells	Hygromycin (50-100 mg/l)	A large number of transgenic plants produced at high efficiency; Mendelian transmission to R_2 generation; T-DNA detection by Southern analysis; sequencing of T-DNA boundaries; 'super-binary' vector was very efficient	Hiei et al. 1994

Indica Basmati 370 Basmati 385	EHA101	CaMV35S-*hpt* CaMV35S-*int-gus*	Calli derived from scutella	Hygromycin (50 mg/l)	Efficient transformation of Basmati cultivars of Indica rice	Rashid et al. 1996
Japonica Taipei 309 Javanica Gulfmont Jefferson	LBA4404 EHA101	CaMV35S-*hpt* CaMV35S-*int-gus*	Calli derived from scutella	Hygromycin (50 mg/l)	'Super-binary' vector was definitely efficient in transformation of Javanica rice; Mendelian transmission to R_2 generation	Dong et al. 1996
Japonica Radon Indica TCS10, IR72	LBA4404	CaMV35S-*hpt* CaMV35S-*int-gus*	Immature embryos	Hygromycin (30-50 mg/l)	Efficient transformation of Japonica rice using immature embryos; transgenic plants from 'group I' varieties of Indica rice; 'super-binary' vector was very efficient	Aldemita and Hodges 1996
Tropical Japonica Maybelle	EHA101	rice *actin1-int-bar*	Isolated shoot apices	PPT (0.5 mg/l)	PPT-resistant transgenic plants were obtained at low frequency; inactivation of the transgene was observed in R_2 progeny	Park et al. 1996
Japonica Nipponbare Kitaake	EHA101	CaMV35S-*hpt* CaMV35S-*int-gus* nos-*hpt* maize *ubi1-bar*	Calli derived from scutella	Hygromycin (50 mg/l) bialaphos (5 mg/l)	Continuous illumination at 30 °C was effective for rapid callus induction and rapid selection of transformed cells	Toki 1997

* The various abbreviations are explained in the text.

9 Future prospects

Transformation of rice that does not depend on tissue culture techniques appears to be a fantasy at present. If such transformation could be achieved, the time for production of transgenic plants would be shortened dramatically and problems associated with tissue culture, such as somaclonal variation, would be eliminated. Furthermore, dependence on particular genotypes might be significantly reduced. Tissue culture-free transformation, mediated by *Agrobacterium*, might be possible using actively dividing cells, such as cells in meristems of intact plants. At this time, efficient transformation without tissue culture has been achieved only in the so-called "*in planta* transformation" of *Arabidopsis thaliana* (Bechtold et al. 1993). Transformants can be recovered among the progeny at a reasonably high frequency. Development of similar systems would, of course, be highly desirable.

One of the advantages in *Agrobacterium*-mediated transformation is that large segments of DNA can be transferred to plants. Typical lengths of T-DNAs in wild-type *Agrobacterium* strains range from 10 to 23 kb. Working with tobacco, Hamilton et al. (1996) demonstrated that DNA segments of as much as 150 kb in length can be transferred to the plant genome by *Agrobacterium*. They used specially designed vectors, known as binary bacterial artificial chromosome (BIBAC) vectors, that allow stable maintenance of large fragments of DNA both in *E. coli* and in *Agrobacterium*. If such vectors would be further improved, larger fragments, such as mega-base-pair fragments might be transferable. Such technology would certainly open up novel opportunities for the genetic engineering of rice and the efficient identification of useful genes in rice.

10 Conclusion

Agrobacerium tumefaciens-mediated transformation of rice has many advantages over transformation methods that involve the direct uptake of DNA. Fewer copies of transgenes are integrated into the genome, with associated high-fidelity inheritance of the transgene. The procedure is less time-consuming, and should be easier to adopt in laboratories in developing countries. *Agrobacterium*-mediated transformation of Japonica rice varieties has become routine in a growing number of laboratories. Additionally, recent studies suggest that, in rice, this methodology is likely to be genotype independent. At present, rice is now unique among food crops, having a small genome that has been well characterized. We predict that the functions of many plant genes will soon be tested in rice using this technique before they are tested in any other plant. *Agrobacterium*-mediated transformation of agronomically important dicotyledonous species is proving to be a commercially viable approach for the development of crops with new traits. Rice is one of the most important food crops in the world, and use of this transformation method is likely to accelerate greatly the application of molecular breeding approaches to improvement of this agriculturally important plant.

Acknowledgments. The authors thank Yuji Ishida, Hideaki Saito, Shozo Ohta, and Takashi Kumashiro for their valuable advice and for helpful discussions.

References

Abdullah R, Thompson J, Khush GS, Kaushik RP, Cocking EC (1989) Protoclonal variation in the seed progeny of plants regenerated from rice protoplasts. Plant Sci 65:97-101

Aldemita RR, Hodges TK (1996) *Agrobacterium tumefaciens*-mediated transformation of *japonica* and *indica* rice varieties. Planta 199:612-617

Alt-Moerbe J, Neddermann P, Von Lintig J, Weiler EW, Schröder J (1988) Temperature-sensitive step in Ti plasmid *vir* region induction and correlation with cytokinin secretion by *Agrobacteria*. Mol Gen Genet 213:1-8

Alt-Mörbe J, Kühmann H, Schröder J (1989) Differences in induction of Ti-plasmid virulence genes *virG* and *virD*, and continued control of *virD* expression by four external factors. Mol Plant-Microbe Interact 2:301-308

An G, Evert PR, Mitra A, Ha SB (1988) Binary vectors. In: Gelvin SB, Schilperoort RA (eds) Plant Molecular Biology Manual A3. Kluwer Academic, Dordrecht, pp 1-19

Ayres NM, Park WD (1994) Genetic transformation of rice. Crit Rev Plant Sci 13:219-239

Battraw M, Hall TC (1992) Expression of a chimeric neomycin phosphotransferase II gene in first and second generation transgenic rice plants. Plant Sci 86:191-202

Bechtold N, Ellis J, Pelletier G (1993) *In planta Agrobacterium*-mediated gene transfer by infiltration of adult *Arabidopsis thaliana* plants. C R Acad Sci Paris Ser III Life Sci 316:1194-1199

Bevan M (1984) Binary *Agrobacterium* vectors for plant transformation. Nucleic Acids Res 12: 8711-8721

Binns AN, Thomashow MF (1988) Cell biology of *Agrobacterium* infection and transformation of plants. Annu Rev Microbiol 42:575-606

Bolton GW, Nester EW, Gordon MP (1986) Plant phenolic compounds induce expression of the *Agrobacterium tumefaciens* loci needed for virulence. Science 232:983-984

Bytebier B, Deboeck F, De Greve H, Van Montagu M, Hernalsteens J-P (1987) T-DNA organization in tumor cultures and transgenic plants of monocotyledon *Asparagus officinalis*. Proc Natl Acad Sci USA 84:5345-5349

Cangelosi GA, Ankenbauer RG, Nester EW (1990) Sugars induce the *Agrobacterium* virulence genes through a periplasmic binding protein and transmembrane signal protein. Proc Natl Acad Sci USA 87:6708-6712

Cao J, Duan X, McElroy D, Wu R (1992) Regeneration of herbicide-resistant transgenic rice plants following microprojectile-mediated transformation of suspension culture cells. Plant Cell Rep 11:586-591

Chan M-T, Chang H-H, Ho S-L, Tong W-F, Yu S-M (1993) *Agrobacterium*-mediated production of transgenic rice plants expressing a chimeric α-amylase promoter/β-glucuronidase gene. Plant Mol Biol 22:491-506

Chan M-T, Lee T-M, Chang H-H (1992) Transformation of indica rice (*Oryza sativa* L.) mediated by *Agrobacterium tumefaciens*. Plant Cell Physiol 33:577-583

Cheng M, Fry JE, Pang S, Zhou H, Hironaka CM, Duncan DR (1997) Genetic transformation of wheat mediated by *Agrobacterium tumefaciens*. Plant Physiol 115:971-980

Chilton M-D, Currier TC, Farrand SK, Bendich AJ, Gordon MP, Nester EW (1974) *Agrobacterium tumefaciens* DNA and PS8 bacteriophage DNA not detected in crown gall tumors. Proc Natl Acad Sci USA 71:3672-3676

Christou P (1997) Rice transformation: bombardment. Plant Mol Biol 35:197-203

Christou P, Ford TL, Kofron M (1991) Production of transgenic rice (*Oryza sativa* L.) from agronomically important indica and japonica varieties via electric discharge particle acceleration of exogenous DNA into immature zygotic embryos. Biotechnology 9:957-962

Christou P, Ford TL, Kofron M (1992) The development of a variety-independent gene-transfer method for rice. Trends Biotechnol 10:239-246

Christou P, Swain WF, Yang N-S, McCabe DF (1989) Inheritance and expression of foreign genes in transgenic soybean plants. Proc Natl Acad Sci USA 86:7500-7504

Cooley J, Ford T, Christou P (1995) Molecular and genetic characterization of elite transgenic rice plants produced by electric-discharge particle acceleration. Theor Appl Genet 90:97-104

Cornejo M-J, Luth D, Blankenship KM, Anderson OD, Blechl AE (1993) Activity of maize ubiquitin promoter in transgenic rice. Plant Mol Biol 23:567-581

Datta SK, Datta K, Soltanifer N, Donn G, Potrykus I (1992) Herbicide-resistant Indica rice plants from IRRI breeding line IR72 after PEG-mediated transformation of protoplasts. Plant Mol Biol 20:619-627

Datta SK, Peterhans A, Datta K, Potrykus I (1990) Genetically engineered fertile indica rice recovered from protoplasts. Biotechnology 8:736-740

De Cleene M, De Ley J (1976) The host range of crown gall. Bot Rev 42:389-466

Dong J, Teng W, Buchholz WG, Hall TC (1996) *Agrobacterium*-mediated transformation of Javanica rice. Mol Breed 2:267-276

Finnegan J, McElroy D (1994) Transgenic inactivation: plants fight back! Biotechnology 12:883-888

Flavell RB (1994) Inactivation of gene expression in plants as a consequence of specific sequence duplication. Proc Natl Acad Sci USA 91:3490-3496

Fraley RT, Rogers SG, Horsch RB, Sanders PR, Flick JS, Adams SP, Bittner ML, Brand LA, Fink CL, Fry JS, Galluppi GR, Goldberg SB, Hoffmann NL, Woo SC (1983) Expression of bacterial genes in plant cells. Proc Natl Acad Sci USA 80:4803-4807

Fromm ME, Taylor LP, Walbot V (1985) Expression of genes transferred into monocot and dicot plant cells by electroporation. Proc Natl Acad Sci USA 882: 5824-5828

Fullner KJ, Nester EW (1996) Temperature affects the T-DNA transfer machinery of *Agrobacterium tumefaciens*. J Bacteriol 178:1498-1504

Glaszmann JC (1987) Isozymes and classification of Asian rice varieties. Theor Appl Genet 74:21-30

Grimsley N, Hohn B, Hohn T, Walden R (1986) Agroinfection, an alternative route for viral infection of plants by using the Ti plasmid. Proc Natl Acad Sci USA 83:3282-3286

Grimsley N, Hohn T, Davies JW, Hohn B (1987) *Agrobacterium*-mediated delivery of infectious maize streak virus into maize plants. Nature 325:177-179

Hamilton CM, Frary A, Lewis C, Tanksley SD (1996) Stable transfer of intact high molecular weight DNA into plant chromosomes. Proc Natl Acad Sci USA 93:9975-9979

Hauptmann RM, Ozias-Akins P, Vasil V, Tabaeizadeh Z, Rogers SG, Horsch RB, Vasil IK, Fraley RT (1987) Transient expression of electroporated DNA in monocotyledonous and dicotyledonous species. Plant Cell Rep 6:265-270

Hayashimoto A, Li Z, Murai N (1990) A polyethylene glycol-mediated protoplast transformation system for production of fertile transgenic rice plants. Plant Physiol 93:857-863

Hernalsteens J-P, Thia-Toong L, Schell J, Van Montagu M (1984) An *Agrobacterium*-transformed cell culture from the monocot *Asparagus officinalis*. EMBO J 13:3039-3041

Hiei Y, Komari T (1996) Stable inheritance of transgenes in rice plants transformed by *Agrobacterium tumefaciens*. In: Khush GS (ed) Rice Genetics III. Proceedings of the Third International Rice Genetics Symposium, Manila (Philippines). International Rice Research Institute, Los Baños, Philippines, pp 131-142

Hiei Y, Ohta S, Komari T, Kumashiro T (1994) Efficient transformation of rice (*Oryza sativa* L.) mediated by *Agrobacterium* and sequence analysis of the boundaries of the T-DNA. Plant J 6: 271-282

Hoekema A, Hirsch PR, Hooykaas PJJ, Schilperoort RA (1983) A binary plant vector strategy based on separation of *vir*- and T-region of the *Agrobacterium tumefaciens* Ti-plasmid. Nature 303:179-180

Hood EE, Fraley RT, Chilton M-D (1987) Virulence of *Agrobacterium tumefaciens* strain A281 on legumes. Plant Physiol 83:529-534

Hood EE, Gelvin SB, Melchers LS, Hoekema A (1993) New *Agrobacterium* helper plasmids for gene transfer to plants. Transgenic Res 2:208-218

Hood EE, Helmer GL, Fraley RT, Chilton M-D (1986) The hypervirulence of *Agrobacterium tumefaciens* A281 is encoded in a region of pTiBo542 outside of T-DNA. J. Bacteriol 168:1291-1301

Hooykaas PJJ, Beijersbergen AGM (1994) The virulence system of *Agrobacterium tumefaciens*. Annu Rev Phytopathol 32:157-179

Hooykaas PJJ, Schilperoort RA (1992) *Agrobacterium* and plant genetic engineering. Plant Mol Biol 19:15-38

Ishida Y, Saito H, Ohta S, Hiei Y, Komari T, Kumashiro T (1996) High-efficiency transformation of maize (*Zea mays* L.) mediated by *Agrobacterium tumefaciens*. Nat Biotechnol 14:745-750

Jefferson RA (1987) Assaying chimeric genes in plants: the GUS gene fusion system. Plant Mol Biol Rep 5:387-405

Jian RK, Jian S, Wang B, Wu R (1996) Optimization of biolistic method for transient gene expression and production of agronomically useful transgenic Basmati rice plants. Plant Cell Rep 15:963-968

Kahl G (1982) Molecular biology of wound healing: the conditioning phenomenon. In: Kahl G, Schell J (eds) Molecular Biology of Plant Tumors. Academic Press, New York, pp 211-267

Klee HJ, Horsch R, Rogers S (1987) *Agrobacterium*-mediated plant transformation and its further applications to plant biology. Annu Rev Plant Physiol 38:467-486

Kohli A, Ghareyazie B, Kim HS, Kush GS, Bennett J (1996) Cytosine methylation implicated in silencing of β-glucuronidase genes in transgenic rice. In: Khush GS (ed) Rice Genetics III. Proceedings of the Third International Rice Genetics Symposium, Manila (Philippines). International Rice Research Institute, Los Baños, Philippines, pp 825-828

Komari T (1989) Transformation of callus cultures of nine plant species mediated by *Agrobacterium*. Plant Sci 60:223-229

Komari T (1990) Transformation of cultured cells of *Chenopodium quinoa* by binary vectors that carry a fragment of DNA from the virulence region of pTiBo542. Plant Cell Rep 9:303-306

Komari T, Hiei Y, Saito Y, Murai N, Kumashiro T (1996) Vectors carrying two separate T-DNAs for co-transformation of higher plants mediated by *Agrobacterium tumefaciens* and segregation of transformants free from selection markers. Plant J 10:165-174

Kumpatla SP, Teng W, Buchholz WG, Hall TC (1997) Epigenic transcriptional silencing and 5-azacytidine-mediated reactivation of a complex transgene in rice. Plant Physiol 115:361-373

Langridge P, Brettschneider R, Lazzeri P, Lörz H (1992) Transformation of cereals via *Agrobacterium* and the pollen pathway: a critical assessment. Plant J 2:631-638

Li L, Rongda Q, Kochko A, Fauquet C, Beachy RN (1993) An improved rice transformation system using the biolistic method. Plant Cell Rep 12:250-255

Li X-Q, Liu C-N, Ritchie SW, Peng J-Y, Gelvin SB, Hodges TK (1992) Factors influencing *Agrobacterium*-mediated transient expression of *gusA* in rice. Plant Mol Biol 20:1037-1048

Li Z, Upadhyaya NM, Meena S, Gibbs AJ, Waterhouse PM (1997) Comparison of promoters and selectable marker genes for use in Indica rice transformation. Mol Breed 3:1-14

Mooney PA, Goodwin PB, Dennis ES, Llewellyn DJ (1991) *Agrobacterium tumefaciens* gene transfer into wheat tissues. Plant Cell Tissue Organ Cult 25:209-218

Oard J, Linscombe SD, Braverman MP, Jodari F, Bollich PK, Blouin DC, Cooley JC, Leech M, Kholi A, Vain P, Christou P (1996) Development, field evaluation and agronomic performance of transgenic herbicide-resistant rice. Mol Breed 2:359-368

Ohta S, Mita S, Hattori T, Nakamura K (1990) Construction and expression in tobacco of a β-glucuronidase (GUS) reporter gene containing an intron within the coding sequence. Plant Cell Physiol 31:805-813

Park SH, Pinson SRM, Smith RH (1996) T-DNA integration into genomic DNA of rice following *Agrobacterium* inoculation of isolated shoot apices. Plant Mol Biol 32:1135-1148

Peng J, Kononowicz H, Hodges TK (1992) Transgenic indica rice plants. Theor Appl Genet 83:855-863

Potrykus I (1990a) Gene transfer to plants: assessment and perspectives. Physiol Plant 79:125-134

Potrykus I (1990b) Gene transfer to cereals: an assessment. Biotechnology 8:535-542

Potrykus I (1991) Gene transfer to plants: assessment of published approaches and results. Annu Rev Plant Physiol Plant Mol Biol 42:205-225

Raineri DM, Bottino P, Gordon MP, Nester EW (1990) *Agrobacterium*-mediated transformation of rice (*Oryza sativa* L.). Biotechnology 8:33-38

Rashid H, Yokoi S, Toriyama K, Hinata K (1996) Transgenic plant production mediated by *Agrobacterium* in Indica rice. Plant Cell Rep 15:727-730

Rathore KS, Chowdhury VK, Hodges TK (1993) Use of *bar* as a selectable marker gene and for the production of herbicide-resistant rice plants from protoplasts. Plant Mol Biol 21:871-884

Schäfer W, Görz A, Kahl G (1987) T-DNA integration and expression in a monocot crop plant after induction of *Agrobacterium*. Nature 327:529-532

Shimamoto K, Terada R, Izawa T, Fujimoto H (1989) Fertile transgenic rice plants regenerated from transformed protoplasts. Nature 338:274-276

Shimoda N, Toyoda-Yamamoto A, Nagamine J, Usami S, Katayama M, Sakagami Y, Machida Y (1990) Control of expression of *Agrobacterium vir* genes by synergistic actions of phenolic signal molecules and monosaccharides. Proc Natl Acad Sci USA 87:6684-6688

Sivamani E, Shen P, Opalka N, Beachy RN, Fauquet CM (1996) Selection of large quantities of embryogenic calli from indica rice seeds for production of fertile transgenic plants using the biolistic method. Plant Cell Rep 15:322-327

Spencer TM, O'Brien JV, Start WG, Adams TR, Gordon-Kamm WJ, Lemaux PG (1992) Segregation of transgenes in maize. Plant Mol Biol 18:201-210

Stachel SE, Messens E, Van Montagu M, Zambryski P (1985) Identification of the signal molecules produced by wounded plant cells that activate T-DNA transfer in *Agrobacterium tumefaciens*. Nature 318:624-628

Tada Y, Sakamoto M, Fujimura T (1990) Efficient gene introduction into rice by electroporation and analysis of transgenic plants: use of electroporation buffer lacking chloride ions. Theor Appl Genet 80:475-480

Tingay S, McElroy D, Kalla R, Fieg S, Wang M, Thornton S, Brettell R (1997) *Agrobacterium tumefaciens*-mediated barley transformation. Plant J 11:1369-1376

Toki S (1997) Rapid and efficient *Agrobacterium*-mediated transformation in rice. Plant Mol Biol Rep 15:16-21

Toriyama K, Arimoto Y, Uchimiya H, Hinata K (1988) Transgenic plants after direct gene transfer into protoplasts. Biotechnology 6:1072-1074

Turk SCHJ, Melchers LS, Den Dulk-Ras H, Regensburg-Tuïnk AJG, Hooykaas PJJ (1991) Environmental conditions differentially affect *vir* gene induction in different *Agrobacterium* strains. Role of the VirA sensor protein. Plant Mol Biol 16:1051-1059

Usami S, Okamoto S, Takebe I, Machida Y (1988) Factor inducing *Agrobacterium tumefaciens vir* gene expression is present in monocotyledonous plants. Proc Natl Acad Sci USA 85:3748-3752

Van Wordragen MF, Dons HJM (1992) *Agrobacterium tumefaciens*-mediated transformation of recalcitrant crops. Plant Mol Biol Rep 10:12-36

Vancanneyt G, Schmidt R, O'Connor-Sanchez A, Willmtzer L, Rocha-Sosa M (1990) Construction of an intron-containing marker gene: splicing of the intron in transgenic plants and its use in monitoring early events in *Agrobacterium*-mediated plant transformation. Mol Gen Genet 220:245-250

Vasil IK (1994) Molecular improvement of cereals. Plant Mol Biol 25:925-937

Veluthambi K, Krishnan M, Gould JH, Smith RH, Gelvin SB (1989) Opines stimulate induction of the *vir* genes of the *Agrobacterium tumefaciens* Ti plasmid. J Bacteriol 171:3696-3703

Vijayachandra K, Palanichelvam K, Veluthambi K (1995) Rice scutellum induces *Agrobacterium tumefaciens vir* genes and T-strand generation. Plant Mol Biol 29:125-133

Watson B, Currier TC, Gordon MP, Chilton M-D, Nester EW (1975) Plasmid required for virulence of *Agrobacterium tumefaciens*. J Bacteriol 123:255-264

Xu X, Li B (1994) Fertile transgenic rice plants obtained by electroporation of the seed embryo cells. Plant Cell Rep 13:237-242

Yamagishi M, Otani M, Shimada T (1996) A comparison of somaclonal variation in rice plants derived and not derived from protoplasts. Plant Breed 115:289-294

Yoder JI, Goldsbrough AP (1994) Transformation systems for generating marker-free transgenic plants. Biotechnology 12:263-267

Zambryski P (1988) Basic processes underlying *Agrobacterium*-mediated DNA transfer to plant cells. Annu Rev Genet 22:1-30

Zambryski P (1992) Chronicles from the *Agrobacterium*-plant cell DNA transfer story. Annu Rev Plant Physiol 43:465-490

Zambryski P, Joos H, Genetello C, Leemans J, Van Montagu M, Schell J (1983) Ti plasmid vector for the introduction of DNA into plant cells without alteration of their normal regeneration capacity. EMBO J 2:2143-2150

Zhang HM, Yang H, Rech EL, Gold TJ, Davis AS, Mulligan BJ, Cocking EC, Davey MR (1988) Transgenic rice plants produced by electroporation-mediated plasmid uptake into protoplasts. Plant Cell Rep 7:379-383

Zhang S, Chen L, Qu R, Marmey P, Beachy R, Fauquet C (1996) Regeneration of fertile indica (group 1) rice plants following microprojectile transformation of embryogenic suspension-cultured cells. Plant Cell Rep 15:465-469

Zhang W, McElroy D, Wu R (1991) Analysis of rice *Act1* 5' region activity in transgenic rice plants. Plant Cell 3:1155-1165

Zhang W, Wu R (1988) Efficient regeneration of transgenic plants from rice protoplasts and correctly regulated expression of the foreign gene in the plants. Theor Appl Genet 76:835-840

Zupan JR, Zambryski P (1995) Transfer of T-DNA from *Agrobacterium* to the plant cell. Plant Physiol 107:1041-10475-840

13
Genetic engineering of virus resistance

Peter M. Waterhouse and Narayana M. Upadhyaya

CSIRO Plant Industry, GPO Box 1600, Canberra, ACT 2601, Australia

1 Introduction

Twenty-three different viruses are known to infect rice (Brunt et al. 1998), and 15 cause significant yield loses in the field. Rice tungro virus, for example, is considered to be one of the most serious economic constraints on rice production in Asia (Herdt 1991). There are resistance genes in rice or its wild relatives against some viruses and against some of the insects that vector viruses. However, for many viruses, there are no resistance genes, the resistance genes are difficult to transfer into domestic rice, or there are single resistance genes that are under threat of being overcome by evolving virus strains.

In 1986, it was shown that introduction of the coat protein gene of tobacco mosaic virus (TMV) as a transgene into tobacco gave the plant TMV resistance (Powell-Abel et al. 1986). Since then, there have been many examples of the introduction of virus-derived transgenes into dicotyledonous plants for virus resistance. The application of similar pathogen-derived resistance (PDR) in rice has been prevented, until recently, by the difficulty of transforming rice and lack of molecular information about rice-infecting viruses. However, in the past 5 years rice transformation has become more routine (see chapters 12 and 14), and a number of important rice-infecting viruses have been characterized at a molecular level. These advances have facilitated the application of PDR in rice. In this chapter, we review natural virus resistance genes available for domestic rice, strategies of PDR that have been successful in protecting dicots against viruses, the characterization of important rice-infecting viruses, and the progress and prospect of genetically engineered virus-resistant rice.

Key words. pathogen derived resistance, movement protein, dysfunctional virus protein, replicase, ribozyme, cosuppression, antisense RNA, antiviral protein

2 Rice viruses and their economic importance

Among the 23 different viruses known to infect rice, 15 are known to have had serious effects on rice crops around the world (Table 1). Thirteen are transmitted by insect vectors, and two have fungal vectors. One of the first records of a virus epidemic in rice was in 1897 when plants were infected with rice dwarf virus (RDV) in Japan (Katsura 1936). Since then there have been sporadic or cyclic virus outbreaks wherever cultivated rice is grown. Some of the history of virus outbreaks in rice is shown in Table 2.

The most economically important rice viruses are insect borne; ten of them occur in Asia and are vectored either by leafhopper or planthopper (Table 1). The only important rice-infecting virus on the American continent is rice hoja blanca (RHBV), which is planthopper borne. The major rice-infecting virus in Africa is rice yellow mottle virus (RYMV), which is transmitted by beetles (Bakker 1975). Rice giallume virus is a strain of barley yellow dwarf virus (Waterhouse et al. 1986); it is aphid transmitted and has only been reported as a problem in rice in northwest Italy (Osler 1984).

The impact of viruses on rice production has been assessed on national (Setboonsarng 1996; Jatilesono 1996; Dey et al. 1996; Hossain et al. 1996) and international scales (Herdt 1991; Ramasamy and Jatileksono 1996) and although there are some inconsistencies, it is possible to give some ranking to the importance of the viruses. In India and Southeast Asia, rice tungro (RTSV and RTBV) is generally the most serious virus disease; in China and Japan rice stripe virus (RSV) is probably the most important; in Africa and Latin America, RYMV and RHBV, respectively, are the most important. Rice ragged stunt (RRSV), rice grassy stunt (RGSV), rice gall dwarf (RGDV), rice transitory yellowing (RTYV), rice dwarf (RDV), and rice black-streaked dwarf (RBSDV) are all important viruses but in different locations and at different times in Asia. It has been assessed that RTV, RRSV, and RGSV in Southeast Asia cause yield losses per annum of about US$ 700 million, US$40 million, and US$10 million, respectively (Herdt 1991).

The important rice viruses are insect borne, so their spread may be controlled by spraying with insecticides. However, major losses occur in epidemic years that are hard to predict, making control of rice losses due to virus infection difficult. For consistent control of virus vectors the crop must be protected by a regular spraying regime, which is expensive, environmentally harmful, and in some years may be unnecessary. If the applications are too early or too late they will be ineffective at controlling insect populations and virus spread. The most cost-effective, lowest risk, and environmentally benign method of protection of rice from viruses is by the use of resistance genes.

Table 1. Viruses of rice

Virus	Group	Vector	Distribution
Maize rough dwarf (MRDV)	Reo – fijivirus	Planthopper	
Rice black streaked dwarf (RBSDV)	Reo – fijivirus	Planthopper	China, Japan, Korea
Echinochloa ragged stunt (ERSV)	Reo – oryzavirus	Planthopper	Taiwan
Rice ragged stunt (RRSV)	Reo – oryzavirus	Planthopper	S. and S.E. Asia, China, Japan, Taiwan
Rice bunchy stunt (RBSV)	Reo – phytoreovirus	Leafhopper	China
Rice dwarf (RDV)	Reo – phytoreovirus	Leafhopper	China, Japan, Korea, Nepal
Rice gall dwarf (RGDV)	Reo – phytoreovirus	Leafhopper	China, Malaysia, Thailand
Rice grassy stunt (RGSV)	Tenuivirus	Planthopper	S. and S.E. Asia, China, Japan, Taiwan
Rice hoja blanca (RHBV)	Tenuivirus	Planthopper	South and Central America
Rice stripe (RTV)	Tenuivirus	Planthopper	China, Japan, Korea, Taiwan, Russia
Rice tungro spherical (RTSV)	Waikavirus	Leafhopper	S. and S.E. Asia, Japan, China
Rice tungro bacilliform (RTBV)	Badnavirus	Leafhopper	S. and SE Asia, China
Rice yellow mottle (RYMV)	Sobemovirus	Beetle	Africa
Rice transitory yellowing (RTYV)	Nucleorhabdovirus	Leafhopper	China, Japan, Taiwan, Thailand
Rice necrosis mosaic (RNMV)	Furovirus	Fungus	
Rice guiallume (RGV)	Luteovirus	Aphid	Italy
Barley yellow dwarf (RYDV)	Luteovirus	Aphid	Experimental
Barley stripe mosaic (RSMV)	Hordeivirus	Fungus	Experimental
Cymbidium mosaic (CMV)	Potexvirus		Experimental
Foxtail mosaic (FMV)	Potexvirus		Experimental
Sugarcane mosaic (SMV)	Potyvirus	Aphid	Experimental
Maize dwarf mosaic (MDMV)	Potyvirus	Aphid	Experimental
Ryegrass mosaic (RMV)	Rymovirus		Experimental

Data from Brunt et al. (1998) and Hibino (1989).

Table 2. Outbreaks of viruses in cultivated rice

Date	Country	Virus	Reference
1897	Japan	Rice dwarf	Ishii et al. 1970
1903	Japan	Rice stripe	Chen et al. 1980
1941	Japan	Rice black streaked dwarf	Kuribayashi and Shinkai 1952
1956	Latin America	Rice hoja blanca	Galvez 1969
1957	Philippines	Rice tungro	Hibino et al. 1978 Omura et al. 1983
1960-1962	Taiwan	Rice transitory yellowing	Su 1969
1964	South China	Rice transitory yellowing	Faan et al. 1965
1965	Thailand	Rice tungro	Ling 1972
1966	Kenya	Rice yellow mottle	Bakker 1975
1969	Bangladesh India Indonesia Malaysia	Rice tungro	Ling 1972 Ou 1985
1970-1973	India Indonesia Philippines	Rice grassy stunt	Ou 1985
1976	Sierra Leone Ivory Coast Nigeria	Rice yellow mottle	Bakker 1975 Rossel 1986
1977	Southeast Asia	Rice ragged stunt	Hibino et al. 1977 Ling et al. 1978
1981	China	Rice gall dwarf	Faan et al. 1983

3 Natural resistance genes

Protection of rice from viruses can be mediated by natural resistance genes that inhibit the virus replicating (or spreading) in the plant and by genes which repel virus-vectoring insects. The resistance can range from complete immunity to moderate tolerance, and sources have been identified in *indica, japonica* and wild species of rice (Kobayashi et al. 1993; Cuevas-Perez et al. 1993). About 40,000 germplasm accessions have been screened for resistance to rice tungro disease using viruliferous green leafhoppers (*Nephotettix virescens*), and 119 lines showed good levels of resistance. Some of these lines were resistant to the green leafhopper (GLH), 40 lines inhibited the replication of RTSV, 16 lines were tolerant of RTBV, and 9 lines appeared resistant to both viruses (Hibino et al. 1990). Sebastian and others (1996) have mapped the chromosomal location of the GLH and RTSV resistance gene(s). Their study shows this dominant gene(s) is located within 5.5 cM of RFLP marker RZ 262 on chromosome 4. A collection of 9766 rice lines screened for resistance to rice yellow dwarf virus (RYDV) yielded 1161 moderately resistant lines, 34 highly resistant lines, and 64 immune lines (Li et al. 1990). Screening 362 accessions for RGSV resistance has identified 30 lines with moderate resistance (Devika et al. 1990). Sources of natural resistance have also been found for RSV, RDV, RRSV, and RYMV (Li et al. 1990; Xing et al. 1985; Nemato et al. 1994; Ikeda and Kaneda 1982; Fomba 1988). Many of the resistance sources found for rice viruses are polygenic, and in most cases their mechanism of action is unknown.

4 Engineered resistance

Recent advances in molecular biology make it possible to take genes from any organism, modify them so that they can be expressed in plants, and then deliver them into the genomes of plants. This has expanded the potential repertoire of genes for crop plants over that possible by conventional breeding and has facilitated pathogen-derived (originally termed parasite-derived) resistance (PDR). The concept of PDR is that genes from a pathogen genome can be expressed inappropriately in its host, thereby interrupting the infection cycle (Sanford and Johnston 1985). Embracing this concept, virus genes have been introduced into plant genomes and in many cases virus-resistant plants were obtained. To use PDR effectively and responsibly, it is important to understand the life cycle and life style of the target virus at a molecular level.

4.1 Life of a virus

Most plant viruses have genomes composed of single-stranded translatable RNA that are packaged in coat proteins to form virus particles. Almost all these viruses have genomes encoded on one, two, or three RNA molecules and have between five and ten genes. All the pathogen-derived resistance research described in Section 4.2 deals with viruses with this type of genome organization and life cycle. Three genes common to all viruses are essential to the life cycle of a plant virus (Fig. 1). They encode a replicase, a coat protein, and a movement protein. The replicase replicates the invading viral genome to produce many new viral RNA molecules, the coat protein packages the newly produced RNA to form virus particles, and the movement protein interacts with the viral RNA and the plant cell plasmodesmata to facilitate the spread of the viral RNA from cell to cell. Interfering with any one of these genes will interrupt the virus life cycle. If the replicase is inhibited the virus cannot multiply; if the coat protein is interfered with, the virus RNA cannot be packaged and is vulnerable to ribonucleases; and without a functional movement protein the virus cannot spread within the plant. These three genes have been the most common targets for PDR against plant viruses.

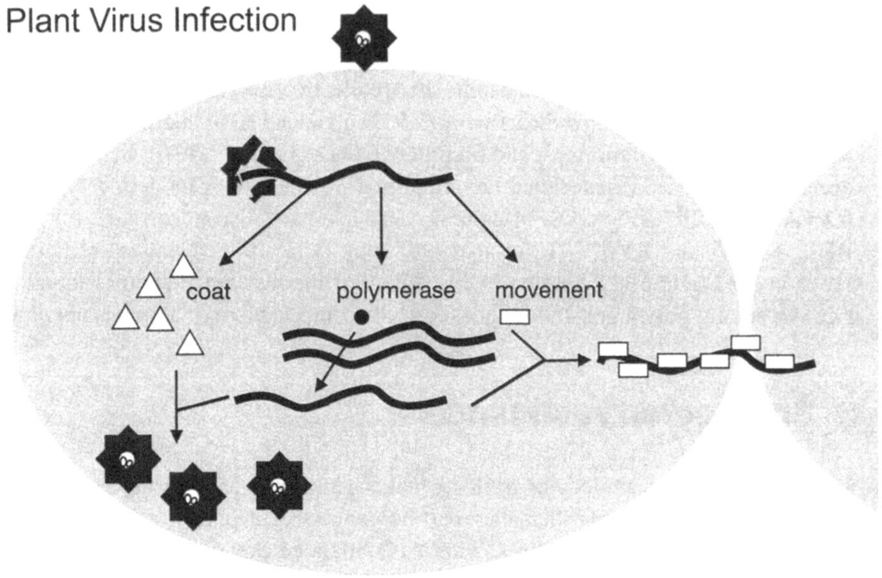

Fig. 1. Life cycle of a virus. Once the virus particle enters the cell, it is disassembled to release the virus RNA genome, which is translated to give at least three proteins: the polymerase, movement, and coat proteins. The polymerase makes new copies of the virus RNA genome, the coat protein packages the newly synthesized RNA into particles, and the movement protein enables the virus RNA to spread into new cells.

4.2 Pathogen-derived resistance (PDR)

As yet, PDR has been applied in rice in a very limited way. However, it has been used very successfully for a large number of viruses in dicotyledonous crops. It is likely that many of the strategies used in dicotyledonous plants are also applicable in rice. In this section we describe these approaches. PDR for virus resistance can be divided into three categories: the expression in plants of wild-type virus proteins, the expression of dysfunctional virus proteins, and the expression of virus RNA or virus-associated RNA. Each approach has variations and has met with some degree of success.

4.2.1 PDR using wild-type virus proteins

It might be expected that expressing a wild-type virus protein in a plant would enhance viral infection rather than inhibit it, as there are examples where deliberately crippled viruses can be complemented by transgenic expression of the wild-type gene (e.g., Osbourn et al. 1990). However in most cases, viruses inoculated onto plants expressing a viral transgene are either inhibited or unaffected. The reasons for this inhibition are not clear, but it may be that the viral protein is expressed inappropriately in terms of the regulated expression of viral genes during a normal infection.

Coat protein-mediated resistance

Coat protein-mediated resistance has been the most widely used form of genetically engineered resistance against plant virus infection (Fitchden and Beachy 1993). It is achieved by incorporating a gene sequence encoding the viral coat protein into a plant gene expression cassette and transforming the cassette into the plant genome. Expression of the virus coat protein in the plant usually gives resistance against not only the virus from which the coat protein was isolated, but also gives some resistance to other related viruses.

The first report of this type of resistance was for tobacco mosaic virus (Powell-Abel et al. 1986). Transgenic tobacco plants expressing a tobacco mosaic virus (TMV) coat protein transgene showed delayed symptom development when inoculated with relatively low levels of TMV. The transgenic plant lines expressing higher levels of coat protein, in general, showed more resistance to the virus than lines with lower levels of coat protein expression (Powell et al. 1990). It was demonstrated that the resistance was mediated by the coat protein, not its mRNA, by making transgenic plants containing the coat protein gene in which the initiation codon was disrupted. These plants, which expressed the mRNA but not the protein, showed no resistance to TMV.

Since this first report, coat protein-mediated resistance has been reported for most of the 23 known plant virus groups. Although this approach has been so widely effective, the mechanism(s) by which it operates are not well understood. However, the resistance seems to act by interfering with virus particle uncoating, virus replication, and virus spread (Register and Beachy 1988; Osbourn et al. 1989; Wisniewski et al. 1990). For coat protein-mediated protection against TMV, there is good evidence for prevention of virus particle uncoating following invasion of the cell (Wilson and Watkins 1986; Register and Beachy 1988; Osbourn et al. 1989). Also for TMV, there is evidence that the coat protein is blocking systemic movement (Wisniewski et al. 1990). When a stem piece from a coat protein expressing transgenic plant was grafted into the stem of a nontransgenic plant and TMV was inoculated below the graft junction, the virus was unable to pass through the section to the nontransgenic stem and leaves above. However, there are inconsistencies between the TMV coat protein-mediated resistance attributes and those of other viruses. Indeed, many of the examples of "coat protein-mediated resistance" have attributes more consistent with RNA-mediated mechanisms, as discussed later.

Replicase-mediated resistance

Replicase-mediated resistance was first described for TMV (Golemboski et al. 1990). Tobacco transformed with the C-terminal portion (54-kDa protein) of the TMV replicase were highly resistant to TMV infection. Introduction of a mutation into the initiation codon of this gene negated its effectiveness (Carr and Zaitlin 1991; Carr et al. 1992), showing that the resistance was protein mediated. Similar C-terminal replicase-mediated resistance also has been reported for pea early browning virus (MacFarlane and Davies 1992) and cucumber mosaic virus (Carr et al. 1994). Full-length replicase-mediated resistance has been reported for potato virus X (Braun and Hemenway 1992) and cymbidium ringspot virus (Rubino et al. 1993). However, in plants carrying a transgene derived from the replicase genes of potato virus X (Mueller et al. 1995), cowpea mosaic virus (Sijen et al. 1996), and pepper mild mottle virus (Tenllado et al. 1995), there is clear evidence that the resistance is mediated at an RNA level.

Movement proteins

Tobacco (Nicotiana tabaccum) is a host of TMV but not of brome mosaic virus (BMV). Tobacco was transformed with a gene encoding the movement protein of brome mosaic virus and the resulting plants inoculated with TMV (Malyshenko et al. 1993). These plants were resistant to TMV infection. This suggests that the expression of a heterologous movement protein that is not adapted to a particular plant species can interfere with the processes mediated by the movement protein of the infecting virus.

4.2.2 PDR using dysfunctional virus proteins

The hypothesis for this approach is that a mutated protein may be able to compete with, and disrupt, the function of a wild-type protein. It has proved to be successful for plant virus protection using mutated movement protein and mutated replicase.

Mutant replicase

The replicases of almost all RNA viruses have a conserved domain which includes the amino acid motif glycine-aspartate-aspartate. In mutant replicase genes used for plant virus protection, this domain is either deleted or modified by amino acid substitution. A deleted version of the cucumber mosaic virus (CMV) replicase has been reported to be effective for protection against CMV (Anderson et al. 1992) and modified replicase has been shown to be effective for potato virus X (Longstaff et al. 1993) and alfalfa mosaic virus (Brederode et al. 1995). An accidentally mutated (during the process of cloning) replicase gene for TMV has been reported to be very effective in contrast to use of the wild-type gene (Donson et al. 1993). In most of these cases, the resistance has been effective only against the homologous virus or its close relatives.

Mutant movement protein

Although the resistance conferred by most mutant replicase genes appears to be restricted to the homologous virus or close relatives, mutant movement proteins seem to confer broader resistance. Four virus families, including the potexvirus and carlavirus groups, have a common organization of movement protein genes termed the "triple block" arrangement. In all the viruses there are three closely located open reading frames (ORFs), and the central ORF encodes a small (~13-kDa) protein with conserved structural motifs of two hydrophobic domains separated by a hydrophilic domain. A transgene derived from the central ORF was made from the potexvirus white clover mosaic virus (WCMV) with the modification that six conserved hydrophilic amino acids in the central domain were converted to hydrophobic amino acids. *Nicotiana benthamiana* plants transformed with this gene showed not only resistance to WCMV but also to other potexviruses and a carlavirus (Beck et al. 1994). Similarly, broad resistance has been found using mutant movement proteins from TMV (Cooper et al. 1995) and potato leaf roll virus (Tacke et al. 1996). In both cases the plants produced showed resistance to viruses outside their respective families.

4.2.3 PDR using RNA

It has been long recognized that expression of a nucleic acid, not a protein, might interfere with the virus infection cycle. There are three different RNA-mediated mechanisms that have been successful in transgenic plants. One is the expression of an RNA that competes against viral RNA for viral proteins; another is the production of ribozymes, which are RNAs that can be designed to cleave viral RNAs at a specific sequence; and the third, which is the least well understood, is involved with antisense and cosuppression and seems to induce the plant to specifically degrade sequences of viral RNA.

Competitive RNA

Satellite RNAs are small single-stranded RNAs which occur naturally in association with some viruses. A satellite RNA does not encode replicase or coat proteins but is replicated and packaged by its associated virus. It also usually ameliorates the virus symptoms. Not all viruses are capable of supporting satellite replication, and satellites have little or no sequence homology with their helper virus. Transgenic plants expressing cloned copies of satellite of tobacco rattle virus (Gerlach et al. 1987) or CMV (Harrison et al. 1987) have been reported. When the plants were challenged with their respective helper viruses, they showed reduced levels of virus replication and amelioration of symptom development. It is thought that this attenuation may be caused by the satellite competing with the viral RNA for replicase or coat protein. However, there are certain virus/satellite/host plant combinations that give enhanced, rather than reduced, virus symptoms (Tien and Wu 1991).

Defective interfering (DI) RNAs are similar to satellites in that they can ameliorate symptoms caused by their helper virus and probably by the same competitive mechanism. However, DIs are small RNAs derived from the helper virus. Transgenic plants expressing a DI from cymbidium ringspot virus showed reduced virus symptoms when challenged with the virus (Koller et al. 1993).

Ribozymes

Many satellite RNAs are replicated from circular templates as multimers within long RNA transcripts. To produce unit-size satellite RNAs, the transcripts undergo self-cleavage. Haseloff and Gerlach (1988) identified the cleavage and the specificity regions and showed that they could be used to make RNA cleaving (in cis or trans) molecules with any desired specificity. The only constraint is that the target RNA must have the sequence GUC and the cleavage occurs immediately after the C of this motif. There has been one report of a ribozyme conferring protection in transgenic plants. Tobacco plants expressing a ribozyme targeted against potato spindle tuber viroid (a virus-like pathogen with an RNA genome) showed a very high level of resistance to infection by the viroid (Yang et al. 1997).

Antisense and cosuppression RNA

Expression of antisense RNA in transgenic plants has been a much used method of inhibiting gene expression (see reviews by van den Elzen et al. 1989; Finnegan and McElroy 1994). It has also been used successfully in transgenic plants to give resistance against PVX (Hemenway et al. 1988), CMV (Cuozzo et al. 1988), TMV (Powell et al. 1989), and tobacco etch virus (Lindbo and Dougherty 1992a, 1992b). In animal systems, antisense has been reported to operate by either preventing the processing or transport of RNA in the nucleus or by hybridizing to the target mRNA in the cytoplasm and preventing translation (Green et al. 1988). However, there is increasing evidence that antisense in plants is inducing a posttranslational silencing mechanism, the same mechanism that is induced by cosuppression (Baulcombe 1996).

Posttranscriptional cosuppression (also called gene silencing; see reviews by Finnegan and McElroy 1994 and Flavell 1994) was first proposed as a mechanism for induced virus resistance in plants by Dougherty and colleagues (Lindbo and Dougherty 1992a,1992b; Lindbo et al. 1993a, 1993b; Smith et al. 1994). They were examining the resistance to two potyviruses (tobacco etch virus and PVY) in plants transformed with the coat protein genes and found that the protein was not required for resistance. They also observed that the plants showing high levels of resistance had low levels of coat protein mRNA in the cytoplasm yet their transgenes were being highly transcribed in the nucleus. Their conclusions, which are steadily gaining widespread acceptance, were that very high levels of transcription of the transgene (usually associated with high transgene copy number) induces a mechanism in the plant which specifically degrades the mRNA of that transgene and, as the transgene mRNA has the same sequence as part of the viral genome, the virus genome is also degraded by the induced mechanism (see reviews by Baulcombe 1996 and Stam et al. 1997).

4.2.4 Antiviral protein-mediated resistance

In addition to the approaches just described, virus resistance has been achieved in transgenic plants by expressing antiviral proteins from other organisms. These include expression of an antiviral protein from pokeweed (Lodge et al. 1993), an interferon-associated protein 2'-5'-oligoadenylate synthetase from rat (Truve et al. 1993), a human cysteine-protease inhibitor (Garcia et al. 1993), a single-chain antibody against a virus coat protein (Tavladoraki et al. 1993), and a double-stranded RNA-specific ribonuclease (Watanabe et al. 1995).

5 Rice viruses

The basic life cycle and genome organization of the majority of plant viruses is described in Section 4.1. However, 11 rice-infecting viruses have major differences from the common theme. These fall into three virus families/genera: the reoviruses, the badnaviruses, and the tenuiviruses.

5.1.1 Waikaviruses and badnaviruses

Although rice tungro disease was originally thought to be caused by one virus and is often still referred to in this way, it is actually caused by a complex of two viruses, rice tungro bacilliform badnavirus (RTBV) and rice tungro spherical waikavirus (RTSV) (Hibino 1983; Jones et al. 1991). RTBV in not insect transmitted without the help of RTSV, and RTBV causes the majority of the disease symptoms. RTSV has an single-stranded (plus sense) linear RNA genome similar to those described in Section 4.1 (with the deviation that it has three coat protein subunits; Fig. 2). However, RTBV is a badnavirus and has a circular double-stranded DNA genome (Fig. 2). A large part of its replication cycle occurs in the nucleus with its genes expressed from a single RNA transcript containing four ORFs. The functions of proteins encoded by ORFs 1, 2, and 4 are unknown. ORF3 is much the largest of the four ORFs and by sequence analysis has been shown to encode a polyprotein containing the coat protein, viral proteinase, reverse transcriptase, and ribonuclease H.

5.1.2 Reoviruses

Seven different reoviruses infect rice (Table 1). Reoviruses are unusual plant viruses as they replicate in their insect vectors, have multisegmented genomes (10-12 segments) of double-stranded (ds)RNA, carry replicase molecules within their particles, and have very complex double-shelled isometric virus particles. The organization of the genes in reoviruses is relatively simple: nearly all the genome segments encode a single protein (e.g., rice ragged stunt virus; Fig. 2). From work on animal-infecting reoviruses (Estes 1990; Gorman et al. 1983; Schiff and Fields 1990), it is possible to present a scenario for the plant reovirus replication cycle. Almost all steps of the life cycle (Fig. 3) occur within the particle. On infection of the plant cell, the outer shell is removed and the replicase within the inner particle produces mRNA transcripts (from some of the dsRNA genome segments) which exit the particle via channels. The proteins translated from these "early" transcripts

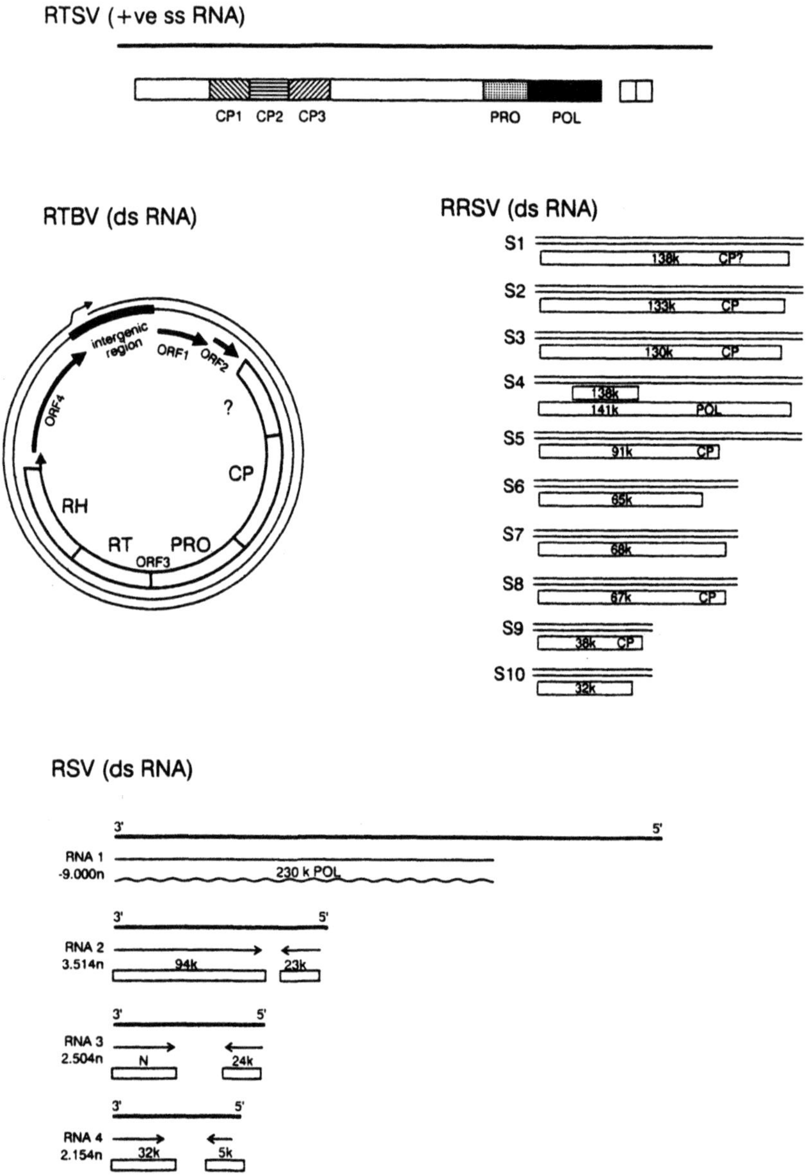
RTSV (+ve ss RNA)
CP1 CP2 CP3
PRO POL
RTBV (ds RNA)
intergenic region
ORF1
ORF2
ORF4
?
CP
RH
RT
PRO
ORF3
RRSV (ds RNA)
S1 138k CP?
S2 133k CP
S3 130k CP
S4 138k 141k POL
S5 91k CP
S6 65k
S7 68k
S8 67k CP
S9 38k CP
S10 32k
RSV (ds RNA)
3' 5'
RNA 1 -9.000n 230 k POL
RNA 2 3.514n 94k 23k
RNA 3 2.504n N 24k
RNA 4 2.154n 32k 5k

form new core particles and specifically package one copy of each of all the "late" RNA transcripts (all 10 or 12 segments), then produce the complementary RNA strand within the particle. These particles are then either coated with the outer shell and become inactive or remain single shelled and repeat the replication process. An important feature of reoviruses is the need to package one copy of each of the genome segments per particle. It seems that the sequences at the termini of each segment form "panhandle" secondary structures that allow their specific recognition during the core packaging process.

During the past 7 years we have sequenced the entire genome (10 segments) of RRSV and identified four genes encoding structural proteins (S3, S5, S8, and S9), one polymerase gene (S4), and two genes encoding nonstructural proteins (S7 and S10) (Li et al. 1996; Upadhyaya et al. 1995, 1996, 1997, 1998). S8 encodes a 43-kDa major capsid protein by an autocatalytic cleavage of a 68-kDa polypeptide (Upadhyaya et al. 1996). One other plant-infecting reovirus, RDV, has been sequenced completely (Suzuki et al. 1994; Murao et al. 1994), and the functions of the genes encoded on the 12 genomic segments have been largely determined. The virus particle consists of six polypeptides, including P8, the major outer capsid protein (Omura et al., 1989); P2, the cell attachment protein (Nakata et al. 1978); P3, the major core protein (Suzuki et al. 1992); P1, the RNA-dependent RNA polymerase (Suzuki et al. 1992); P5, the guanyltransferase (Suzuki et al. 1996), and P7, a major core component (Omura 1995).

5.1.3 Tenuiviruses

Rice stripe, rice grassy stunt, and rice hoja blanca viruses are tenuiviruses and have some unusual properties. Like the reoviruses, they replicate in their insect vectors as well as their plant hosts; they package dsRNA, the termini of the genome segments can form panhandle structures, and their particles contain replicase molecules. Unlike reoviruses, they have simple filamentous virus particles, package both single- (negative) stranded and double-stranded forms of their four genome

---

Fig. 2. Genome organization of rice-infecting viruses. Rice tungro spherical waikavirus (RTSV) has a genome of single-stranded RNA that is translated into one polyprotein which is cleaved by the viral protease (PRO). The polyprotein contains three coat proteins (CP1, CP2, and CP3) and an RNA-dependent RNA polymerase (POL). Rice tungro bacilliform badnavirus (RTBV) has a genome of double-stranded DNA which is transcribed to give a mRNA containing four open reading frames (ORFs). ORF3 encodes a polyprotein containing coat protein (CP), protease (PRO), reverse transcriptase (RT), and Rnase H domains. Rice ragged stunt reovirus (RRSV) has a genome of double-stranded RNA divided into ten segments. Each segment contains a single translated ORF, except S4, which contains two ORFs one of which encodes an RNA-dependent RNA polymerase (POL). Many of the ORFs encode coat proteins (CP), which make up the complex virus particle. Rice stripe tenuivirus (RSV) has a genome of RNA which is packaged in double-stranded or negative-sense single-stranded form. The genome is made up of four segments of which the largest (RNA 1) encodes an RNA-dependent RNA polymerase (POL). RNAs 2, 3, and 4 contain ORFs in both their sense and negative-sense forms.

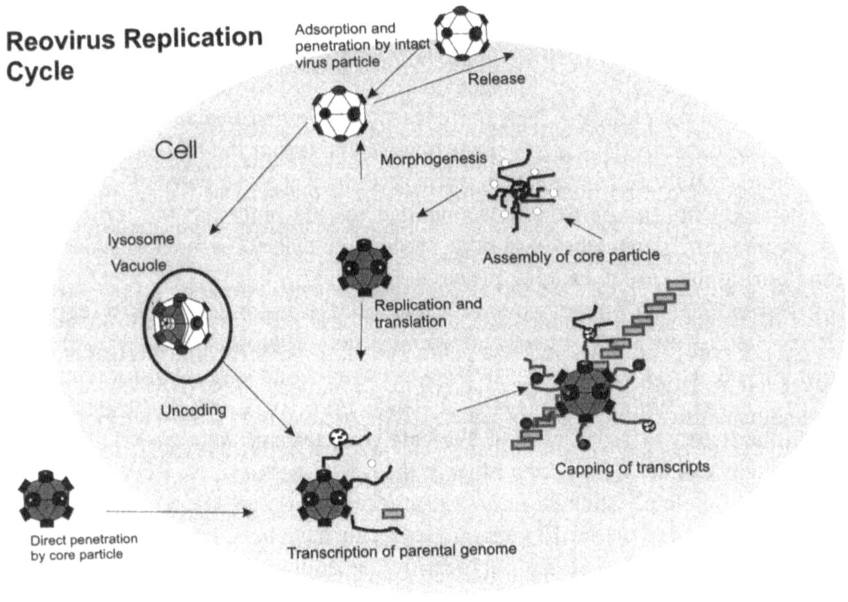

Fig. 3. Reovirus replication cycle. Reoviruses have complex virus particles in which the virus carries out most of its life cycle. Once the particle invades the cell, the outer shell is removed, opening channels within the particle spikes. Single-stranded RNA is transcribed from the double-stranded RNA within the particle and escapes through the channels. The RNAs are either translated to generate particle proteins or are selectively packaged into new core particles. Each new core particle contains only one copy of each virus RNA segment. The complementary RNA strands are synthesized within the particle, and the particle is either coated with the outer protein layer or remains in its core state to undergo another round of replication and translation.

segments, and express their genes using an ambisense coding strategy (e.g., rice stripe virus; Fig 2). Genome segment 1 encodes the replicase. Segments 2, 3, and 4 use ambisense expression to produce six proteins including a nucleoprotein (N protein), which is associated with the virus particle and sometimes referred to as the coat protein.

5.2 Engineered resistance to rice viruses

There is a wealth of approaches that have conferred virus resistance on dicotyledonous plants using transgenesis (see Section 4). Some or all of these approaches may also be useful for protecting rice and other monocotyledonous plants against viruses. However, as illustrated in Section 5.1, a number of the viruses that are a problem in rice differ significantly from the type of viruses shown to be responsive to the approaches described in Section 4. Their differences may make them intractable to some of these approaches but might also provide opportunities for alternative resistance strategies. To our knowledge, there is only one published report of virus resistance in rice conferred by transgenesis (Haykawa et al. 1992). However, we know of at least three research groups (including our own) that are working on different viruses and have transgenic rice plants which by preliminary analysis appear to be showing virus resistance.

5.2.1 Waikaviruses and badnaviruses

Rice tungro spherical waikavirus

Rice tungro disease is such an important disease in Asia that the Rockefeller Foundation rice biotechnology program has simultaneously supported at least three different laboratories (Scripps, USA; John Innes, UK; ITH, Switzerland) in their endeavors to produce transgenic rice with tungro resistance. This has led to a molecular understanding of the genomes of both RTSV (Shen et al. 1993) and RTBV (Qu et al. 1991) and, recently, to preliminary data suggesting that resistance has been achieved for RTSV. RTBV causes the majority of the rice tungro disease symptoms but cannot be vectored by insects without RTSV. Providing rice with RTSV resistance may prevent tungro epidemics by limiting the spread of the two-virus complex.

RTSV has a single-stranded RNA genome and a life cycle strategy similar to the majority of plant viruses. Its isometric virus particles, however, are composed of three different types of coat protein subunits. Researchers at the Scripps Research Institute have recently introduced the three coat protein (cp1, cp2, and cp3) genes, each under the control of the maize ubiquitin promoter, individually into rice using the particle bombardment method of transformation. The resulting transgenic plants were tested for RTSV resistance as primary transformants and as T1 and T2 progeny. Some of the plants with cp1, cp2, or cp3 genes showed moderate levels of resistance. Interestingly, the plants showing this resistance had detectable levels of mRNA for the transgenes but undetectable levels of the coat proteins. Very recent results using RTSV replicase seems to be giving a high level of resistance (Huett, personal communication).

Rice tungro bacilliform badnavirus

RTBV has a genome composed of dsDNA. This virus and other members of the badnavirus group have only recently been identified and characterized. To date there are no reports of PDR to RTBV.

5.2.2 Reoviruses

We are using PDR approaches for RRSV. We have placed the genes (in both sense and antisense orientations) encoded by genome segments S5, S7, S8, S9, and S10 into plant expression vectors and transformed *japonica* (cultivar Taipei 309) and *indica* (cultivar Chinsurah Boro II) rice with them. Plants containing the sense transgenes under the control of the maize *Ubiquitin* or rice *Actin1* promoters produced detectable levels of transgene protein. Preliminary glasshouse tests (in China) indicate that resistance to RRSV has been achieved with transgenes derived from genome segments S5, S7, S8, S9, and S10. In some cases the effective transgenes were in the sense orientation, and in others they were in the antisense orientation. Interestingly, RRSV resistance seems to be more frequent with transgenes driven by the vascular specific promoter *rolC*. The level of resistance varied from complete immunity to delayed symptom development.

Transgenes derived from RDV S8 or RRSV S9 genes have been placed under the control of the CaMV35S promoter and transformed into rice (Matsumura and Tabayashi 1995). Unfortunately, these plants showed no increased resistance to RDV.

5.2.3 Tenuiviruses

Rice stripe virus

The only published report of virus-resistant rice obtained by PDR, to date, is that for RSV (Haykawa et al. 1992). Two *japonica* rice varieties (Kinuhikari and Nipponbare) were transformed with the N protein (referred to by the authors as the coat protein) gene of RSV under the control of the cauliflower mosaic virus 35S promoter with the castor bean catalase intron in the 5' untranslated region of the transcript. The gene construct was electroporated into protoplasts which were subsequently regenerated into 33 independent plant lines. Nineteen of these lines were found to be expressing detectable levels of N protein at a level of ~0.5% total protein. When 5 of the 33 lines were inoculated (3 to 4 vegetative clones of the primary transformant per line) with RSV, plants from 2 lines showed virus symptoms and plants from 3 lines were symptomless (e.g., Fig. 4). The transgene-encoded N protein was detectable in the three symptomless lines but not in the two susceptible lines.

Other primary transformed lines were selfed and their progeny tested for N-protein expression. The progeny of two lines showed 3:1 segregation of detectable N protein and none of the progeny of another line had detectable N-protein levels. When these three lines and a nontransgenic line were challenged with RSV, 80%-90% of the nontransgenic or nonexpressing plants (~20 plants per line) showed virus symptoms whereas only 20%-40% of the plants from the two N-protein-expressing lines showed virus symptoms. The authors concluded that expression of the RSV N protein was needed to give resistance and noted that the level of ~0.5% of total protein was similar to the level found to be effective for TMV coat protein protection against TMV infection (Powell-Abel et al. 1986). They also suggested that the mechanism causing the resistance was similar to that demonstrated for TMV, the prevention of viral uncoating early in the infection process.

RSV is a tenuivirus and has the unusual properties described in Section 5.1.3. This demonstration of N-protein-mediated resistance for RSV suggests that it will be a successful approach for resistance to other tenuiviruses such as rice grassy stunt and rice hoja blanca viruses. Because this mechanism seems to be acting before viral replication commences, the unusual nature of the tenuivirus replication

Fig. 4. Transgenic rice (*left*) protected against rice stripe virus (provided by Ko Shimamoto).

strategy is perhaps irrelevant for this mechanism of resistance.

Rice hoja blanca virus

Encouraged by the N-protein-mediated resistance in rice against RSV (Haykawa et al. 1992), a research group at CIAT (Z. Lentini, L. Calvert, I. Loranzo, E. Tabares, M. Cuervo, and E. Restrepo) are investigating virus protection by expression of the RHBV nucleoprotein (N) gene in rice. They have transformed Latin American indica rice variety Circa-8, using biolistic delivery, with a construct encoding the N gene under the control of the CaMV 35S promoter. The plant lines obtained show extreme resistance to RHBV, Mendelian inheritance of this trait, and unaltered agronomic performance characteristics (in glasshouse tests) in the absence of the virus. Their best lines hold great promise for breeding with rice containing conventional RHBV resistance to produce a superior rice variety for use in Latin America. The CIAT team has found that some of their resistant lines contain N genes with partial deletions. They have also found that their most resistant lines had no detectable levels of N protein and very low levels of mRNA. This suggests that, in contrast to the RSV results, the mechanism is operating at the RNA level, is probably posttranscriptional degradation, as described in Section 4.2.3, and that the mechanism can operate against viruses utilizing ambisense gene expression.

6 Problems and prospects

We have seen that resistance to viruses has been achieved in dicotyledonous plants using a number of PDR approaches and that one of these approaches (although probably operating by two different mechanisms) has produced rice with resistance to RSV and RHBV. There are also encouraging preliminary data suggesting that rice with pathogen-derived resistance for RTSV and RRSV will soon be available. It seems very likely that rice with resistance to RGSV could be readily achieved using either N gene as this approach has worked well for the other tenuiviruses, RSV and RHBV. Resistance to RYMV may also be simply achieved using standard PDR approaches as it has a genome structure and life cycle similar to many of the dicot viruses for which PDR has been effective.

If the preliminary data for RRSV are confirmed, the expectation is that the other six rice-infecting reoviruses will also be amenable to such strategies. Reoviruses spend most of their life within a particle core, and it therefore seemed that this might make them less vulnerable to straightforward PDR strategies. A number of strategies, special to reoviruses, could be pursued, such as overexpressing the protein that covers the channels of the core particle through which viral RNA is exported. Another approach might be to produce RNAs which contain prematurely terminated reading frames but have the "panhandle" sequences that label them as specific segments. Such molecules should compete with viral RNAs for packaging into new core particles but not produce wild-type protein in the next round of translation/replication. A number of other strategies could be postulated based on the exceptional reovirus life cycle. However, this may not be necessary as preliminary RRSV results suggest that an RNA-mediated resistance (since antisense works) is effective. It may be that this mechanism is acting on the viral RNA, which is briefly exposed for translation before the new core particles are synthesized.

RTBV is probably the most difficult virus to protect rice against and is one of the most devastating to the rice crop. Controlling RTSV may control RTBV as it will not be able to spread within the rice crop without its helper virus. However, resistance to RTSV will not protect rice from RTBV coming from external reservoirs. It would be more satisfactory to have direct resistance to RTBV. RTBV has a dsDNA genome and in dicots there has been little success for PDR against DNA viruses. Nevertheless, we believe that a cosuppression approach may be successful as the RTBV RNA transcripts may be vulnerable early in infection.

An advantage of using a PDR transgene is that it can be followed in breeding programs using perfect markers, whereas natural resistance genes are often polygenic and usually have, at best, linked markers. However, polygenic resistance is likely to be more durable in the field. We will probably see, in the future, the use of combinations of natural virus and insect resistance genes with virus and insect resistance transgenes. In cases in which there is no effective or easily transferred natural rice virus resistance gene, one might be made using PDR and then combined in a rice background containing natural virus resistance to other viruses.

The use of virus resistance transgenes pose concerns other than their effectiveness. In the late 1980s, when transgenic plants containing viral genes were first produced and shown to have virus resistance, little concern was raised over the possible environmental impact of these plants. However, there have since been many discussions (e.g., AIBS 1995; Tepfer and Balazs 1997; McLean et al. 1997) about the wisdom of using such transgenic plants in broad-acre agriculture. Some scientists believe the benefits of transgenic virus resistant plants greatly outweigh the minimal risks (Falk and Breuning 1994; Miller et al. 1997) while others encourage a cautious approach (Gibbs et al. 1997). The potential risks posed by plants expressing rice virus genes are similar to those expressing other plant virus genes and can be categorized into four scenarios: (1) interaction of the transgene protein with an invading virus causing a new disease; (2) selection by the transgene-mediated resistance for the creation of a new resistance-breaking virus strain; (3) escape of the transgene into weed· species, conferring on them greater fitness; and (4) recombination between a transgene mRNA and the genome of an invading viruses to create a novel virus and a new disease.

All these possibilities are worst case scenarios and seem to have very low chances of occurring. However, each of these scenarios needs to be addressed. Indeed it may be necessary to modify or abandon some of the PDR approaches to satisfy governments, regulators, and the general public of the complete safety of such transgenic crops. We believe that careful and responsible use of PDR transgenes in rice will minimize yield losses from viruses and that within a decade transgenic virus-resistant rice plants will be in widespread use to help feed the world's growing population.

Acknowledgments. We thank Elumalai Sivamani, Ping Shen, Ching Ang Ong, Herve Huet, Alexandre de Kochko, Claude Fauquet, Roger Beachy, Z. Lentini, Lee Calvert, I. Lozano, E. Tabres, M. Cuervo, and E. Restrepo for access to their unpublished information and Ko Shimamoto for the RSV photograph used in Figure 4. We would also like to thank Garry Toenissen and the Rockefeller Foundation for their support throughout our RRSV work.

References

AIBS (1995) Proceedings of the workshop on transgenic virus-resistant plants and new plant viruses. American Institute of Biological Sciences, Beltsville, MD. http://www.resto.com/aibs/transgenic/virus/html

Anderson JM, Palukaitis P, Zaitlin M (1992) A defective replicase gene induces resistance to cucumber mosaic virus in transgenic tobacco plants. Proc Natl Acad Sci USA 89:8759-8763

Bakker W (1975) Rice yellow mottle virus. Descriptions of plant viruses. No 149. Commonwealth Mycological Institute, UK

Baulcombe DC (1996) Mechanisms of pathogen-derived resistance to viruses in transgenic plants. Plant Cell 8:1833-1844

Beck DL, Van Dolleweerd CJ, Lough TJ, et al (1994) Disruption of virus movement confers broad spectrum resistance against systemic infection by plant viruses with a triple gene block. Proc Natl Acad Sci USA 91:10310-10314

Braun CJ, Hemenway CL (1992) Expression of amino-terminal portions of full-length viral replicase genes in transgenic plants confers resistance to potato virus X infection. Plant Cell 4:735-744

Brederode FT, Taschner PEM, Posthumus E, et al (1995) Replicase-mediated resistance to alfalfa mosaic virus. Virology 207:467-474

Brunt A, Crabtree K, Dallwitz M et al (1998) VIDE Plant Viruses Online. http://biology.anu.edu.au/Groups/MES/vide/refs.htm#intro

Carr JP, Amit GO, Palukaitis P, et al (1994) Replicase-mediated resistance to cucumber mosaic virus in transgenic plants involves suppression of both virus replication in the inoculated leaves and long-distance movement. Virology 199:439-447

Carr JP, Marsh LE, Lomonossoff GP, et al (1992) Resistance to tobacco mosaic virus induced by the 54-kDa gene sequence requires expression of the 54-kDa protein. Mol Plant-Microbe Interact 5:397-404

Carr JP, Zaitlin M (1991) Resistance in transgenic tobacco plants expressing a nonstructural gene sequence of tobacco mosaic virus is a consequence of markedly reduced virus replication. Mol Plant-Microbe Interact 4:579-585

Chen CC, Chiu RJ (1980) Factors affecting transmission of rice transitory yellowing virus by green leafhoppers. Plant Prot Bull 22:297-306

Cooper B, Lapidot M, Heick JA, et al (1995) A defective movement protein of TMV in transgenic plants confers resistance to multiple viruses whereas the functional analog increases susceptibility. Virology 206:307-313

Cuevas-Perez F, Ventura-Flores C, Correa-Victoria FJ (1993) Using pedigree analysis to identify sources of resistance to rice hoja blanca virus (RHBV). Int Rice Res Newslett 18:12-13

Cuozzo M O'Connell KM, Kaniewski W, et al (1988) Viral protection in transgenic tobacco plants expressing the cucumber mosaic virus coat protein gene or its antisense RNA. Biotechnology 6:549-557

Devika R, Bai NR, Regina A, et al (1990) Evaluation of rice germplasm for resistance to grassy stunt virus (GSV). Int Rice Res Newslett 15:11

Dey MM, Miah MNI, Mustsfi BAA, et al (1996) Rice production constraints in Bangladesh: implications for further research priorities. In: Evenson RE, Herdt RW, Hossain M (eds) Rice research in Asia: progress and priorities. CAB International, Wallingford, UK, pp 179-192

Donson J, Kearney CM, Turpen TH, et al (1993) Broad resistance to tobacco viruses is mediated by a modified tobacco mosaic virus replicase transgene. Mol Plant-Microbe Interact 6:635-642

Estes MK (1990) Rotaviruses and their replication. In: Fields BN, Knipe DM, (eds) Virology. Second edn. Raven Press, New York, pp 1329-1352

Faan HG, Li YG, Fei WY, et al (1965) Preliminary studies on rice yellow stunt in Kwangtung (in Chinese). Zhi Wu Baa Fu (Plant Protection) 3:143-145

Faan S, Liut C, Zhuo L, et al (1983) Rice gall dwarf - a new disease epidemic in the west of Guangdong Province of South China. Acta Phytopathol Sin 13:1-6

Falk BW, Bruening G (1994) Will transgenic crops generate new viruses and new disease? Science 263:1395-1396

Flavell RB (1994) Inactivation of gene expression in plants as a consequence of specific sequence duplications. Proc Natl Acad Sci USA 91:3490-3496

Finnegan J, McElroy D (1994) Transgene inactivation:plants fight back! Biotechnology 12:883-888

Fitchden JH, Beachy RN (1993) Genetically engineered protection against viruses in transgenic plants. Annu Rev Microbiol 47:739-763

Fomba SN (1988) Screening for seedling resistance to rice yellow mottle virus in some rice cultivars in Sierra Leone. Plant Dis 72:641-642

Galvez GE (1969) Hoja blanca disease of rice. In: Hopkins J (ed) The virus disease of rice plant. Baltimore/IRRI, Philippines, pp 35-49

Garcìa JA, Cervera MT, Riechmann JL, et al (1993) Inhibitory effects of human cystatin-Con plum pox potyvirus proteases. Plant Mol Biol 22:697-701

Gerlach WL, Llewellyn D, Haseloff J (1987) Construction of a plant disease resistance gene from the satellite RNA of tobacco ringspot virus. Nature 328:802-805

Gibbs MJ, Waterhouse PM, Weiller GF (1997) Analysis of natural viral recombination may assist the design of new virus resistance traits. In: McLean GD, Waterhouse PM, Evans G, Gibbs MJ (eds) Risk, benefit and trade consideration. Australian Government Publishing, Canberra

Golemboski DB, Lomonossoff GP, Zaitlin M (1990) Plants transformed with a tobacco mosaic virus nonstructural gene sequence are resistant to the virus. Proc Natl Acad Sci USA 87:6311-6315

Gorman BM, Taylor J, Walker PJ (1983) Orbiviruses. In: Joklik WK (ed) The reoviridae. Plenum Press, New York, pp 287-357

Green PJ, Pines O, Inouye M (1988) The role of antisense in gene regulation. Annu Rev Biochem 55:569-597

Harrison BD, Mayo MA, Baulcombe DC (1987) Virus resistance in transgenic plants that express cucumber mosaic virus satellite RNA. Nature 328:799-801

Haseloff J, Gerlach WL (1988) Simple RNA enzymes with new and highly specific endoribonuclease activities. Nature 334:585-591

Haykawa T, Zhu Y, Itoh K, et al (1992) Genetically engineered rice resistant to rice stripe virus, an insect-transmitted virus. Proc Natl Acad Sci USA 89:9865-9869

Hemenway C, Fang RF, Kaniewski WK, et al (1988) Analysis of the mechanism of protection in transgenic plants expressing the potato virus X coat protein or its antisense RNA. EMBO J 7:1273-1280

Herdt RW (1991) Research priorities for rice biotechnology. In: Khush GS, Toenniessen GH (eds) Rice biotechnology. CAB International, Wallingford, UK, pp 19-54

Hibino H (1983) Relations of rice tungro bacilliform and rice tungro spherical viruses with their vector *Nephotettix virescens*. Annu Phytopathol Soc Jpn 49:545-553

Hibino H (1989) Insect-borne viruses of rice. In: Harris KF (ed) Advances in disease vector research. Springer-Verlag, New York, pp 209-241

Hibino H, Daquioag RD, Mesina EM, et al (1990) Resistances in rice to tungro associated viruses. Plant Dis 74:923-926

Hibino H, Rocchan M, Sudarisman S (1978) Association of two type of virus particles with penyaki habang (tungro disease) of rice in Indonesia. Phytopathology 68:1412-1416

Hibino H, Roechan M, Sudarsiman S, et al (1977) A virus disease of rice (kerdil hampa) transmitted by brown planthopper, *Nilaparvata lugens* Stal. in Indonesia. Cent Res Inst Agric Bogor 35:15

Hossain M, Gascon FB, Revilla IM (1996) Constraints to growth in rice production in the Philippines. In: Evenson RE, Herdt RW, Hossain M (eds) Rice research in Asia: progress and priorities. CAB International, Wallingford, UK, pp 239-250

Ikeda R, Kaneda C (1982) Genetic relationships of brown planthopper resistance to dwarf disease and stripe disease resistance in rice. Jpn J Breed 32:177-185

Jatilesono K (1996) Rice production losses from pests in Indonesia. In: Evenson RE, Herdt RW, Hossain M (eds) Rice research in Asia: progress and priorities. CAB International, Wallingford, UK, pp 251-260

Jones MC, Gough K, Dasgupta I, et al (1991) Rice tungro virus is caused by an RNA and a DNA virus. J Gen Virol 72:757-761

Katsura S (1936) The stunt disease of Japanese rice, the first plant virosis shown to be transmitted by an insect vector. Phytopathology 26:887-895

Kobayashi N, Ikeda R, Vaughan DA, et al (1993) Resistance to rice tungro in wild species of rice (*Oryza* spp.) Jpn J Breed 43:247-255

Koller A, Dalmay T, Burgyan J (1993) Defective interfering RNA-mediated resistance against cymbidium ringspot tombusvirus in transgenic plants. Virology 193:313-318

Li XK, Huang FY, Xiao LR, et al (1990) Evaluation of rice germplasm for resistance to yellow dwarf virus. Crop Genet Resourc 1:16-18

Li Z, Upadhyaya NM, Kositratana W et al (1996) Genome segment 5 of rice ragged stunt virus encodes a virion protein. J Gen Virol 77:3155-3160

Lindbo JA, Dougherty WG (1992a) Pathogen-derived resistance to a potyvirus: immune and resistance phenotypes in transgenic tobacco expressing altered forms of a potyvirus coat protein nucleotde sequence. Mol Plant-Microbe Interact 5:144-153

Lindbo JA, Dougherty WG (1992b) Untranslatable transcripts of the tobacco etch virus coat protein gene sequence can interfere with tobacco etch virus replication in transgenic plants and protoplasts. Virology 189:725-733

Lindbo JA, Silva-Rosales L, Dougherty WG (1993a) Pathogen derived resistance to potyviruses: working but why? Semin Virol 4:369-379

Lindbo JA, Silva-Rosales L, Proebsting WM, et al (1993b) Induction of a highly specific antiviral state in transgenic plants: implications for regulation of gene expression and virus resistance. Plant Cell 5:1749-1759

Ling KC (1972) Rice virus diseases. International Rice Research Institute, Los Banos, Philippines, p 142

Ling KC, Tiongco ER, Aguiero VM (1978) Rice ragged stunt , a new virus disease. Plant Dis Rep 3:8

Lodge JK, Kaniewski WK, Tumer NE (1993) Broad-spectrum virus resistance in transgenic plants expressing pokeweed antiviral protein. Proc Natl Acad Sci USA 90:7089-7093

Longstaff M, Brigneti G, Boccard F, et al (1993) Extreme resistance to potato virus X infection in plants expressing a modified component of the putative viral replicase. EMBO J 12:379-386

MacFarlane SA, Davies JW (1992) Plants transformed with a region of the 201-kilodalton replicase gene from early browning virus RNA 1 are resistant to virus infection. Proc Natl Acad Sci USA 89:5829-5833

Malyshenko SI, Kondakova OA, Nazarova JV (1993) Reduction of tobacco mosaic virus accumulation in transgenic plants producing non-functional viral transport proteins. J Gen Virol 74:1149-1156

Matsumura T, Tabayashi N (1995) Transformation of rice plants by plant reovirus genes. Semin Virology 6:133-139

McLean GD, Waterhouse PM, Evans G, Gibbs MJ (eds) (1997) Commercialisation of transgenic crops: risk, benefit and trade considerations. Australian Government Publishing, Canberra.

Miller WA, Koev G, Mohan BR (1997) Are there risks associated with transgenic resistance to luteoviruses. Plant Dis 81:700-710

Mueller E, Gilbert J, Devenport G, et al (1995) Homology-dependent resistance: transgenic virus resistance in plants related to homology-dependent gene silencing. Plant J 7:1001-1013

Murao K, Suda N, Uyeda I, et al (1994) Genomic heterogeneity of rice dwarf phytoreovirus field isolates and nucleotide sequences of variants of genome segment 12. J Gen Virol 75:1843-48

Nakata M, Fukunaga K, Suzuki N (1978) Polypeptide components of rice dwarf virus. Ann Phytopathol Soc Jpn 44:288-296

Nemato H, Ishikawa K, Shimura E (1994) The resistances to rice stripe virus and small brown planthopper in rice variety,IR 50. Breed Sci 44:13-18

Omura T (1995) Genomes and primary protein structures of phytoreoviruses. Semin Virol 6:97-102

Omura T, Ishikawa K, Hirano H, et al (1989) The outer capsid protein of rice dwarf virus is encoded by genome segment S8. J Gen Virol 70:2759-2764

Omura T, Saito Y, Usugi T et al (1983) Purification and serology of rice tungro spherical and rice tungro baciliform viruses. Ann Phytopathol Soc Jpn 49:73-76

Osbourn JK Watts JW, Beachy RN, et al (1989) Evidence that nucleocaspid disassembly and a later step in virus replication are inhibited in transgenic tobacco expressing TMV coat protein. Virology 172:370-373

Osbourn JK, Sarkar, S Wilson TMA (1990) Complementation of coat protein-defective TMV mutants in transgenic tobacco plants expressing TMV coat protein. Virology 179:921-925.

Osler R (1984) Rice Guiallume, a disease related to barley yellow dwarf in Italy. In: Burnett PA (ed) Barley yellow dwarf, a workshop. CIMMYT, Mexico, pp 125-131.

Ou SH (1985) Rice diseases. 2nd edn. Commonwealth Mycological Institute, UK, p 380

Powell PA, Start DM, Sanders PR (1989) Protection against tobacco mosaic virus in transgenic plants that express tobacco mosaic virus antisense RNA. Proc Natl Acad Sci USA 86:6949-6952

Powell PA, Sanders PR, Tumer N, et al (1990) Protection against tobacco mosaic virus infection in transgenic plants requires accumulation of coat protein rather than coat protein RNA sequences. Virology 175:124-130

Powell-Abel PA, Nelson RS, De B, et al (1986) Delay of disease development in transgenic plants that express the tobacco mosaic virus coat protein gene. Science 232:738-743

Qu RD, Bhattacharya M, Laco GS, et al (1991) Characterization of the genome of rice tungro bacilliform virus:comparison with Commelina yellow mottle virus and caulimoviruses. Virology 185:354-364

Ramasamy C, Jatileksono T (1996) Intercountry comparison of insect and disease losses. In: Evenson RE, Herdt RW, Hossain M (eds) Rice research in Asia: progress and priorities. CAB International, Wallingford, UK, pp 305-316

Register JC, Beachy RN (1988) Resistance to TMV in transgenic plants results from interference with an early event in infection. Virology 166:524-532

Rossel HW (1986) Rice yellow mottle and African soybean dwarf, newly discovered virus diseases of economic importance in West Africa. Trop Agric Res Ser 19:146-153

Rubino L, Lupo R, Russo M (1993) Resistance to cymbidium ringspot tombusvirus infection in transgenic *Nicotiana benthamiana* plants expressing a full-length replicase gene. Mol Plant-Microbe Interact 6:729-734

Sanford JC, Johnston SA (1985) The concept of parasite-derived resistance: deriving resistance genes from the parasite's own genome. J Theor Biol 113:395-405
Schiff LA, Fields BN (1990) Reoviruses and their replication. In: Fields BN, Knipe DM (eds) Virology. Second edn. Raven Press, New York, pp 1275-1306
Sebastian LS, Ikeda R, Huang N, et al (1996) Molecular mapping of resistance to rice tungro spherical virus and green leafhopper. Phytopathology 86:25-30
Setboonsarng S (1996) Rice research priorities in Thailand. In: Evenson RE, Herdt RW, Hossain M (eds) Rice research in Asia: progress and priorities. CAB International, Wallingford, UK, pp 217-238
Shen P, Kaniewska M, Smith, et al (1993) Nucleotide sequence and genomic organization of rice tungro spherical virus. Virology 193:621-630
Sijen T, Wellink J, Hiriart JB, et al (1996) RNA-mediated virus resistance: role of repeated transgenes and delineation of targeted regions. Plant Cell 8:2277-2294
Smith HA, Swaney SL, Parks TD, et al (1994) Transgenic plant virus resistance mediated by untranslatable sense RNAs: expression, regulation, and fate of nonessential RNAs. Plant Cell 6:1441-1453
Stam M, Mol JNM, Kooter JM (1997) The silence of genes in transgenic plants. Ann Bot 79:3-12
Su HJ (1969) Transitory yellowing of rice in Taiwan. In: Hopkins J (ed) The virus diseases of the rice plant. Baltimore/IRRI, Philippines, pp 13-21
Suzuki N, Kusano T, Matsuura Y, et al (1996) Novel NTP binding property of rice dwarf phytoreovirus minor core protein P5. Virology 219:471-474
Suzuki N, Sugawara M, Kusano T, et al (1994) Immunodetection of rice dwarf phytoreoviral proteins in both insect and plant hosts. Virology 202:41-48
Suzuki N, Tanimura M, Watanabe Y, et al (1992) Molecular analysis of rice dwarf phytoreovirus segment S1: interviral homology of the putative RNA-dependent RNA polymerase between plant- and animal-infecting reoviruses. Virology 190:240-247
Tacke E, Salamini F, Rohde W (1996) Genetic engineering of potato for broad spectrum protection against virus infection. Nat Biotechnol 14:1597-1601
Tavladoraki P, Benvenuto E, Trinca S, et al (1993) Transgenic plants expressing a functional single-chain Fv antibody are specifically protected from virus attack. Nature 366:469-472
Tenllado F, Garcia-Luque I, Serra MT, et al (1995) *Nicotiana benthamiana* plants transformed with the 54-kDa region of the pepper mild mottle tobravirus replicase gene exhibit two types of resistance responses against viral infection. Virology 211:170-183
Tepfer M, Balazs E (eds) (1997) Virus resistant transgenic plants: potential ecological impact. Springer, Heidelberg, p 126
Tien P, Wu G (1991) Satellite RNA for the biocontrol of plant disease. Adv Virus Res 39:321-339
Truve E, Aaspôllu A, Honkanen J, et al (1993) Transgenic potato plants expressing mammalian 2'-5' oligoadenylate synthetase are protected from potato virus X infection under field conditions. Biotechnology 11:1048-1052
Upadhyaya NM, Ramm K, Gellatly JA, et al (1997) Rice ragged stunt oryzavirus genome segments S7 and S10 encode non-structural proteins of *Mr* 68,025 (Pns7) and *Mr* 32,364 (Pns10). Arch Virol 142:1719-1726
Upadhyaya NM, Ramm K, Kositratana W, et al (1998) Rice ragged stunt oryzavirus genome segment S4 could encode an RNA dependent RNA polymerase and a second protein of unknown function. Arch Virol 143:1815-1822

Upadhyaya NM, Yang M, Kositratana W, et al (1995) Molecular analysis of rice ragged stunt oryzavirus segment S9 and sequence conservation among isolates from Thailand and India. Arch Virol 140:1945-1956

Upadhyaya NM, Zinkowsky E, Li Z, et al (1996) The Mr 43K major capsid protein of rice ragged stunt oryzavirus is a post-translationally processed product of a Mr 67348 polypeptide encoded by genome segment 8. Arch Virol 141:1689-1701

Van den Elzen PJM, Huisman MJ, Posthumus L, et al (1989) Engineering virus resistance in agricultural crops. Plant Mol Biol 13:337-346

Watanabe Y, Ogawa T, Takahashi H, et al (1995) Resistance against multiple plant viruses in plants mediated by double stranded-RNA specific ribunuclease. FEBS Lett 372:165-168

Waterhouse PM, Gildow FE, Johnstone GR (1988) The luteovirus group. AAB descriptions of plant viruses. No 339. Institute of Horticultural Research, Wellesbourne, Warwick

Wilson TMA, Watkins PAC (1986) Influence of exogenous viral coat protein on the cotranslational disassembly of tobacco mosaic virus (TMV) particles *in vitro*. Virology 149:132-135

Wisniewski LA, Powell PA, Nelson RS, et al (1990) Local and systemic spread of tobacco mosaic virus in transgenic tobacco. Plant Cell 2:559-567

Xing ZY, He JQ, Liu ZW, et al (1985) Studies on the hybridization of indica and japonica rice. II. Breeding for resistance to rice stripe. Acta Agron Sin 11:1-7

Yang X, Yie Y, Zhu F, et al (1997) Ribozyme-mediated high resistance against potato spindle tuber viroid in transgenic potatoes. Proc Natl Acad Sci USA 94:4861-4865

14
Rice improvement by genetic transformation

Andreas Klöti and Ingo Potrykus*

ETH Zürich, Institute for Plant Sciences, Universitätsstr. 2, 8092 Zürich, Switzerland
*Present address: Paradigm Genetics, Inc., North Carolina, USA

1 Introduction

Genetic engineering is a powerful tool for the introduction of new traits to rice. It can support the rice breeder in two ways: first, compared with conventional breeding, by relatively quick and precise transfer of desired genes from one variety to another without bringing in undesirable qualities as a backpack, and second, by the transfer to rice of genes that are not available so far in the rice gene pool.

Rice genetic engineering is now looking back on its first decade. The first reports on successful production of transgenic rice plants were published in 1988, and since then, enormous progress has been made in the development of different transformation protocols. Techniques for transformation of rice are most advanced among all cereals and are used in many laboratories all over the world. The International Program in Rice Biotechnology from the Rockefeller Foundation has been a strong promoter of rice genetic engineering. The aim of the Foundation's program is to "increase crop yields of smallholder farmers in developing countries profitably and without degrading natural resources." Since 1983, the program has funded research in rice biotechnology with around US$75 million, funding numerous research institutions all over the world (Herdt 1996; Rockefeller Foundation 1997).

Most of the ongoing projects in rice genetic engineering aim at the improvement of agronomic traits such as resistances to diseases, pests, abiotic stresses, and herbicides or at quality improvement in terms of micronutrients, vitamins, or amino acids. Many traits can be engineered by the transfer of single genes to rice; others require the combined transfer and expression of several genes. The strategies rely

Key words. cereals, genetic engineering, crop improvement, transgenic plants, promoter, quality improvement, insect resistance, disease resistance, wheat, maize, barley

on overexpression of endogenous genes, the expression of heterologous genes, or inhibition of gene expression.

2 Gene transfer to rice

2.1 Methods

The first reports on transgenic rice were published in 1988 by Toriyama et al., Zhang et al., and Zhang and Wu. All three groups transformed *Japonica* cultivars by direct gene transfer to protoplasts, mediated by polyethylenglycol (PEG) or electroporation. Two years later, Datta et al. (1990) transformed *Indica* rice using the same transformation method. The protoplasts were isolated from embryogenic cell suspensions (ECS), which were derived from calli grown from the scutellum cells of rice embryos. Up to now, the scutellum cells have been the most common source for totipotent cells for rice transformation since a reproducible protocol for plant regeneration from leaf cells has not been presented so far. Direct gene transfer to protoplasts was quite successful and even led to the transformation of modern *Indica* varieties like IR72 (Datta et al. 1992). However, this method has a few drawbacks. First, the establishment of ECS is expensive, time- and labor consuming, and extremely genotype dependent; second, somaclonal variations accumulate in ECS; and third, multiple, rearranged gene copies are integrated into the genome.

A very promising alternative was the development of a protocol for DNA delivery to rice cells by microprojectile bombardment (Christou et al. 1991). By using the scutellum cells of freshly isolated immature embryos as target tissue, ECS were no longer required for transformation of many *Japonica* and *Indica* varieties (Christou et al. 1991; Wünn et al. 1996). However, very efficient protocols for production of phenotypically normal, fertile transgenic *Japonica* and *Indica* plants by particle bombardment were also developed using callus or ECS as target material (Li et al. 1993; Zhang et al. 1996; Jain et al. 1996). Alternatively, transgenic rice was obtained after transfer of the DNA into the scutellum of mature embryos by tissue electroporation (Xu and Li 1994).

The most recent milestone in the field of rice transformation was the development of protocols for *Agrobacterium tumefaciens*-mediated transformation of scutellum-derived callus material. The protocols are simple and allow efficient production of transgenic plants from *Japonica* (Hiei et al. 1994) and *Indica* varieties (Rashid et al. 1996). *Agrobacterium*-mediated rice transformation is described in detail in the chapter 12.

To date, particle bombardment and *Agrobacterium*-mediated transformation of scutellum-derived cells are both very efficient and reproducible methods for the production of large quantities of transgenic lines. So far, neither of the methods is clearly superior to the other, and some problems still remain with both methods. First, using scutellum-derived cells as targets for gene transfer and going through tissue culture steps, both methods are still genotype dependent. Certain *Indica* varieties therefore remain recalcitrant. The development of an efficient protocol for in vivo transformation of the rice germline, analogous to the *Agrobacterium*-mediated transformation of *Arabidopsis* flowers (Bechtold et al. 1993), would therefore be desireable. Second, integration of the transgenes into the rice nuclear genome basically occurs at random. Thereby, using either method, multiple, often rearranged transgene copies get integrated somewhere into the chromosomes, leading to unpredictable and often unstable transgene expression levels due to position effects and gene silencing.

Transgene silencing in rice has been reported by Oard et al. (1996) in plants transformed by particle bombardment, and has been described in detail, by Kumpatla et al. (1997) and Itoh et al. (1997), in plants transformed by direct gene transfer to protoplasts. Also, large numbers of lines with silenced transgenes are routinely obtained from *Agrobacterium*-mediated transformation (A. Klöti 1997, unpublished observations). This problem is overcome by production of large numbers of transgenic lines and subsequent selection and further cultivation of the best performers. However, development of a protocol for site-directed transgene integration into the rice nuclear genome, as was shown in the moss *Physcomitrella patens* (Schaefer and Zyrd 1997) and in *Arabidopsis* (Kempin et al. 1997), would be desirable. An alternative is the introduction of the transgenes into the genome of plastids, as has been shown for tobacco (Svab et al. 1990). Incorporation of DNA sequences into the circular plastid genome occurs by homologous recombination and can be directed by border sequences with homologies to the integration site.

So far no mechanism similar to gene silencing is known in plastids. Additionally, transgene expression levels in plastids can be expected to be very high because many genome copies are present in a single cell (McBride et al. 1995), and the prokaryotic genome organization of the plastids may allow expression of multiple genes from bacterial operons. Therefore, although enormous progress has been made in the first decade of rice genetic engineering, major technology improvements are still possible.

2.2 Promoters

The most frequently used constitutive promoters to drive chimeric gene expression in rice are the cauliflower mosaic virus (CaMV) 35S promoter, the rice actin 1 promoter (McElroy et al. 1990), and the maize ubiquitin 1 promoter (Toki et al. 1992), both including the first exon and the first intron of the respective genes. The rice actin promoter is very strong and active in almost all tissues (Wen et al. 1993), whereas the maize ubiquitin 1 promoter is also very strong in rice, being active in most tissues but showing variable expression during plant development (Cornejo et al. 1993). The CaMV 35S promoter is rather weak (Wen et al. 1993), active mainly in the vascular tissues (Terada and Shimamoto 1990) but also in mesophyll and parenchyma cells of leaves and roots (Battraw and Hall 1990). Introduction of an intron into a transgene sequence can increase gene expression (discussed in Fütterer and Potrykus 1995). In combination with an intron and promoter downstream sequences from rice tungro bacilliform virus (RTBV), the CaMV 35S promoter is very strong in rice, active in almost all young tissues of leaves and roots (A. Klöti, 1997, unpublished results). Other promoters which confer tissue-specific and/or inducible transgene expression have been used for controlling transgene expression in rice. Examples are the maize pyruvate, orthophosphate dikinase promoter, which is light regulated and is active in photosynthetic organs exclusively (Matsuoka et al. 1993), the rice glutelin 1 promoter, which confers specific expression to the endosperm (Kim et al. 1993), and the rice phenylalanine ammonia-lyase (PAL) promoter, which is wound inducible and specific for root apical tips and vascular and epidermal tissues of stems and roots (Zhu et al. 1995).

2.3 Marker genes

Different antibiotic or herbicide resistance genes have been used for positive selection of transformed rice cells. Of 70 articles about rice transformation published between 1988 and 1997, 32 reported the use of a hygromycin phosphotransferase gene (*hptI* or *aph4*) from *E. coli* conferring resistance to hygromycin B (Gritz and Davies 1983), 18 used the neomycin phosphotransferase gene (*nptII*) from the *E. coli* transposon Tn5, conferring resistance to kanamycin and related compounds like geneticin (Bevan et al. 1983), and 15 used phosphinothricin acetyl transferase genes from *Streptomyces viridiochromogenes* (*pat*) or *S. hygroscopicus* (*bar*) (Wohlleben et al. 1988; Murakami et al. 1986), both conferring resistance to phosphinothricin, glufosinate, and bialaphos. Other selectable marker genes that have been used in rice transformation are a mutant acetolactate synthase gene from *Arabidopsis*, which confers resistance to chlorsulfuron (Zhijian et al. 1992), and a mouse dihydrofolate reductase gene, which confers resistance to methotrexate (Meijer et al. 1991). The presence of these marker genes is no longer required in the regenerated transgenic plants and they can cause problems, as repeated transformations using the same marker gene are not possible. Komari et al. (1996) demonstrated that transgenic rice containing only the gene of interest but not the selectable marker gene can be produced efficiently by transferring the genes by *Agrobacterium* with two separate T-DNA molecules. More than half of the plants cotransformed with two separate T-DNAs contained at least one locus of the nonselectable T-DNA that was unlinked to the selectable marker gene, yielding marker gene-free plants in the T1 generation. By the use of transposable genetic elements (Ow 1996), it should be possible to remove the marker gene already in the primary regenerant, as demonstrated by Ebinuma et al. (1997) in transgenic tobacco.

3 Application of rice genetic engineering

3.1 Constraints to rice production

In 1997, 572 million metric tons of rice were produced worldwide (FAO 1998). Yields ranged from as little as around 1 t/ha in many countries of Africa and more than 6 t/ha in China and Japan to more than 9 t/ha in Australia (FAO 1998). Major yield determinants are the potential of the grown cultivar, abiotic conditions such as water and nutrient availability, and soil toxicity as well as crop protection measures against weeds, pathogens, and animals. Estimations about losses due to specific stress conditions vary considerably from source to source, the importance of specific deficiencies, pests, and diseases varying greatly from region to region. However, any abiotic or biotic stress alone can cause significant loss to rice farmers.

Twelve bacterial, 17 viral or mycoplasma-like, and 58 fungal pathogens are known to cause diseases of rice (Oerke et al. 1994). Among the most serious rice pathogens are bacterial blight (*Xanthomonas oryzae* pv. *oryzae*) and the fungal pathogens rice blast (*Pyricularia oryzae* or *Magnaporthe grisea*) and sheath blight (*Rhizoctonia solani*). Serious viral diseases are, among others, the rice tungro virus disease, a composite disease caused by rice tungro spherical virus (RTSV) and rice tungro bacilliform virus (RTBV), and the diseases caused by rice grassy stunt virus (RGSV), rice dwarf virus (RDV), rice ragged stunt virus (RRSV), rice stripe virus (RSV), rice yellow mottle virus (RYMV), and rice hoja blanca virus (RHBV) (Hibino 1989).

Rice animal pests include 100 insect species alone in Southeast Asia (Oerke et al. 1994). Despite the use of insecticides at US $1.2 billion in 1993, an estimated 145 million tons of rice were lost to insects (Krattiger 1997). Among the major rice insect pests are the brown planthopper (BPH; *Nilaparvata lugens*), the striped stem borer (SSB; *Chilo suppressalis*), the yellow stem borer (YSB; *Scirpophaga incertulas*), the pink stem borer (PSB; *Sesamia inferens*), and the leaffolder (*Cnaphalocrocis medinalis*) (Oerke et al. 1994). Further damage is caused by various nematodes, rats, and birds. Weed competition is also a big threat to rice production. The major weeds of rice include *Echinochloa*, *Cyperus*, and *Fimbristylis* species. Particularly in the first month, it is essential to eliminate weed competition. Although much weeding in traditional rice production is still done manually, in 1988 US $885 millions were spent worldwide for herbicides in rice production (Oerke et al. 1994).

Severe yield loss is caused by a range of abiotic stress conditions, which include drought, flooding, heat, cold, salinity, toxicity from aluminum, boron, iron, or manganese, and deficiencies of nitrogen, phosphorus, potassium, silicon, sulfur, or zinc (IRRI 1983). With the increasing number of genes discovered in plants and other organisms and the increasing knowledge of plant–pathogen interactions, genetic engineering can contribute more and more to the introduction of traits for resistance to biotic and abiotic stress.

3.1.1 Insect resistance

Various proteins with insecticidal activity are known that can be used for engineering plants with intrinsic pest resistance, including protease inhibitors, α-amylase inhibitor, and the crystal δ-endotoxins (Cry proteins) from *Bacillus thuringiensis* (*Bt*). The Cry proteins are highly specific against a narrow spectrum of insect pests. Three major insect families respond to the more than 50 insecticidal Cry proteins that are known so far: *Lepidoptera* (caterpillars), *Coleoptera* (beetles), and *Diptera* (flies and mosquitoes). *Bt* genes have been transferred to many different crops (see Krattiger 1997). The bacterial *Bt* genes are expressed poorly in transgenic plants, but synthetic genes with a crop-adapted codon usage and a lack of improper splice sites and polyadenylation signals are expressed well (Koziel et al. 1993; Fujimoto et al. 1993; Sardana et al. 1996).

Synthetic *cryIA(b)* and *cryIA(c)* genes under control of different promoters have been transferred to rice (Table 1). Cry protein levels in leaf material (as percentage of total soluble protein) ranged from 0.01% (Wünn et al. 1996) to 0.1% (Ghareyazie et al. 1997). Insecticidal activity of the Cry proteins was scored with in vitro bioassays by feeding plant material to larvae of stemborers or leaffolders, assaying survival rate and development of the larvae. Alternatively, whole plants were infested with larvae in the greenhouse and plants were screened for surviving insects and "white heads" or folded leaves, the typical damage caused by stemborers or

Table 1. Insect-resistant transgenic rice

Promoter::*gene*	Variety	Pest	Effect	Source
CaMV 35S::*cryIA(b)*	Nipponbare	SSB, leaffolder	50%-55% mortality; larvae on seedlings	Fujimoto et al. 1993
CaMV 35S::*cryIA(b)*	IR58	SSB,YSB	100% mortality; larvae on leaves and on plants	Wünn et al. 1996
		leaffolders	feeding inhibition; larvae on plants	
maize PEPC::*cryIA(b)*	Tarom Molaii	SSB, YSB	100% mortality; larvae on plants	Ghareyazie et al. 1997
maize ubi 1::*cryIA(c)*	IR64	YSB	92% mortality; larvae on nodal stem tissue	Nayak et al. 1997
pin2::*pin2*	Tainung 67 Pi4	PSB	Increased plant resistance	Duan et al. 1996
rice actin 1::*CpTi*	Taipei 309	PSB, SSB	Increased resistance; field test	Xu et al. 1996b

SSB, striped stem borer; YSB, yellow stem borer; PSB, pink stem borer; ubi, ubiquitin; PEPC, phosphoenol pyruvate carboxylase; pin, potato proteinase inhibitor; CpTi, cowpea trypsin inhibitor.

leaffolders, respectively. As summarized in Table 1, high mortality rates were found with CryIA(b) to SSB and YSB (Fujimoto et al. 1993; Wünn et al. 1996; Ghareyazie et al. 1997) and to leaffolders (Fujimoto et al. 1993; Wünn et al. 1996) and with CryIA(c) to YSB (Nayak et al. 1997). These results with *Bt*-toxin producing plants are encouraging; however, field experiments will have to show how these lines perform under more natural conditions in the field with high insect pressure and variable environmental factors.

In contrast to *Bt* δ-endotoxins, protease inhibitors are antimetabolic against a wide range of insect species. Two genes coding for protease inhibitors have been transferred to rice: the potato proteinase inhibitor II (PINII) gene under control of its own wound-inducible promoter (Duan et al. 1996) and a cowpea trypsin inhibitor (CpTI) gene under control of the rice actin 1 promoter (Xu et al. 1996b). In both cases, high protein levels of more than 2% of total soluble protein accumulated in the leaves, and increased resistance to PSB (PINII) or PSB and SSB (CpTI) was observed in whole plant assays or a small-scale field experiment, respectively.

The range of potentially useful genes for insect control has recently been expanded by the discovery of new proteins with insecticidal activity: vegetative insecticidal proteins (VIPs) from *Bacillus cereus* and *B. thuringiensis* and a cholesterol oxidase from *Streptomyces* (reviewed by Estruch et al. 1997).

3.1.2 Resistance to bacterial and fungal diseases

On attack by a pathogen, specific receptors of a resistant plant (products of the resistance, or *R*-genes) recognize certain elicitors of the pathogen and induce a response for protection of the plant. This response can include the production of phytoalexin antibiotics like stilbene, or pathogenesis-related proteins (PR-proteins) like β-1,3 glucanases, chitinases, thaumatin-like proteins, or ribosome-inactivating proteins (RIPs). The antifungal activity of all these protein classes has been demonstrated in transgenic plants (see Yun et al. 1997). In rice, the transfer of two genes coding for antifungal proteins has been reported so far (Table 2). Transgenic rice overexpressing a rice chitinase gene showed reduced symptoms after infection with *Rhizoctonia solani* (Lin et al. 1995). Plants containing a grapevine stilbene

Table 2. Fungal or bacterial resistant transgenic rice

Promoter::*gene*	Variety	Pathogen	Effect	Source
CaMV 35S:: *chitinase* *(O. sativa)*	Chinsurah Boro II	*Rhizoctonia solani*	Reduced symptoms on plants	Lin et al. 1995
stilbene synthase:: *stilbene synthase* *(Vitis vinifera)*	Nipponbare	*Magnaporthe grisea*	Enhanced resistance	Stark-Lorenzen et al. 1997
Xa21::*Xa21* *(O. longistaminata)*	Taipei 309	*Xanthomonas oryzae* pv. *o.*	Resistance	Wang et al. 1996

synthase gene controlled by its own promoter expressed the transgene on infection with *Magnaportha grisea* and showed reduced disease symptoms (Stark-Lorenzen et al. 1997).

Recently, *Xa21*, an *R*-gene conferring resistance to *Xanthomonas oryzae* pv. *oryzae* and originating from the wild rice species *O. longistaminata,* was cloned and transferred to *Japonica* rice (Song et al. 1995). In transgenic plants, *Xa21* conferred resistance to 29 isolates of *X. oryzae* pv. *oryzae* from eight different countries, similar as in *O. longistaminata* (Wang et al. 1996). Two more cases of protection against fungal pathogens with transgenic plants containing the *bar* gene have been reported. In both, it is not the transgene product which protects the plant, but a bialaphos treatment of the plants. The plants were protected against infections with *R. solani* (Uchimiya et al. 1993) or *Magnaporthe grisea* (Tada et al. 1996a), respectively, after bialaphos treatment within 2 days after infection.

3.1.3 Resistance to viral diseases

Sanford and Johnston (1985) postulated the concept of pathogen-derived resistance (PDR), which aims at the disruption of the viral life cycle by the expression of viral sequences within the host plant. The concept has successfully been exploited in different crops. Many virus-resistant plants have been obtained after expression of viral coat proteins (Wilson 1993). Also, the transfer of mutated forms of viral replicase genes (Carr and Zaitlin 1993) and genes for viral RNA polymerases (Müller et al. 1995) led to strong viral resistance. The mechanism involved in these latter

Table 3. Virus-resistant transgenic rice

Gene	Variety	Virus	Effect	Source
Coat protein	Kinuhikari Nipponbare	RSV	Fewer disease symptoms	Hayakawa et al. 1992
Coat protein	*Japonica* cultivar	RYSV	Resistance	Fang et al. 1997
Coat protein	n.s.	RHBV	Attenuated disease symptoms	Lentini et al. 1997
Coat protein	Taipei 309	RTBV, RTSV	Reduced virus accumulation	Fauquet et al. 1997
Polymerase in sense or antisense orientation	Taipei 309	RTSV	Immunity	Fauquet et al. 1997
Different sequences	n.s.	RRSV	Resistance	Upadhyaya et al. 1997

RSV, rice stripe virus; RYSV, rice yellow stunt rhabdovirus; RHBV, rice hoja blanca virus; RTBV, rice tungro bacilliform virus; RTSV, rice tungro spherical virus; RRSV, rice ragged stunt virus; n.s., not specified.

resistances is probably related to gene silencing. Müller et al. (1995) proposed homology-dependent gene silencing with posttranscriptional degradation of viral RNA.

In 1992, Hayakawa and co-workers reported the transfer of the RSV coat protein gene to two *Japonica* rice cultivars (Table 3). Plants which expressed the coat protein at high levels, up to 0.5% of total soluble protein, showed fewer disease symptoms when infected with viruliferous insects. At the General Meeting of the International Program on Rice Biotechnology of the Rockefeller Foundation in September 1997, several groups reported work on engineered virus resistance in rice: resistance to rice yellow stunt rhabdovirus (RYSV; Fang et al. 1997), attenuated disease symptoms after infection with RHBV (Lentini et al. 1997), and reduced accumulation of RTBV and RTSV (Fauquet et al. 1997) after transfer of the respective coat protein gene(s), immunity to RTSV in transgenic plants, containing the RTSV polymerase gene in sense or antisense orientation (Fauquet et al. 1997), and RRSV resistance after transfer of different RRSV sequences to rice (Upadhyaya et al. 1997). However, these claims must be substantiated by further reports. The topic of virus resistance is discussed more extensively in the chapter by P. Waterhouse in this volume.

3.1.4 Resistance to herbicides

Several transgenic herbicide-resistant crops already have been engineered and commercialized. These crops have engineered resistance to one of the following herbicidal compounds: glyphosate, glufosinate, bromoxynil, or acetolactate synthase inhibitors (reviewed by Powles et al. 1997). Transgenic rice plants resistant to high doses of glufosinate under greenhouse conditions were obtained by Datta et al. (1992) and Cao et al. (1992) by constitutive expression of the *bar* gene in the varieties IR72 or TP 309, respectively. In transgenic plants of the variety Gulfmont, constitutively expressing the *bar* gene, foliar application of 2.24 kg/ha glufosinate under field conditions had little or no effect on grain yield (Oard et al. 1996). In another field experiment, the same transgenic lines were tested for the control of the weedy red rice (*Oryza sativa* L.) in the United States by glufosinate application (Sankula et al. 1997). The authors showed that sequential foliar application of glufosinate allowed an effective control of the weedy red rice when the flood levels covered no more than 25% of the leaf surface.

3.1.5 Resistance management

Similar to resistance traits introduced into crops by classical breeding, genetically engineered resistance can be overcome by insects, pathogens, and weeds. Resistant populations can evolve quickly under heavy selection pressure by the selective survival of individuals carrying (mostly recessive) resistance alleles. The field frequency of alleles for resistance to CryIA(c) in the cotton pest *Heliothis virescens* was experimentally determined in insect populations that had not been exposed to *Bt* before and was found to be 1.5 x 10^{-3}; i.e., more than 1 in 1000 insects carried a specific, recessive resistance allele (Gould et al. 1997). Several strategies to avoid a fast development of insect resistance to *Bt*-toxins have been discussed (McGaughey and Whalon 1992; Tabashnik 1994).

The strategy applied by seed companies with genetically engineered, insect-resistant *Bt* crops such as maize and cotton in the United States combines separate refuge areas with high gene expression levels (Krattiger 1997). Refuges where nontransgenic plants are grown maintain a population of nonresistant insects available for mating with the resistant homozygous individuals, thereby producing heterozygous progeny. These individuals, carrying the recessive resistance allels only heterozygously, will be eliminated by high doses of *Bt* protein. As an alternative to the refuge areas, the seeds of transgenic and nontransgenic rice could be mixed, providing refuges within the field (IRRI 1996). Most importantly, pyramiding of several genes with different modes of action or, in case of *Bt*, with different binding sites in the insect midgut, will further reduce the possibility of insect resistance development (IRRI 1996).

A problem that must be considered with all resistance genes is the outcrossing of the gene to weedy relatives. Although most rice cultivars are almost completely autogamic, outcrossing could rapidly lead to herbicide-resistant weedy rice ecotypes. A solution for this problem may be the production of transplastomic rice in which the transgenes are integrated in the plastid genome and thus are not transmitted by pollen but are maternally inherited.

3.1.6 Resistance to abiotic stress

To adapt to various environmental stresses such as drought, salinity, and extreme temperatures, plants have developed physiological and biochemical strategies. These strategies involve changes in cellular chemistry by accumulation of solutes and rapidly induced specific proteins (Skriver and Mundy 1990).

In attempts to engineer rice cultivars which are resistant to abiotic stress, rice has been transformed with several heterologous stress-induced genes. Xu et al. (1996a) reported transgenic rice that constitutively expresses *hva1*, a gene for a barley late embryogenesis abundant protein. In transgenic rice, HVA1 was produced in leaves and roots at concentrations great than 1.0% of total soluble protein. Under water-deficit and salt-stress conditions, transgenic plants performed significantly better than control plants, and performance was correlated with the amount of HVA1 protein produced in the different transgenic lines.

The transfer of several other putative resistance genes to rice was announced at the General Meeting of the International Program on Rice Biotechnology from the Rockefeller Foundation in September 1997. Although the results on the performance of these transgenic plants have not been published yet, some interesting lines may have been developed. The following genes were transferred to rice: *cor47*, encoding the cold-regulated protein COR47 from *Arabidopsis* for cold tolerance; *mtlD*, encoding the mannitol-1-phosphate dehydrogenase from *E. coli* for salt and oxidative stress tolerance; *p5cs*, the first enzyme of the proline biosynthesis pathway from *Arabidopsis* for drought and salt tolerance (Wu et al. 1997); and *codA* from the soil bacterium *Arthrobacter globiformis*, encoding choline oxidase, which converts choline to glycine betaine for tolerance to salt stress and low temperature (Sakamoto et al. 1997).

3.2 Improved endosperm quality

In Asia, rice is responsible for as much as 80% of the caloric and 68% of the protein uptake (IRRI 1993). Rice is a very good energy source and is highly digestible. However, especially milled rice, consisting exclusively of the endosperm, is poor in fat, protein, and a whole range of vitamins and micronutrients (FAO 1993). Therefore, diets based mainly on rice lead to malnutrition with the most severe deficiencies in lysine, (pro)vitamin A, iron, iodine, and zinc (FAO 1993). Classical breeding has not been able to solve these problems so far. Therefore, genetically engineered rice with improved endosperm quality is highly desirable. Projects have been started which aim at the correction of some of the most severe deficiencies in milled rice, and some have already shown partial success (see below).

Next to the correction of deficiencies, rice endosperm starch composition can be modified to fit the needs of the consumer and the industry. Starch consists of two glucose polymers, amylose and amylopectin. The ratio of amylose to amylopectin has a major impact on the physical properties of the grain starch and amylose content is usually lower for *Japonica* rice varieties (17%-22%) than for *Indica* varieties (20%-30%). A low amylose content leads to an opaque, waxy appearance of the endosperm. Natural *waxy* mutants with no or low amylose yield sticky, glutinous rice which is used for many sweet dishes in Asia (IRRI 1993). A transgenic approach for downregulation of an enzyme involved in starch biosynthesis led to genetically engineered low-amylose rice as shown below (Shimada et al. 1993). The same strategy of downregulation was applied for reduction of seed proteins that cause allergic reactions in consumers (Tada et al. 1996b).

3.2.1 Increased lysine content in rice endosperm

To increase lysine content in rice endosperm, Zheng et al. (1995) transformed rice (cv. Nipponbare) with a gene for a seed storage protein, β-phaseolin, from the common bean (*Phaseolus vulgaris* L.). β-Phaseolin has a relatively high lysine content of 6%. The authors reported a positive gene dosage effect with the gene under control of the rice glutelin 1 promoter and a β-phaseolin protein content in transgenic homozygous plants of 4% of total endosperm protein. It still remains to be demonstrated whether this additional lysine content will increase the lysine uptake in a rice diet.

3.2.2 Production of provitamin A in rice endosperm

Insufficient dietary (pro)vitamin A levels lead to severe clinical symptoms like the eye disease xerophthalmia and several childhood diseases (Sommer 1988). Unlike the photosynthetic tissues, rice endosperm contains neither β-carotene (provitamin A) nor its C_{40} carotenoid precursors (Burkhardt et al. 1997). Improved (pro)vitamin

A nutrition for people depending mainly on rice could be expected to prevent 1-2 million deaths annually among children aged 1-4 years, and an additional 0.25-0.5 million deaths during later childhood (Humphrey et al. 1992).

β-Carotene is synthesized in plastids from the general C_{20} isoprenoid precursor geranyl geranyl diphosphate (GGPP) by four specific plant enzymes: phytoene synthase, phytoene desaturase, ζ-carotene desaturase, and lycopene cyclase (Armstrong and Hearst 1996). Since GGPP is abundant in rice endosperm, expression of the four genes whose products are needed for the conversion of GGPP to β-carotene should lead to engineered β-carotene production in rice endosperm; 2 μg β-carotene per gram dry seed weight would cover the daily provitamin A needs of children in 300 g of rice (Burkhardt et al. 1997). This β-carotene content in rice endosperm may lead to a slightly yellowish coloration as is the case in maize. The first gene, a phytoene synthase gene from daffodil (*Narcissus pseudonarcissus*) with its own plastid localization signal and under control of the glutelin 1 promoter has been transferred to rice. The highest phytoene content measured was 0.74 μg phytoene per gram dry seed weight in a heterozygous seed population (Burkhardt et al. 1997). This result shows that it is possible to engineer an isoprenoid pathway in rice endosperm. However, it remains to be demonstrated that the phytoene level can be increased and that it is possible to engineer the remaining three steps on the way to β-carotene production.

3.2.3 Increased iron content in rice endosperm

Iron deficiency is the most prevalent micronutrient deficiency worldwide, and it has been estimated that more than 1 billion people suffer from iron deficiency anemia (McPhail and Bothwell 1992). Rice in its milled form as it is consumed by most people is characterized by a very low content of iron, 0.2 to 2.8 mg/100 g. Absorption of nonheme iron in the intestinal lumen is affected by several factors promoting or impairing iron availability. Depending on these factors, absorption of iron from a meal may be in the range of 1% to 20% (FAO 1988). None of the current intervention strategies, such as iron supplementation in the form of tablets or food fortification with different iron compounds, has been successful in developing countries. Genetic engineering of rice may improve the situation in three ways:

1. Increase of the total iron content of rice by expression of a ferritin gene from *Phaseolus vulgaris* (Lammers et al. 1991). Iron from ferritin can be a dietary source as shown recently by Beard et al. (1996).
2. Decrease of the level of phytic acid by expression of phytase genes from *Aspergillus niger* (van Hartingsveldt et al. 1993) or from *Aspergillus fumigatus* (heat-stable phytase gene (Pasamontes et al. 1997). The absorbability of nonheme iron is reduced by the presence of phytic acid. Although only 0.02% to 0.07% of the phytic acid of the rice seed is present in the endosperm, these small quantities strongly inhibit iron absorption. For a significant increase in iron absorption, phytic acid level in the endosperm must be reduced to less than 0.01% (Hurrell et al. 1992).
3. Addition of absorption-enhancing cysteine by expression of a gene for a cys-

teine-rich, methallothionein-like protein from rice (Hsieh et al. 1995). Cysteine increases the absorption of dietary nonheme iron about twofold, as shown for cysteine-containing peptides from meat (Layrisse et al. 1984).

Experiments are under way to express all these genes in rice under control of the rice glutelin 1 promoter (P. Lucca, 1997, unpublished results). The effect of the transgenes on iron-uptake in the human intestine and a putative additive effect of the genes remains to be demonstrated.

3.2.4 Altered starch composition in rice endosperm

Three enzymes in the biosynthesis of starch have been considered to catalyze key regulatory steps: ADP-glucose pyrophosphorylase, granule-bound starch synthase (GBSS) and branching enzyme. The gene coding for GBSS has been identified as the one affected in the *waxy* mutants and is therefore called *waxy* gene (Wang et al. 1990). Shimada et al. (1993) showed that by expression of a 1.0-kb antisense sequence of the *waxy* gene in transgenic rice (cv Nipponbare), the amylose content of rice starch could be decreased from about 19% to 6%.

3.2.5 Reduction of allergenic protein levels in rice seeds

Antisense strategy was also applied for the downregulation of allergenic proteins in rice seeds (Tada et al. 1996b). Cereal flour can provoke allergic diseases such as asthma upon inhalation or ingestion. A rice seed protein of about 16 kDa (RA17) was isolated with reactivity for IgE of rice-allergic patients. RA17 is encoded by a multigene family and is similar to allergenic proteins of other cereals, like wheat α-amylase inhibitor, barley trypsin inhibitor, and a castor bean storage protein (described by Tada et al. 1996b). To downregulate expression of RA17, rice was transformed with a 550-bp antisense sequence from *ra17* cDNA under control of different seed-specific promoters from rice. Accumulation of RA17 and other related proteins of 14–16 kDa was reduced to one-fifth of the wild-type protein level (Tada et al. 1996b). Although it has yet to be assessed whether such rice with reduced levels of allergenic proteins is tolerable by patients allergic to rice, these results show that antisense regulation can be applied to molecular breeding for downregulation of gene expression in rice grains.

4 Outlook

In the first decade of rice genetic engineering,protocols have been developed for efficient production of a large number of transgenic *Japonica* and *Indica* plants. At the same time, many projects for improvement of different agronomical traits have been started and many show promising results. Further technical improvements like the development of protocols for an in vivo transformation system, site-directed transgene integration, and rice plastid transformation would facilitate the routine production of rice lines that stably express multiple transgenes. In addition, the discovery of further useful genes and expression signals will allow tissue- and developmental stage specific expression of more homologous and heterologous single genes and whole biosynthetic pathways.

Until recently, most of the work on the improvement of transformation techniques in rice and the production of transgenic rice was conducted by public institutions. Now, it seems that more and more companies have gained interest in genetic engineering of rice. This tendency is illustrated in the United States by the permits that have been issued by the USDA for environmental releases for transgenic rice (USDA 1998). From 1990 to 1995, eight permits were issued to universities. Starting from 1996, four different companies have applied for release permits and from the 12 applications from 1996 to 1997, only 5 came from universities. It may become lucrative for private seed companies to market transgenic varieties in Asia, with introduced resistances or improved crop quality, perhaps as hybrid rice. A permit issued by the USDA to a U.S. biotech company in 1997 for a field trial with rice containing the genes for an antithrombin and a serum albumin (USDA 1998) implies that rice may also be used for production of biomolecules for use in therapy, diagnosis, and industrial processes. Rice will also become more and more important as a model plant for other grasses because of the high degree of synteny among grass species (Devos and Gale 1997).

To our knowledge no transgenic rice varieties have been released for consumption to date anywhere in the world. However, a considerable number of field trials with transgenic rice have been performed in the United States (markers, *Bt*, storage proteins, chitinase, glucanase, phosphinothricin resistance, antithrombin, serum albumin), Japan (virus resistance, quality improvement), China (virus resistance), and Mexico (markers) (James and Krattiger 1996; USDA 1998) and it can be expected that some transgenic lines will be released soon.

Acknowledgment. Dr. Martin Schrott is kindly acknowledged for revision of the manuscript.

References

Armstrong GA, Hearst JE (1996) Genetics and molecular biology of carotenoid pigment biosynthesis. FASEB J 10:228-237

Battraw MJ, Hall TC (1990) Histochemical analysis of CaMV 35S promoter-β-glucuronidase gene expression in transgenic rice plants. Plant Mol Biol 15:527-538

Beard JL, Burton JW, Theil EC (1996) Purified ferritin and soybean meal can be sources of iron for treating iron deficiency in rats. J Nutr 126:154-160

Bechtold N, Ellis J, Pelletier G (1993) In planta *Agrobacterium*-mediated gene transfer by infiltration of adult *Arabidopsis thaliana* plants. CR Acad Sci Ser III Life Sci 316:1194-1199

Bevan MW, Flavell RB, Chilton MD (1983) A chimaeric antibiotic resistance gene as a selectable marker for plant cell transformation. Nature 394:184-187

Burkhardt PK, Beyer P, Wünn J, et al (1997) Transgenic rice (*Oryza sativa*) endosperm expressing daffodil (*Narcissus pseudonarcissus*) phytoene synthase accumulates phytoene, a key intermediate of provitamin A biosynthesis. Plant J 11:1071-1078

Cao J, Duan X, McElroy D, et al (1992) Regeneration of herbicide resistant transgenic rice plants following microprojectile-mediated transformation of suspension culture cells. Plant Cell Rep 11:586-591

Carr JP, Zaitlin M (1993) Replicase-mediated resistance. Semin Virol 4:339-347

Christou P, Ford T, Kofron M (1991) Production of transgenic rice (*Oryza sativa* L.) plants from agronomically important indica and japonica varieties via electric discharge particle acceleration of exogenous DNA into immature zygotic embryos. Biotechnology 9:957-962

Cornejo MJ, Luth D, Blankenship KM, et al (1993) Activity of a maize ubiquitin promoter in transgenic rice. Plant Mol Biol 23:567-581

Datta SK, Peterhans A, Datta K, et al (1990) Genetically engineered fertile *Indica*-rice recovered from protoplasts. Biotechnology 8:736-740

Datta SK, Datta K, Soltanifar N, et al (1992) Herbicide-resistant *Indica* rice plants from IRRI breeding line IR72 after PEG-mediated transformation of protoplasts. Plant Mol Biol Interact 20:619-629

Devos KM, Gale MD (1997) Comparative genetics in the grasses. Plant Mol Biol 35:3-15

Duan X, Li X, Xue Q, et al (1996) Transgenic rice plants harboring an introduced potato proteinase inhibitor II gene are insect resistant. Nat Biotechnol 14:494-498

Ebinuma H, Sugita K, Matsunaga E, et al (1997) Selection of marker-free transgenic plants using the isopentenyl transferase gene. Proc Natl Acad Sci USA 94:2117-2121

Estruch JJ, Carozzi NB, Desai N, et al (1997) Transgenic plants: an emerging approach to pest control. Nat Biotechnol 15:137-141

Fang RX, Luo ZL, Chen XY, et al (1997) Analysis of genetically-engineered resistance to RYSV in progeny rice plants. In: General Meeting of the International Program on Rice Biotechnology, September 1997, Malacca, p 60 (abstract)

FAO (1988) Iron. In: Requirements of vitamin A, iron, folate and vitamin B_{12}. FAO Food and Nutrition Series 23. FAO, Rome, pp 17-32

FAO (1993) Rice in human nutrition. FAO, Rome, pp 17-84

FAO (1998) FAOSTAT Agriculture Statistics Database; see <http://www.fao.org/WAICENT/Agricul.htm>

Fauquet CM, Huet H, Ong CA, et al. (1997) Control of the rice tungro disease by genetic engineering is now a reality! In: General Meeting of the International Program on Rice Biotechnology, September 1997, Malacca, pp 59 (abstract)

Fütterer J, Potrykus I (1995) Transformation of *Poaceae* and gene expression in transgenic plants. Agronomie 15:309-319

Fujimoto H, Itoh K, Yamamoto M, et al (1993) Insect resistant rice generated by introduction of a modified δ-endotoxin gene of *Bacillus thuringiensis*. Biotechnology 11:1151-1155

Ghareyazie B, Alinia F, Menguito CA, et al (1997) Enhanced resistance to two stem borers in an aromatic rice containing a synthetic *cry*IA(b) gene. Mol Breed 3:401-414

Gould F, Anderson A, Jones A, et al (1997) Initial frequency of alleles for resistance to *Bacillus thuringiensis* toxins in field populations of *Heliothis virescens*. Proc Natl Acad Sci USA 94:3519-3523

Gritz L, Davies J (1983) Plasmid-encoded hygromycin B resistance: the sequence of hygromycin B phosphotransferase and its expression in *Escherichia coli* and *Saccharomyces cerevisisiae*. Gene 25:179-188

Hayakawa T, Zhu Y, Itoh K, et al (1992) Genetically engineered rice resistant to rice stripe virus, an insect transmitted virus. Proc Natl Acad Sci USA 89:9865-9869

Herdt RW (1996) Establishing the Rockefeller Foundation's priorities for rice biotechnology research in 1995 and beyond. In: Khush GS (ed) Rice Genetics III. Proceedings of the Third International Rice Genetics Symposium. IRRI, Manila, pp 17-29

Hibino H (1989) Insect-borne viruses of rice. In: Advances in disease vector research. Vol 6. Springer, New York, pp 209-241

Hiei Y, Ohta S, Komari T, et al (1994) Efficient transformation of rice (*Oryza sativa* L.) mediated by *Agrobacterium* and sequence analysis of the boundaries of the T-DNA. Plant J 6:271-282

Humphrey JH, West KP Jr, Sommer A (1992) Vitamin A deficiency and attributable mortality among under-5-year-olds. WHO Bull 70:225-232

Hurrell RF, Juillerat MA, Reddy MB, et al (1992) Soy protein, phytate, and iron absorption in humans. Am J Clin Nutr 56:573-578

Hsieh HM, Liu WK, Huang PC (1995) A novel stress-inducible metallothionein-like gene from rice. Plant Mol Biol 28:381-389

IRRI (1983) Field problems of tropical rice. International Rice Research Institute (IRRI), Manila, pp 154-169

IRRI (1993) IRRI Rice almanac. IRRI, Manila, pp 8-9

IRRI (1996) Bt Rice: research and policy issues. IRRI Information Series No. 5. IRRI, Manila, pp 17-18

Itoh K, Nakajima M, Shimamoto K (1997) Silencing of *waxy* genes in rice containing *Wx* transgenes. Mol Gen Genet 255:351-358

Jain RK, Jain S, Wang B, et al (1996) Optimization of biolistic method for transient gene expression and production of agronomically useful transgenic Basmati rice plants. Plant Cell Rep 15:963-968

James C, Krattiger AF (1996) ISAAA Briefs No. 1. Global review of the field testing and commercialization of transgenic plants: 1986 to 1995. The first decade of crop biotechnology. ISAAA, Ithaka, pp 6-22

Kempin SA, Liljegren SJ, Block LM, et al (1997) Targeted disruption in *Arabidopsis*. Nature 389:802-803

Kim WT, Li X, Okita TW (1993) Expression of storage protein multigene families in developing rice endosperm. Plant Cell Physiol 34:595-603

Komari T, Hiei Y, Saito Y, et al (1996) Vectors carrying two separate T-DNAs for co-transformation of higher plants mediated by *Agrobacterium tumefaciens* and segregation of transformants free from selection markers. Plant J 10:165-174

Koziel MG, Beland GL, Bowman C, et al (1993) Field performance of elite transgenic maize plants expressing an insecticidal protein derived form *Bacillus thuringiensis*. Bio/Technology 11:194-200

Krattiger AF (1997) ISAAA Briefs No. 2. Insect resistant crops: A case study of *Bacillus thuringiensis* (*Bt*) and its transfer to developing countries. ISAAA, Ithaka

Kumpatla SP, Teng W, Buchholz WG, et al (1997) Epigenetic transcriptional silencing and 5-Azacytidine-mediated reactivation of a complex transgene in rice. Plant Physiol 115:361-373

Lammers PJ, Spencer MJ, Henzl MT (1991) The structure of a *Phaseolus vulgaris* cDNA encoding the iron storage protein ferritin. Plant Mol Biol 17:499-504

Layrisse M, Martinez-Torres C, Leets I, et al (1984) Effect of histidine, cysteine, glutathione or beef on iron absorption in humans. J Nutr 114:217-223

Lentini Z, Calvert L, Tabares E, et al (1997) The development of resistance to rice hoja blanca virus using the N gene in transgenic rice. In: General Meeting of the International Programm on Rice Biotechnology, September 1997, Malacca, pp 61 (abstract)

Li L, Qu R, de Kochko A, et al (1993) An improved rice transformation system using the biolistic method. Plant Cell Rep 12:250-255

Lin W, Anuratha CS, Datta K, et al (1995) Genetic engineering of rice for resistance to sheath blight. Biotechnology 13:686-691

Matsuoka M, Tada Y, Fujimura T, et al (1993) Tissue-specific light-regulated expression directed by the promoter of a C-4 gene, maize pyruvate, orthophosphate dikinase, in a C-3 plant, rice. Proc Natl Acad Sci USA 90:9586-9590

McBride KE, Svab Z, Schaaf DJ, et al (1995) Amplification of a chimeric *Bacillus* gene in chloroplasts leads to an extraordinary level of an insecticidal protein in tobacco. Biotechnology 13:362-365

McElroy D, Zhang W, Cao J, et al (1990) Isolation of an efficient actin promoter for use in rice transformation. Plant Cell 2:163-172

McGaughey WH, Whalon ME (1992) Managing insect resistance to *Bacillus thuringiensis* toxins. Science 258:1451-1455

McPhail AP, Bothwell JD (1992) The prevalence and causes of nutritional iron deficiency anaemia. In: Formon SJ, Zlotkin S (eds) Nutritional Anaemias. Nestle Nutrition Workshop Series 30. Raven Press, New York, pp 1-12

Meijer EGM, Schilperoort RA, Rueb S, et al (1991) Transgenic rice cell lines and plants: expression of transferred chimeric genes. Plant Mol Biol 16:807-820

Müller E, Gilbert J, Davenport G, et al (1995) Homology-dependent resistance: transgenic virus resistance in plants related to homology-dependent gene silencing. Plant J 7:1001-1013

Murakami T, Anzai H, Imai S, et al (1986) The bialaphos biosynthetic genes of *Streptomyces hygroscopicus*: molecular cloning and characterization of the gene cluster. Mol Gen Genet 221:245-250

Nayak P, Basu D, Das S, et al (1997) Transgenic elite *Indica* rice plants expressing CryIAc δ-endotoxin of *Bacillus thuringiensis* are resistant against yellow stem borer (*Scirpophaga incertulas*). Proc Natl Acad Sci USA 94:2111-2116

Oard JH, Linscombe SD, Braverman MP, et al (1996) Development, field evaluation, and agronomic performance of transgenic herbicide resistant rice. Mol Breed 2:359-368

Oerke EC, Dehne HW, Schönbeck F, et al (1994) Crop losses in rice. In: Crop production and crop protection. Elsevier, Amsterdam, pp 89-179

Ow DW (1996) Recombinase-directed chromosome engineering in plants. Curr Opin Biotechnol 7:81-186

Pasamontes L, Haiker M, Wyss M, et al (1997) Gene cloning, purification and characterisation of a heat-stable phytase from the fungus *Aspergillus fumigatus*. Appl Env Microbiol 63:1696-1700

Powles SB, Preston C, Bryan IB, et al (1997) Herbicide resistance: impact and management. In: Sparks DL (ed) Adv Agron 58:57-93

Rashid H, Yokoi S, Toriyama K, et al (1996) Transgenic plant production mediated by *Agrobacterium* in *Indica* rice. Plant Cell Rep 15:727-730

Rockefeller Foundation (1997) 1996 Annual Report. The Rockefeller Foundation, New York, pp 15, 65-67

Sakamoto A, Murata A, Murata N (1997) Genetic engineering of glycinebetaine accumulation and stress tolerance in rice. In: Book of abstracts, Fifth International Congress of Plant Molecular Biology, abstract 841. SPMB, Singapore

Sanford JC, Johnston SA (1985) The concept of pathogen derived resistance: deriving resistance genes from the pathogen's own genome. Theor Biol 113:395-405

Sankula S, Braverman MP, Jodari F, et al (1997) Evaluation of glufosinate on rice (*Oryza sativa*) transformed with the *bar* gene and red rice (*Oryza sativa*). Weed Technol 11:70-75

Sardana R, Dukiandjiev S, Giband M, et al (1996) Construction and rapid testing of synthetic and modified toxin gene sequences *cry*IA (b&c) by expression in maize endosperm culture. Plant Cell Rep 15:677-681

Schaefer DG, Zyrd JP (1997) Efficient gene targetting in the moss *Physcomitrella patens*. Plant J 11:195-1206

Shimada H, Tada Y, Kawasaki T, et al (1993) Antisense regulation of the rice *waxy* gene expression using a PCR-amplified fragment of the rice genome reduces the amylose content in grain starch. Theor Appl Genet 86:665-672

Skriver K, Mundy J (1990) Gene expression in response to abscisic acid and osmotic stress. Plant Cell 2:503-512

Sommer A (1988) New imperatives for an old vitamin (A). J Nutr 119:96-100

Song WY, Wang GL, Chen LL, et al (1995) A receptor kinase-like protein encoded by the rice disease resistance gene, *Xa21*. Science 270:804-806

Stark-Lorenzen P, Nelke B, Hänssler G, et al (1997) Transfer of a grapevine stilbene synthase gene to rice (*Oryza sativa* L.). Plant Cell Rep 16:668-673

Svab Z, Hajdukiewicz P, Maliga P (1990) Stable transformation of plastids in higher plants. Proc Natl Acad Sci USA 87:8526-8530

Tabashnik BE (1994) Evolution of resistance to *Bacillus thuringiensis*. Annu Rev Entomol 39:47-79

Tada T, Kanzaki H, Norita E, et al (1996a) Decreased symptoms of rice blast disease on leaves of *bar*-expressing transgenic rice plants following treatment with bialaphos. Mol Plant-Microbe Interact 9:758-759

Tada Y, Nakase M, Adachi T, et al (1996b) Reduction of 14-16 kDa allergenic proteins in transgenic rice plants by antisense gene. FEBS Lett 391:341-345

Terada R, Shimamoto K (1990) Expression of CaMV35S-GUS gene in transgenic rice plants. Mol Gen Genet 220:389-392

Toki S, Takamatsu S, Nojiri C, et al (1992) Expression of a maize ubiquitin gene promoter-*bar* chimeric gene in transgenic rice plants. Plant Physiol 100:1503-1507

Toriyama K, Arimoto Y, Uchimiya H, et al (1988) Transgenic rice plants after direct gene transfer into protoplasts. Biotechnology 6:1072-1074

Uchimiya H, Iwata M, Nojiri C, et al (1993) Bialaphos treatment of transgenic rice plants expressing a *bar* gene prevents infection by the sheath blight pathogen (*Rhizoctonia solani*). Biotechnology 11:835-836

Upadhyaya NM, Li Z, Ramm K, et al (1997) Transgenic rice with rice ragged stunt virus derived resistance genes. In: General Meeting of the International Programm on Rice Biotechnology, September 1997, Malacca, pp 98 (abstract)

USDA (1998) Environmental releases database for the U.S. <http://gophisb.biochem.vt.edu/cgi-dos/bioone.exe>, updated 01/28/98

van Hartingsveldt W, Van Zeijl CMJ Harteveld GM, et al (1993) Cloning, characterization and overexpression of the phytase-encoding gene (*phyA*) of *Aspergillus niger*. Gene 127:87-94

Wang GL, Song WY, Ruan DL, et al (1996) The cloned gene, *Xa21*, confers resistance to multiple *Xanthomonas oryzae* pv. *oryzae* isolates in transgenic plants. Mol Plant-Microbe Interact 9:850-855

Wang Z, Wu Z, Xing Y, et al (1990) Nucleotide sequence of rice *waxy* gene. Nucleic Acids Res 18:5898

Wen FJ, Peng JY, Lister RM, et al (1993) Comparative efficiency of promoters for stable expression of the gusA transgene inheritance in transgenic rice. Chin J Bot 5:102-109

Wilson TMA (1993) Strategies to protect crop plants against viruses: pathogen-derived resistance blossoms. Proc Natl Acad Sci USA 90:3134-3141

Wohlleben W, Arnold W, Broer I, et al (1988) Nucleotide sequence of the phosphinothricin *N*-acetyltransferase gene from *Streptomyces viridochromogenes* Tu494 and its expression in *Nicotiana tabacum*. Gene 70:25-37

Wu R, Su J, Cheng W, et al (1997) Produciton of transgenic rice plants that are resistant to insect pests and fungal diseases or to water and salt stress. In: General Meeting of the International Programm on Rice Biotechnology, September 1997, Malacca, pp 113 (abstract)

Wünn J, Klöti A, Burkhardt PK, et al (1996) Transgenic *Indica* rice breeding line IR58 expressing a synthetic *cry*IA(b) gene from *Bacillus thuringiensis* provides effective insect pest control. Biotechnology 14:171-176

Xu D, Duan X, Wang B, et al (1996a) Expression of a late embryogenesis abundant protein gene, *HVA1*, from barley confers tolerance to water deficit and salt stress in transgenic rice. Plant Physiol 110:249-257

Xu D, Xue Q, McElroy D, et al (1996b) Constitutive expression of a cowpea trypsin inhibitor gene, CpTi, in transgenic rice plants confers resistance to two major rice insect pests. Mol Breed 2:167-173

Xu X, Li B (1994) Fertile transgenic *Indica* rice plants obtained by electroporation of the seed embryo cells. Plant Cell Rep 13:237-242

Yang W, Wang S, Liu W, et al (1996) Inhibition of replication of rice dwarf virus in transgenic rice plant expressing antisense ribozyme gene. Virol Sin 11:277-283

Yun DJ, Bressan RA, Hasegawa PM (1997) Plant antifungal proteins. Plant Breed Rev 14:39-88

Zhang HM, Yang H, Rech EL, et al (1988) Transgenic rice plants produced by electroporation-mediated plasmid uptake into protoplasts. Plant Cell Rep 7:379-384

Zhang S, Chen L, Qu R, et al (1996) Regeneration of fertile transgenic indica (group 1) rice plants following microprojectile transformation of embryogenic suspension culture cells. Plant Cell Rep 15:465-469

Zhang W, Wu R (1988) Efficient regeneration of transgenic plants from rice protoplasts and correctly regulated expression of the foreign genes in the plants. Theor Appl Genet 76:835-840

Zheng Z, Sumi K, Tanaka K, et al (1995) The bean seed storage protein β-phaseolin is synthesized, processed and accumulated in the vacuolar type-II protein bodies of transgenic rice endosperm. Plant Physiol 109:777-786

Zhijian L, Hayashimoto A, Murai N (1992) A sulfonylurea herbicide resistance gene from *Arabidopsis thaliana* as a new selectable marker for production of fertile transgenic rice plants. Plant Physiol 100:662-668

Zhu Q, Dabi T, Beeche A, et al (1995) Cloning and properties of a rice gene encoding phenyl-alanine ammonia-lyase. Plant Mol Biol 29:535-550

Key word index

GPSR Compliance

The European Union's (EU) General Product Safety Regulation (GPSR) is a set of rules that requires consumer products to be safe and our obligations to ensure this.

If you have any concerns about our products, you can contact us on ProductSafety@springernature.com

In case Publisher is established outside the EU, the EU authorized representative is:

Springer Nature Customer Service Center GmbH
Europaplatz 3
69115 Heidelberg, Germany

Batch number: 10755329

Printed by Printforce, the Netherlands